SCIENCE of CYCLING

TRANSFORM YOUR RIDE, GAIN STRENGTH, REVOLUTIONIZE YOUR TRAINING

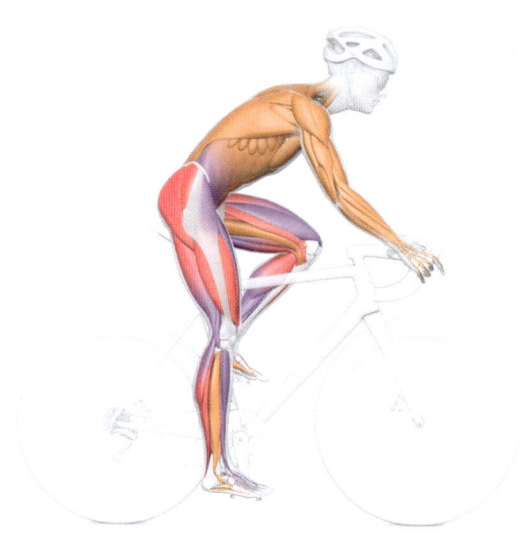

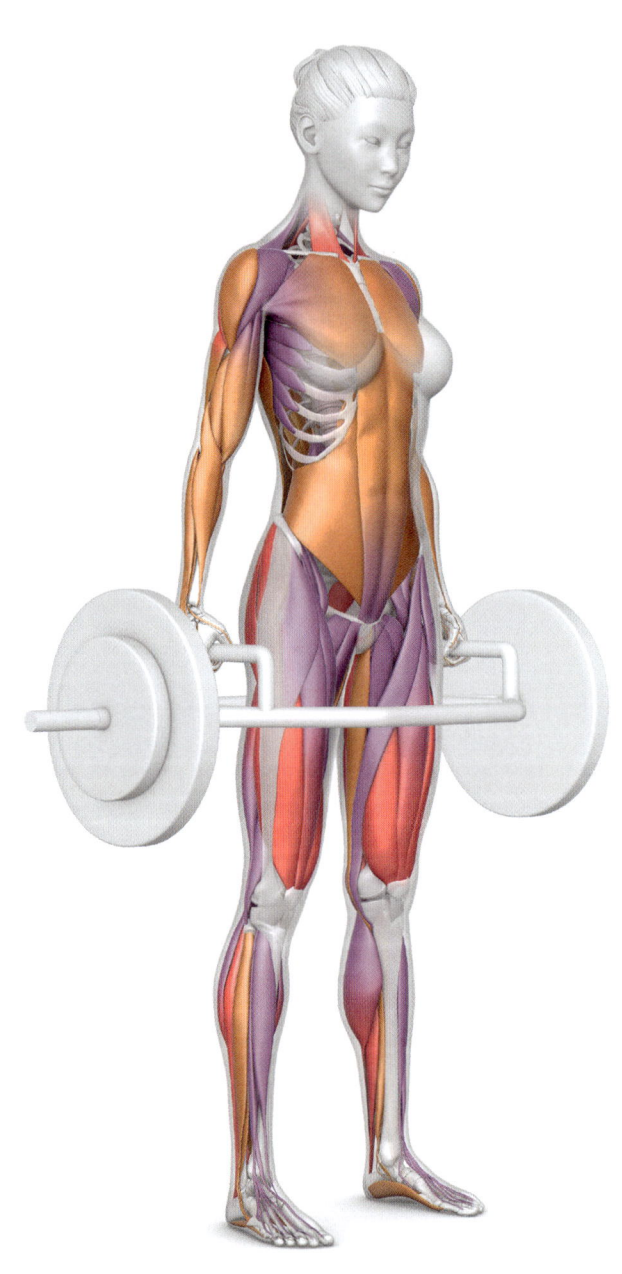

SCIENCE of CYCLING

TRANSFORM YOUR RIDE, GAIN STRENGTH, REVOLUTIONIZE YOUR TRAINING

Dr David Bailey

CONTENTS

Introduction 06
The Evolution of Cycling 08
The Physics of Cycling 10

ANATOMY OF CYCLING 12

Key Muscle Groups 14
Cycling Joint Actions 20
Muscle Recruitment in Different Disciplines 22
Cycling Kinematics 24
Aerodynamic Positioning 26
Pedalling Mechanics 28
Understanding Cadence 30
Cadence and Biomechanics 32
Monitoring Biomechanics 34
Bike Positioning 36
Fit Adjustments 38
Common Injuries 40

PHYSIOLOGY OF CYCLING 44

Physiological Demands 46
Energy Systems 48
Energy Duration 50
The Cardiorespiratory System 52
Muscle Fibre Types 54
Ventilatory Thresholds 56
The Altitude Effect 58
Fuelling Our Energy Systems 60
Timing and Availability of Fuel Sources 62
Fuelling Strategies for Different Events 64
Periodization of Nutrition 66
Nutrition Supplements 68
Common Nutrition Pitfalls 70
Thermoregulation 72
Hydration 74
Fatigue and Recovery 76
Recovery Strategies 78

CYCLING TECHNOLOGY 80

The Bicycle: An Engineering Marvel 82
Design Advances 84
Efficient Setup 88
Sensor Technology and Data Analytics 92
The Future of Data 94

COMPETITIVE CYCLING 96

The Road Racing Scene 98
Drafting and the Peloton 100
Echelons and Breakaways 102
How to Race a Climb 104
Lead-Outs and Sprint Finishes 106
Time Trials 108
Game Theory 112
Mental Performance 114
Managing Anxiety 116

HOW TO TRAIN 118

Quantification of Training	120
Periodization	122
Interval Training	124
Strength Training	126

THE TRAINING PROGRAMMES 128

8-Week Training Plan for a Criterium Race	132
8-Week Training Plan for a Cross-Country Mountain Bike Race	134
8-Week Training Plan for a Cyclocross Race	136
8-Week Training Plan for a Time Trial	138
8-Week Training Plan for a Track Cyclist	140
12-Week Training Plan for a Gran Fondo/Cyclosportive	142
12-Week Training Plan for a Road Cyclist	145

STRENGTH AND STRETCH EXERCISES 148

Why Work Out Off-Bike?	150
Barbell Front Squat	154
Split Squat	156
Trap Bar Deadlift	160
Step Up with Dumbbells	162
Hip Thrust	164
Leg Press	166
Calf Raise with Dumbbells	168
Box Jump	172
Pull Up	176
Push Up	178
Bent-Over Row with Dumbbells	180
Dumbbell Shoulder Press	182
Plank with Shoulder Taps	184
Russian Twist with Medicine Ball	188
Deadbug	190
Glute Bridge	192
Hanging Knee Raise	194
Cat Cow	196
Child's Pose	198
Thread the Needle	200
Figure 4 Stretch	202
Half-Kneel Hip Flexor Stretch	204
Quad Stretch with Foot Raised	206
Static Hamstring Stretch	208
Gastrocnemius Stretch	210

Index	212
Bibliography	219
Acknowledgements	222
About the Author	224

INTRODUCTION

At its core, cycling is simple: you push the pedals, generate forward motion, and the bicycle does the rest. Yet beneath this simplicity lies an extraordinarily complex interaction of human physiology, biomechanics, nutrition, psychology, and technology. Each of these components plays a critical role in shaping how we perform on the bike, how we adapt to training, and how we maintain health over time.

Whether you're racing in the professional peloton, pursuing personal bests, or riding purely for fitness and enjoyment, understanding the science behind cycling offers clear advantages. Informed cyclists train more effectively, recover more efficiently, and reduce the risk of injury, allowing them to perform better, ride longer, and enjoy the sport to its fullest.

WHY CYCLING?

Cycling is one of the most rewarding and accessible sports available. It challenges both aerobic and anaerobic fitness, strengthens key muscle groups, and improves cardiovascular capacity, while offering the added benefit of being low impact on joints and bones. This makes it suitable for people across the lifespan, from young athletes to older adults seeking to maintain fitness and mobility.

Scientific research consistently highlights the broad health benefits of regular cycling: improved metabolic health, reduced risk of cardiovascular disease, better mental wellbeing, and even protection against cognitive decline. For those looking to optimize physical performance, cycling develops both endurance and muscular power, while the psychological benefits of outdoor exercise, stress reduction, and social connection add to its appeal.

However, cycling also brings its own set of physiological challenges. Repetitive motion, positional demands, and prolonged workloads can increase the risk of overuse injuries, fatigue, and underperformance if not managed properly. This is where scientific understanding becomes essential.

HOW THIS BOOK WORKS

The *Science of Cycling* is designed to translate current scientific knowledge into practical application for cyclists of all levels. The first sections explore the anatomy and physiology underpinning cycling performance, explaining how the body moves and generates energy, how muscular and cardiovascular systems respond to different types of training, and how adaptation occurs over time. Topics such as cycling biomechanics, bike positioning, pedalling mechanics, and energy metabolism are introduced in a way that is based on sound scientific principles and easy to apply.

Further chapters offer evidence-based guidance on endurance development, interval training, strength and conditioning, and recovery with strategies for nutrition, hydration, and injury prevention. The emphasis throughout is on helping you understand not just what to do, but why it works, empowering you to make informed decisions.

The book also explores the role of technology in cycling, covering everything from the intricacies of bike design to the growing range of tools available for monitoring training, and provides guidance on how to interpret and apply these insights to optimize performance while reducing the risk of overtraining.

WHO THIS BOOK IS FOR

This book is for a broad spectrum of cyclists: from beginners seeking a solid foundation, to dedicated amateurs aiming for structured progression, to competitive cyclists striving for peak performance. It is also intended as a resource for coaches, sports scientists, and other practitioners who support cyclists and want to ground their practice in sound scientific principles.

WHAT I HOPE TO ACHIEVE

Throughout my career working with some of the world's top professional cyclists, I've seen how an understanding of exercise physiology, combined with smart, individualized training can unlock extraordinary athletic potential.

My aim with this book is to provide a comprehensive yet accessible guide to the science that underpins cycling success. I want to help cyclists train more intelligently, perform more consistently, and sustain their love for the sport over the long term.

Ultimately, cycling is not just about speed or distance, it's about resilience, growth, and the pursuit of human potential. My hope is that this book equips you with the knowledge and confidence to take control of your performance, enjoy your time on the bike, and reach whatever goals you set for yourself.

Dr David Bailey
Sports Scientist, Professional Cycling

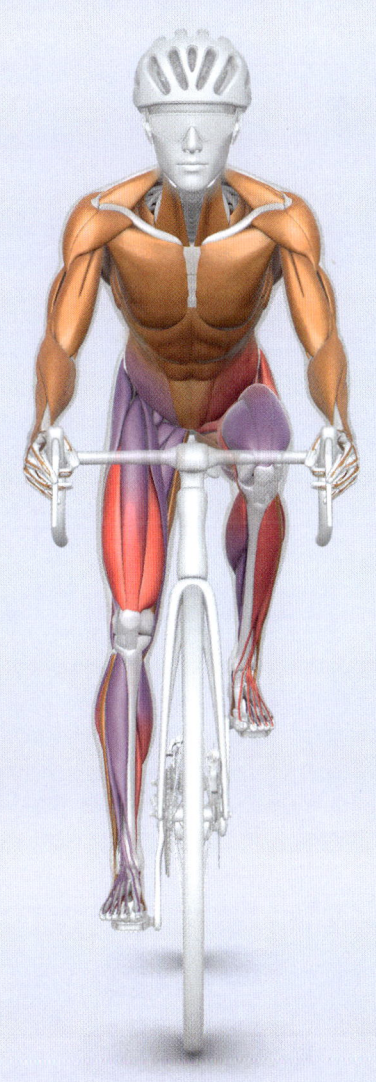

THE EVOLUTION OF CYCLING

Since the earliest bicycles were developed in the nineteenth century, cycling's evolution has been marked by numerous advancements in technology, training, tactics, and physiology. The bikes we see today are the lighter, more aerodynamic, and more specialized than ever before.

In the late 19th century, races were endurance challenges, often covering hundreds of kilometres on basic steel-frame bikes with fixed gears, poor brakes, and heavy tyres. Riders relied on sheer stamina to navigate rough, unpaved roads.

20TH CENTURY

As the sport gained popularity in the early 20th century, technological improvements began to impact performance. The introduction of multi-speed gearing allowed riders to adjust to different gradients, a critical development for navigating mountain passes. Tyres became lighter and more durable, while frame designs evolved to improve aerodynamics and rider efficiency. Team-based tactics emerged, with teams beginning to adopt structured strategies. Riders were assigned roles, such as climbers, sprinters, and domestiques, designed to protect leaders and control the race intensity.

MID-20TH CENTURY

By the mid-20th century, racing became more professionalized, with greater attention to training methodologies. Cyclists moved from a reliance on natural talent and long-distance riding to more targeted preparation. Structured interval training, often guided by training zones, started to shape performance programmes. Nutrition also became more important, with cyclists and teams experimenting with carbohydrate loading and hydration strategies to optimize performance and recovery. This era saw greater use of lightweight materials such as alloy components, further reducing bike weight and improving handling.

1980S

In the 1980s, sports science revolutionized performance in cycling. Coaches implemented periodization in training, focussing on endurance, power, and peak conditioning phases. Heart rate monitors and glycogen replenishment became essential. Equipment innovations like aero handlebars, disc wheels, and wind

19TH CENTURY

Basic steel-frame bikes with fixed gears, poor brakes, and heavy tyres meant that success depended on endurance more than tactics.

20TH CENTURY

Bikes had multi-speed gearing, lighter, more durable tyres and more aerodynamic frame designs. More professionalized training and nutrition programmes developed.

21ST CENTURY

Carbon fibre frames offer exceptional strength-to-weight ratios, electronic shifting enables quick gear changes with ease, and tubeless tyres lower rolling resistance and puncture risks.

INTRODUCTION | The Evolution of Cycling

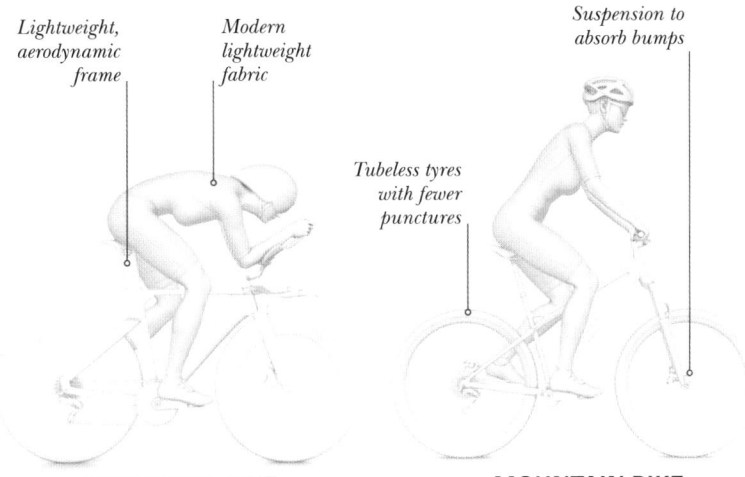

TIME TRIAL BIKE
- Lightweight, aerodynamic frame
- Modern lightweight fabric

MOUNTAIN BIKE
- Suspension to absorb bumps
- Tubeless tyres with fewer punctures

tunnel testing improved time trial performance. However, the sport faced ethical challenges with the rise of blood doping and erythropoietin (EPO) in the 1990s, enhancing riders' oxygen delivery and recovery. Doping scandals overshadowed performance improvements. Stricter testing protocols were implemented in the 2000s to uphold the sport's integrity.

RECENT DECADES

In the early decades of the 21st century the use of power meters significantly influenced training and racing strategies. Riders now train based on specific power outputs to target physiological adaptations for climbing, sprinting, and time trialling. GPS devices offer real-time information on distance, speed, and elevation, providing riders and coaches with performance insights. Online training platforms have enabled more data collection and analysis, allowing riders and their teams to track metrics such as training stress, load, and adaption.

MODERN CYCLING

Professional cycling teams now function as performance ecosystems, involving experts in sports science, biomechanics, nutrition, and mental performance. Nutrition plans are customized, focussing on periodized carbohydrate intake, recovery protein ingestion timing, and race-day fuelling. Recovery methods like compression therapy, cold water immersion, and sleep optimization are essential for maintaining performance. Mental resilience training is also becoming crucial for sustaining peak form and managing competition pressures. Equipment continues to evolve with carbon-fibre frames and aerodynamic designs now standard in road and time trial bikes as teams seek performance gains.

Race tactics have advanced significantly. Teams use detailed reports, simulations, and data analysis to plan races meticulously. Riders specialize in roles: climbers optimize power-to-weight ratios, sprinters focus on anaerobic power, and time triallists perfect sustained power output.

This evolution is not confined to professional road cycling, as the other competitive disciplines have benefited from the collective development in the sport. Mountain biking has pioneered suspension technology and was the first discipline to use disc brakes and tubeless tyres. Many of the aerodynamic advances were first implemented in track cycling where this component of cycling performance predominates. Finally, training techniques for cycling cross, BMX racing, and gravel racing have led the other disciplines to drive practices further.

In cycling, small improvements can make a big difference. Teams innovate in bike fit, clothing materials, and aerodynamic positioning. Sustainability and diversity are shaping the sport to reduce environmental impact and attract more people.

> *With advances in performance science and technology, cycling continues to push the limits of human potential.*

THE PHYSICS OF CYCLING

As a form of human-powered movement, cycling is fundamentally governed by Newtonian mechanics, energy transfer, and biomechanical efficiency. At its core, cycling involves the conversion of chemical energy into mechanical energy through muscle contractions, which then generate force on the pedals, drive the cranks, and ultimately rotate the wheels to propel the cyclist forward.

$$F_{resistance} = F_{drag} + F_{gravity} + F_{rolling} + F_{drivetrain}$$

RESISTIVE FORCES

The propulsive force from pedalling is required to overcome resistive forces associated with cycling. These include aerodynamic drag, gravitational force, rolling resistance, and drivetrain resistance and are expressed in the formula shown above. The cyclist produces power through pedalling, transferring energy to the drivetrain (gears and chain), and then to the wheels. This power output, measured in watts (W), influences the cyclist's ability to accelerate, maintain speed, and climb efficiently as against these resistive forces.

DRAG

Aerodynamic drag is the most significant resistive force at higher speeds, especially above 40kph (25mph). It is influenced by factors such as the cyclist's body position, clothing, equipment, and environmental conditions including wind and air density (temperature and humidity). The force of drag increases rapidly with speed, making aerodynamic efficiency crucial to cycling performance.

Cyclists can reduce drag by adopting a low frontal position (see pp.26–27), using streamlined equipment, and drafting behind others to benefit from the slipstream, which can reduce air resistance by 30–50 per cent (see pp.100–101).

$$F = \tfrac{1}{2} C_d A \rho v^2$$
DRAG

WHERE:
C_d is the coefficient of drag (depends on rider position and bike design)
A is the frontal area of the cyclist and their bike (m²)
ρ is the air density (kg/m³)
v is the bike's velocity (m/s)

$$F = mg\sin(\theta)$$
GRAVITY

WHERE:
m is the mass of the cyclist and their bike (kg)
g is the gravitational acceleration constant (9.81m/s)
θ is the angle of the slop the cyclist is climbing

$$F = C_r mg$$
ROLLING

WHERE:
C_r is the coefficient of rolling resistance dictated by the type of surface cycled on
m is the mass of the cyclist and their bike (kg)
g is the gravitational acceleration constant (9.81m/s)

INTRODUCTION | The Physics of Cycling

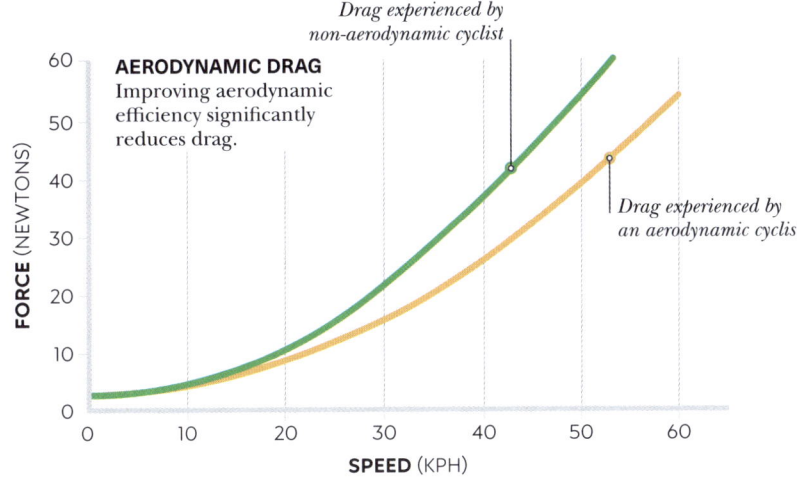

AERODYNAMIC DRAG
Improving aerodynamic efficiency significantly reduces drag.

Drag experienced by non-aerodynamic cyclist

Drag experienced by an aerodynamic cyclist

GRAVITY

The gravitational force is the component of weight of the cyclist and their bicycle, acting parallel to the road surface on an incline. As the gradient increases this resistive force increases resulting in escalating power demands. On inclines, gravitational force becomes the dominant resistive factor, making power-to-weight ratio a key determinant of climbing ability. Lighter cyclists with high power output tend to perform better on steep gradients, as the force of gravity acting against them is lower relative to their power production.

ROLLING

Rolling resistance is another important resistive factor, caused by the deformation of the tyres against the road surface. It depends on the tyre's material, pressure, and the texture of the road's surface. Lower rolling resistance from smooth, properly inflated tyres improves efficiency, allowing the cyclist to maintain speed with less energy expenditure..

DRIVECHAIN

The final resistive force is drivetrain efficiency. This mechanical resistive force is the percentage of power from the cyclists that is successfully transferred to the rear wheel. It is not a constant and depends on several factors including the specific gearing (chainring and cassettes) used as well as the age, material, and lubrication of the drivetrain. It can be improved through keeping the chain clean and lubricated.

$$P_{DRIVETRAIN} = \eta P_{INPUT}$$

WHERE:
η is efficiency, typically 94–98 per cent

ENERGY EFFICIENCY

The human body converts only about 20–25 per cent of ingested food energy into mechanical work, with the remainder lost as heat. This makes cooling mechanisms, such as sweating and airflow, vital for sustained performance. Cyclists must manage their energy expenditure carefully, balancing effort levels to optimize endurance and performance over long distances.

The physics of stability and control also play a role in cycling dynamics. The gyroscopic effect of spinning wheels contributes to balance, while counter steering helps initiate turns. Effective weight distribution and bike handling skills influence cornering ability and braking efficiency, especially in technical racing situations.

Elite cycling is a constant interplay between power output, resistive forces, energy efficiency, and aerodynamics. Cyclists and teams optimize these factors through training, equipment selection, and race tactics to gain a competitive advantage.

2–5% of *total power output* can be lost through the *drivetrain*

ANATOMY OF **CYCLING**

Cycling is a sport that necessitates a high degree of coordination among muscle groups, biomechanics, and energy systems. Unlike high-impact activities, it involves a continuous and repetitive motion that requires muscular endurance, neuromuscular coordination, and efficient energy transfer. An understanding of the anatomical characteristics of cycling enables cyclists to enhance their performance, mitigate fatigue, and prevent injuries.

KEY MUSCLE GROUPS

Cycling predominantly relies on lower body muscles for power, core muscles for stability and power transfer, and upper body muscles for control and stability.

Knowing how each muscle group works can enhance performance, and efficiency, and prevent injuries. The quadriceps, glutes, hamstrings, and calves provide power, while core muscles stabilize the torso and the upper body aids control. Incorporating strength, flexibility, and endurance training for these muscles enhances a cyclist's efficiency, power, and longevity in the sport.

A zoomed-in view shows myofibrils lined up with one another

Visible stripes (striations) reflect the arrangement of muscle proteins

Skeletal muscle fibres
These muscle fibres consist of long, multinucleated, cylindrical cells made up of thousands of myofibrils. They contain proteins that enable muscle contraction.

Pectorals
Pectoralis major
Pectoralis minor

Intercostal muscles

Brachialis

Abdominals
Rectus abdominis
External abdominal obliques
Internal abdominal obliques (deep, not shown)
Transversus abdominis

Hip flexors
Iliopsoas (iliacus and psoas major)
Rectus femoris (see quadriceps)
Sartorius
Adductors (see below)

Elbow flexors
Biceps brachii
Brachialis (deep)
Brachioradialis

Adductors
Adductor longus
Adductor brevis
Adductor magnus
Pectineus
Gracilis

Quadriceps
Rectus femoris
Vastus medialis
Vastus lateralis
Vastus intermedius (deep, not shown)

Ankle dorsiflexors
Tibialis anterior
Extensor digitorum longus
Extensor hallucis longus

SUPERFICIAL   **DEEP**

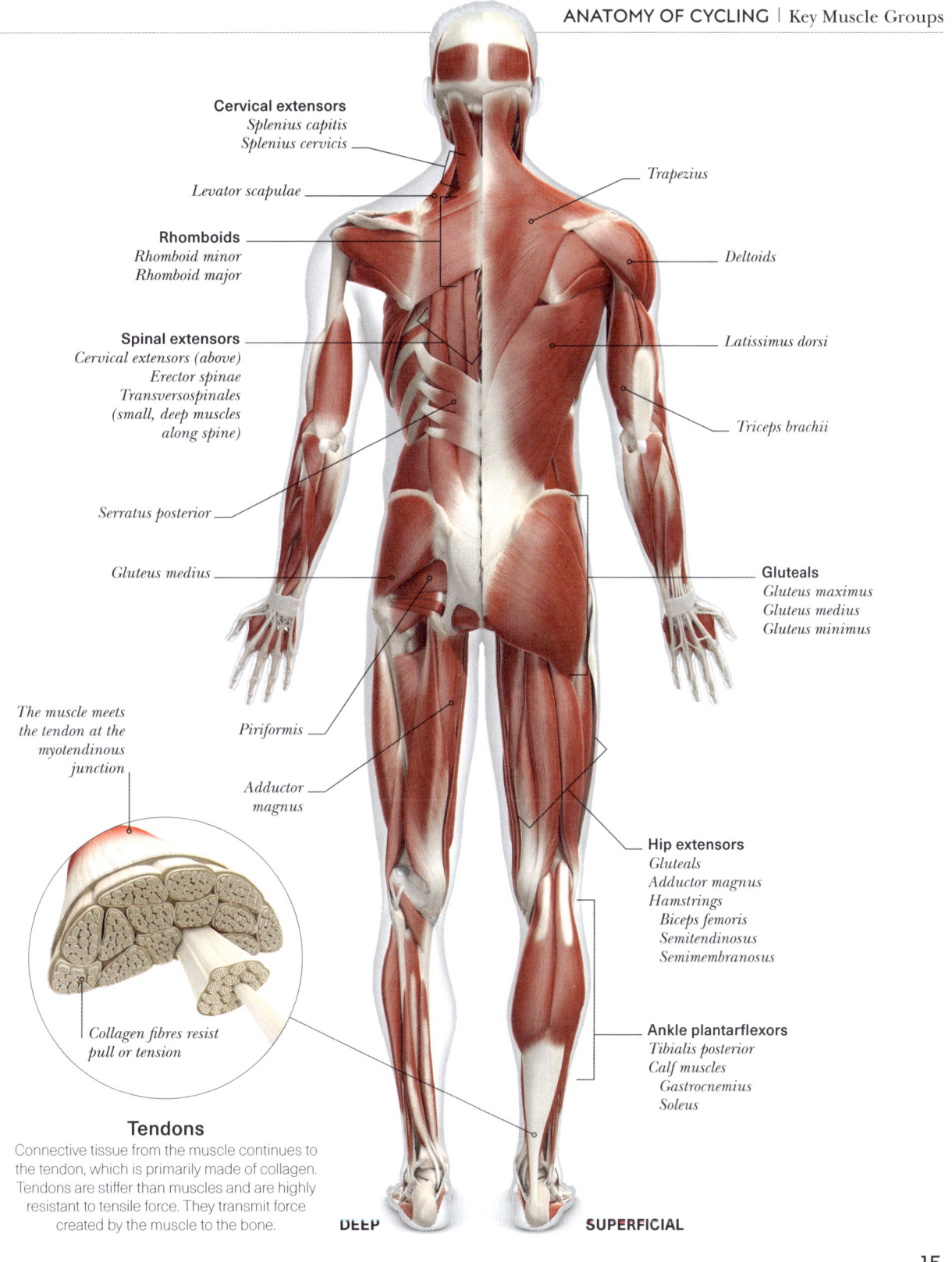

LOWER BODY MUSCLES

The lower body muscles are responsible for transferring force to the pedals and propelling the bike forward, with different muscle groups activating during various phases of the pedal stroke.

Hip flexors
The hip flexors are responsible for lifting the knee and bringing the leg back up during the upstroke phase. While the upstroke contributes less power than the downstroke, strong hip flexors help maintain a smooth pedalling motion, particularly in high-cadence efforts and sprinting. The rectus femoris, part of the quadriceps, plays a dual role by extending the knee and flexing the hip.

Quadriceps muscles
The quadriceps are the dominant force producers in the pedal stroke, particularly during the power phase. They extend the knee and push the pedal downwards, generating a large portion of the forward force. The vastus medialis is essential for knee stability, while the rectus femoris is involved in both hip flexion and knee extension.

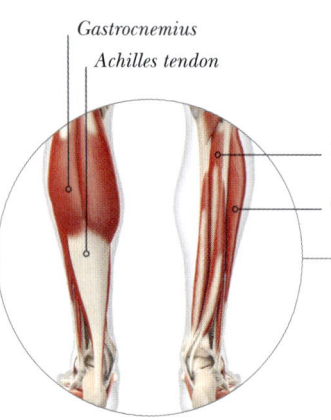

POSTERIOR   ANTERIOR

Calves
The gastrocnemius and soleus muscles contribute to ankle plantarflexion (pointing the toes downwards), helping to smooth out the pedal stroke. These calf muscles are particularly engaged during the bottom of the stroke and help control foot stability throughout the cycle.

ANATOMY OF CYCLING | Key Muscle Groups

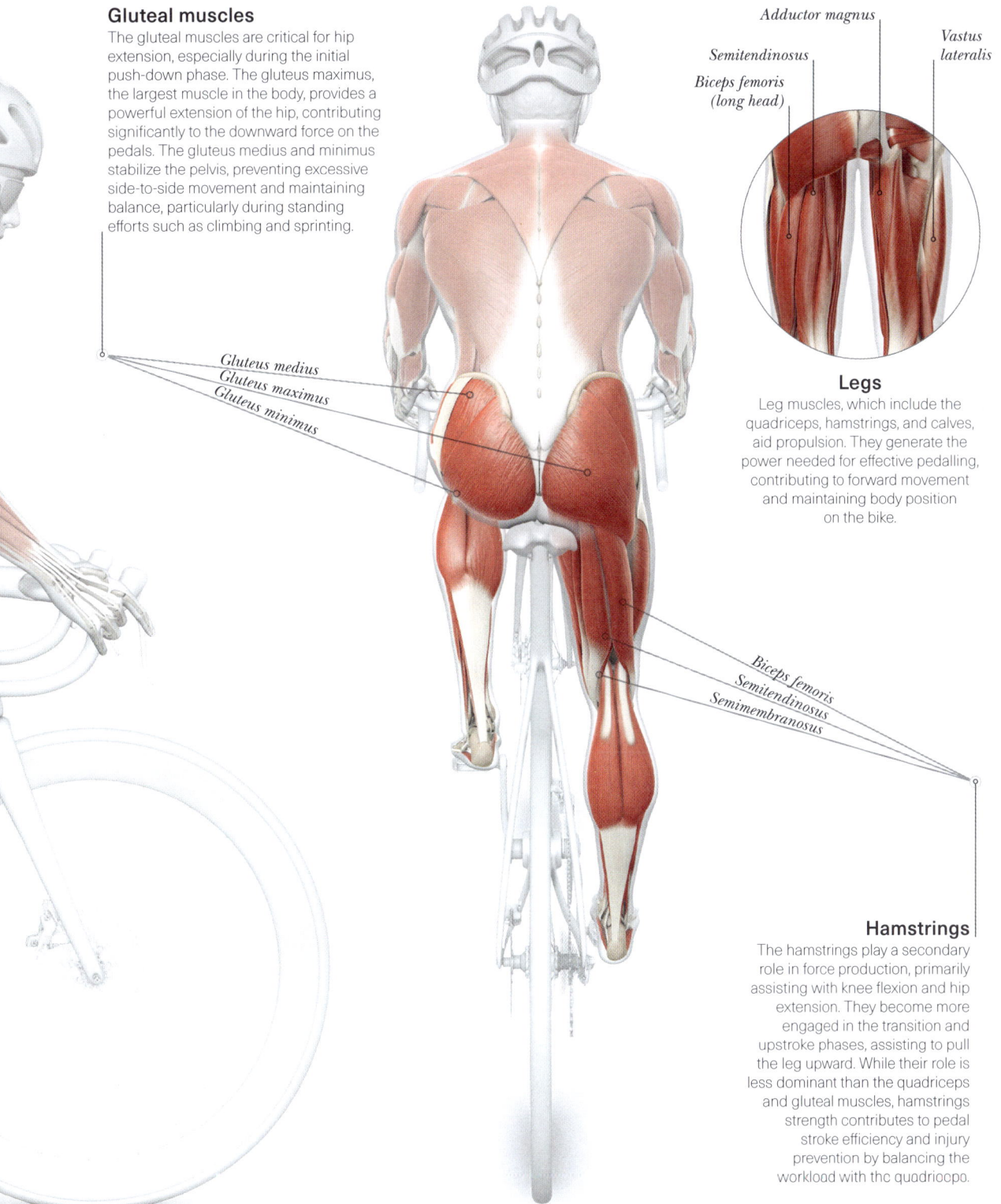

Gluteal muscles
The gluteal muscles are critical for hip extension, especially during the initial push-down phase. The gluteus maximus, the largest muscle in the body, provides a powerful extension of the hip, contributing significantly to the downward force on the pedals. The gluteus medius and minimus stabilize the pelvis, preventing excessive side-to-side movement and maintaining balance, particularly during standing efforts such as climbing and sprinting.

Gluteus medius
Gluteus maximus
Gluteus minimus

Adductor magnus
Semitendinosus
Biceps femoris (long head)
Vastus lateralis

Legs
Leg muscles, which include the quadriceps, hamstrings, and calves, aid propulsion. They generate the power needed for effective pedalling, contributing to forward movement and maintaining body position on the bike.

Biceps femoris
Semitendinosus
Semimembranosus

Hamstrings
The hamstrings play a secondary role in force production, primarily assisting with knee flexion and hip extension. They become more engaged in the transition and upstroke phases, assisting to pull the leg upward. While their role is less dominant than the quadriceps and gluteal muscles, hamstrings strength contributes to pedal stroke efficiency and injury prevention by balancing the workload with the quadriceps.

CORE MUSCLES

The core muscles play a crucial role in maintaining a strong, stable position on the bike, preventing excessive movement that could reduce efficiency or cause discomfort.

Back and spinal erectors
The lower back muscles work in conjunction with the abdominals to maintain an aerodynamic position and support the spine during prolonged riding. They also play a role in absorbing road vibrations and preventing excessive strain on the lumbar region, particularly in aggressive time trial or climbing positions.

Abdominal muscles
The abdominal muscles help stabilize the torso, reducing excess movement and maintaining balance. A strong core minimizes unnecessary rocking of the hips and ensures that power generated by the legs is efficiently transferred to the pedals. The transverse abdominis, the deepest layer of abdominal muscles, provides essential support for spinal stability.

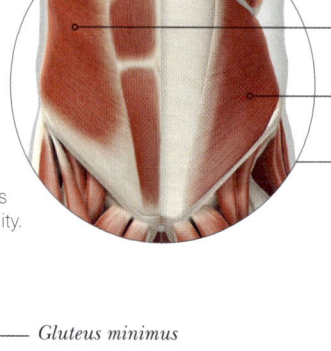

- *Rectus abdominis*
- *Transversus abdominis*
- *External oblique*

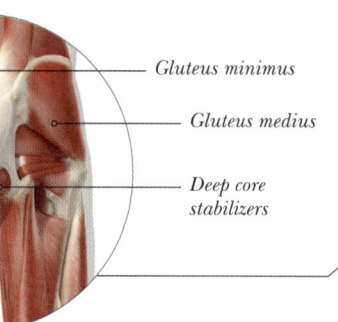

- *Gluteus minimus*
- *Gluteus medius*
- *Deep core stabilizers*

Pelvic stabilizers
These muscles prevent lateral rocking of the hips, which can lead to inefficiencies and imbalances in the pedal stroke. Weak pelvic stabilizers can contribute to knee misalignment and overuse injuries.

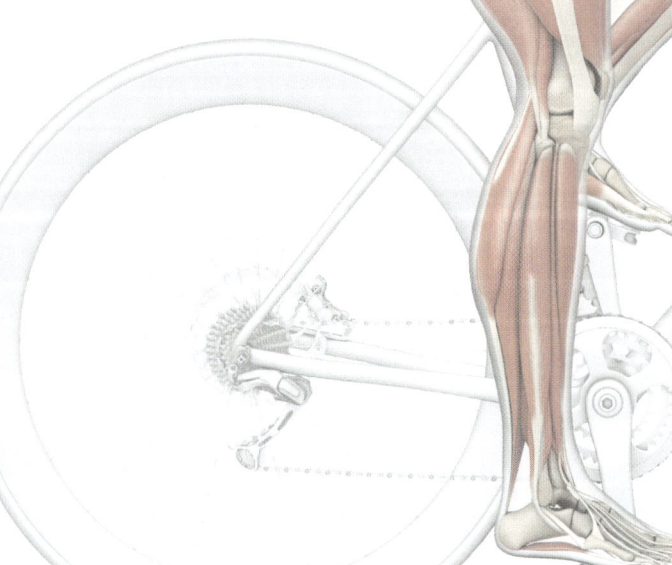

ANATOMY OF CYCLING | Key Muscle Groups

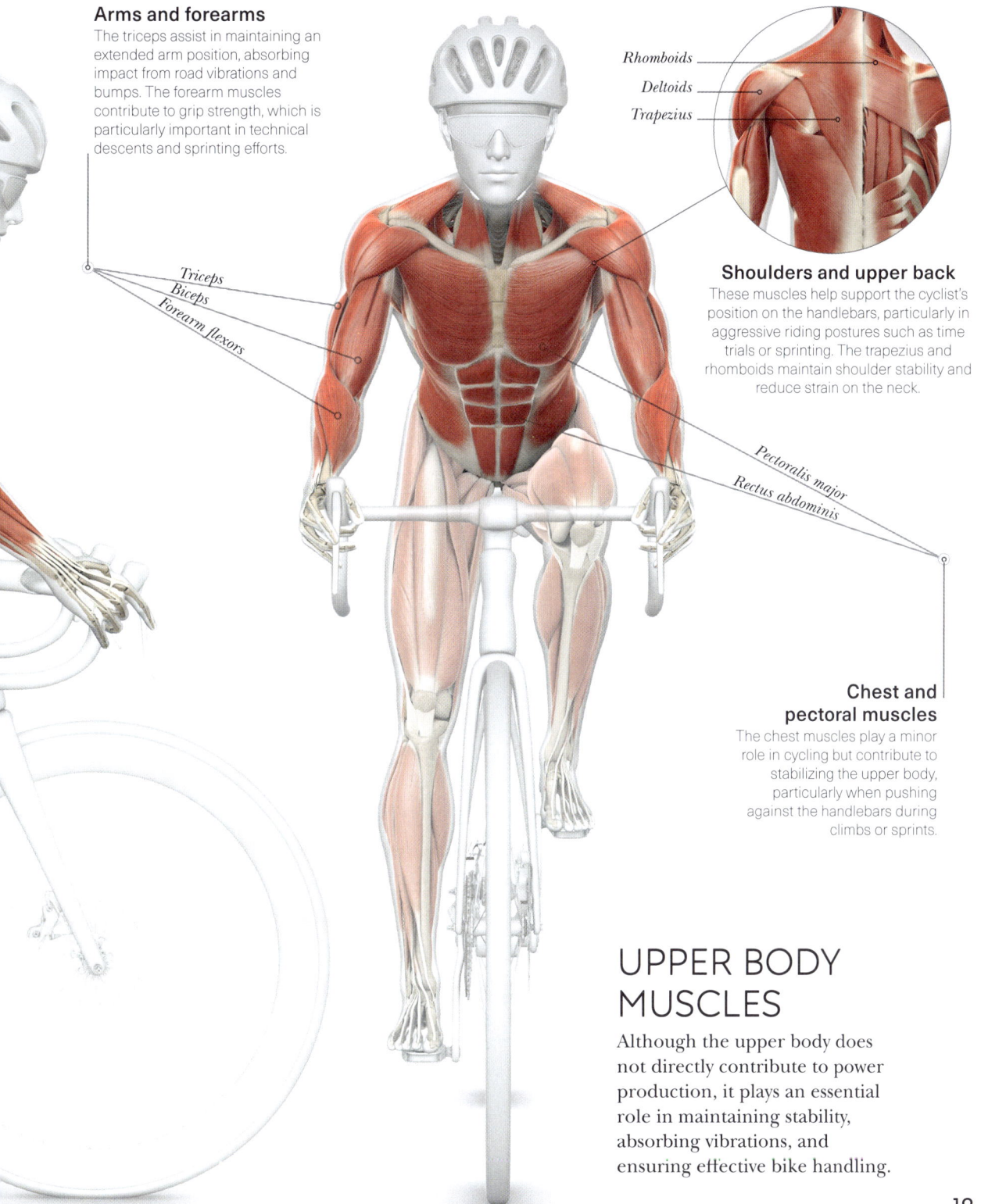

Arms and forearms
The triceps assist in maintaining an extended arm position, absorbing impact from road vibrations and bumps. The forearm muscles contribute to grip strength, which is particularly important in technical descents and sprinting efforts.

Triceps
Biceps
Forearm flexors

Rhomboids
Deltoids
Trapezius

Shoulders and upper back
These muscles help support the cyclist's position on the handlebars, particularly in aggressive riding postures such as time trials or sprinting. The trapezius and rhomboids maintain shoulder stability and reduce strain on the neck.

Pectoralis major
Rectus abdominis

Chest and pectoral muscles
The chest muscles play a minor role in cycling but contribute to stabilizing the upper body, particularly when pushing against the handlebars during climbs or sprints.

UPPER BODY MUSCLES

Although the upper body does not directly contribute to power production, it plays an essential role in maintaining stability, absorbing vibrations, and ensuring effective bike handling.

19

CYCLING JOINT ACTIONS

Cycling involves a complex, repetitive movement pattern that requires coordinated joint actions across both the lower and upper body to achieve efficient and powerful pedalling. Although the main emphasis is on the lower limbs, the contributions of the upper body and trunk are critical for force transfer, stability, and control.

At **90 RPM** (revolutions per minute) each *knee* performs about *5,400 movements* per hour

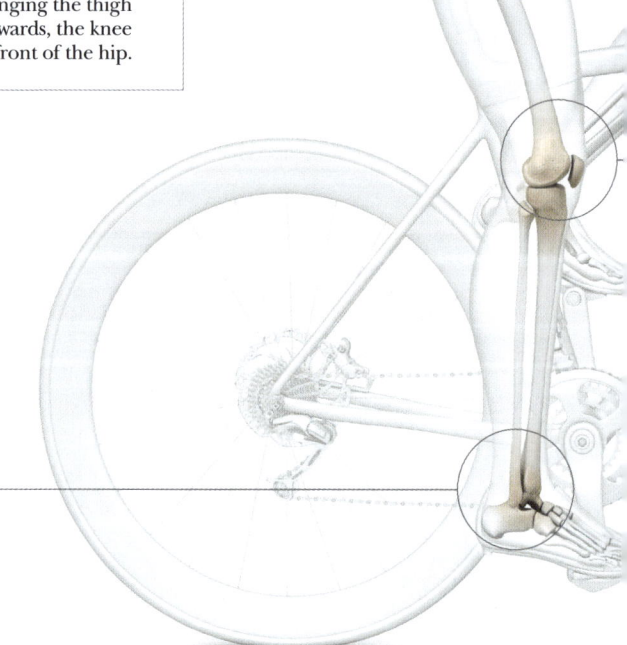

Hip joint actions

The hip joint plays a central role in power production during cycling. The primary movement is hip extension, particularly during the downstroke phase, where the gluteus maximus and hamstrings drive the pedal downwards. During the upstroke, hip flexion occurs, predominantly through the iliopsoas, rectus femoris, and sartorius. Although this contributes less to propulsion, it reduces the drag created by the rising leg and supports a smooth pedal stroke.

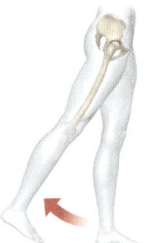

EXTENSION
Extending the thigh backwards, moving the knee behind the hip.

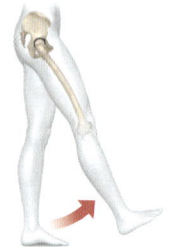

FLEXION
Bringing the thigh forwards, the knee in front of the hip.

Ankle joint actions

Though often underappreciated, the ankle joint contributes to the fine-tuning of force application and pedalling efficiency. During the downstroke, the ankle moves from dorsiflexion to plantarflexion, driven by the gastrocnemius and soleus. This action helps transmit power to the pedal and allows the foot to follow the circular pedal path. During the upstroke, a return to dorsiflexion occurs, primarily through the tibialis anterior, aiding in foot clearance and reducing resistance.

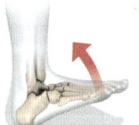

Dorsiflexion
Pointing the foot and toes upwards.

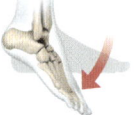

Plantarflexion
Pointing the foot and toes downwards.

ANATOMY OF CYCLING | Cycling Joint Actions

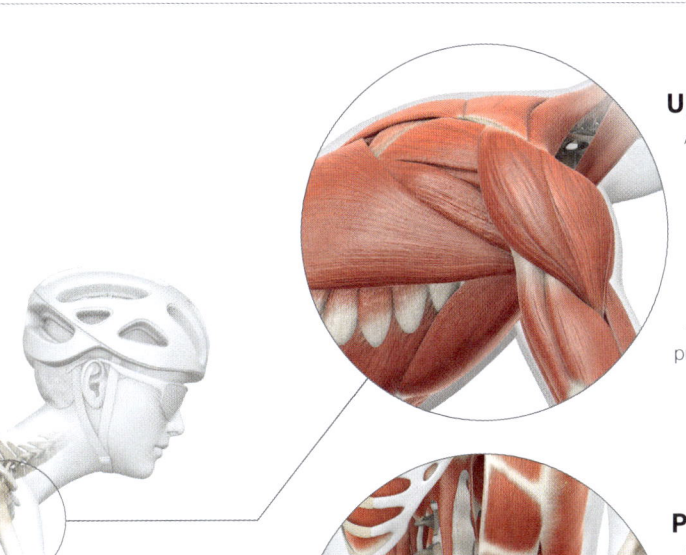

Upper body involvement

Although upper limb motion is minimal in seated cycling, isometric contractions of the shoulders, arms, and hands provide stability and control of the handlebars. The deltoids, trapezius, triceps, and forearm muscles work to stabilize the upper body and counteract forces generated by the legs. During sprinting or climbing when out of the saddle, these muscles play a greater role in force production and body coordination.

Pelvis and trunk stabilization

While there is not a large range of motion at the pelvis and trunk, core stabilization is critical. The abdominals, obliques, erector spinae, and deep spinal stabilizers work isometrically to maintain a stable torso, allowing effective transfer of power from the hips to the legs without energy loss through unnecessary movement. A stable trunk also reduces the risk of lower back discomfort and compensatory movements.

Knee joint actions

The knee joint undergoes significant extension and flexion throughout the pedal cycle. Knee extension, powered by the quadriceps (vastus lateralis, vastus medialis, and rectus femoris), is dominant during the downstroke and contributes directly to pedal force. As the pedal moves through the bottom of the stroke and begins the recovery phase, knee flexion occurs, involving the hamstrings and gastrocnemius. Coordination of these actions aids smooth pedalling and energy conservation.

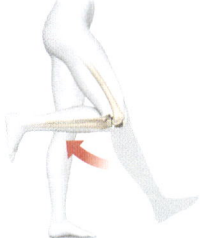

Flexion
Bending at the knee, bringing the foot closer to the thigh.

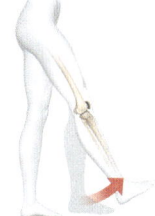

Extension
Straightening at the knee, bringing the foot forwards.

MUSCLE RECRUITMENT IN DIFFERENT DISCIPLINES

Muscle recruitment is the process by which your nervous system activates specific muscle fibres to produce force or movement. It determines which muscles and how many muscle fibres are used during an activity. The order and amount of muscle recruitment in different cycling disciplines is based on posture, intensity, duration, and terrain.

THE DISCIPLINES

ROAD RACING
In road racing, the body adapts to varying terrain, pacing, and race dynamics. On long climbs, the glutes and quadriceps are primarily engaged, especially during seated efforts. When the gradient increases or the pace intensifies, cyclists often stand up, activating the hamstrings, calves, and upper body for more powerful efforts.

During a sprint, the cyclist uses the triceps, shoulders, and upper back to pull on the handlebars while generating lower-body power. Throughout a 4–6-hour race, slow-twitch (Type I) muscle fibres (see pp.54–55) handle most of the workload, whereas fast-twitch fibres are used during accelerations in critical moments.

TIME TRIALS
Time trials involve steady state, high-intensity efforts in a fully optimized aerodynamic position, which can impact specific muscle use. Glutes and quadriceps remain the major muscle delivering force to the pedals, but hamstrings and hip flexors may be less activated compared to an upright road position. Core stability is critical for maintaining aerodynamic position, engaging spinal erectors and deep abdominal muscles as fatigue sets in.

The upper body helps maintain posture and handling with recruitment of shoulder, upper

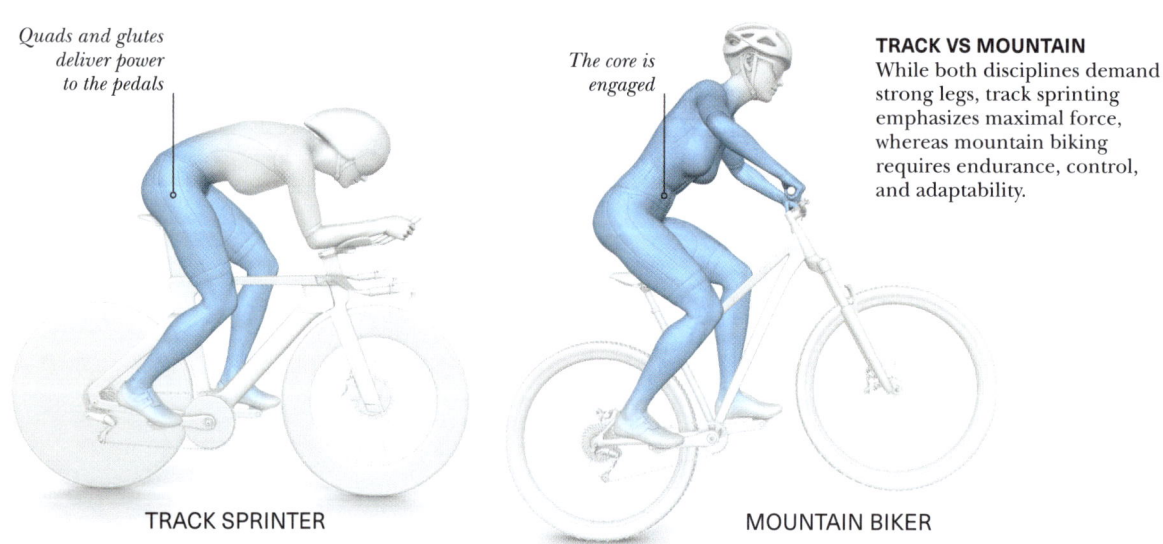

Quads and glutes deliver power to the pedals

TRACK SPRINTER

The core is engaged

MOUNTAIN BIKER

TRACK VS MOUNTAIN
While both disciplines demand strong legs, track sprinting emphasizes maximal force, whereas mountain biking requires endurance, control, and adaptability.

back, and arm muscles to stabilize the bike and counter the forces generated in the lower body guiding the bike, all while holding the upper body in the most aerodynamic position.

TRACK CYCLING

On the track, especially in sprint events, the focus shifts to maximal force and speed. Whether it's a standing start in the 1km time trial or a high-speed match sprint, track cyclists recruit nearly all available fast-twitch (Type IIx) fibres (power muscle fibres that contract quickly). The quadriceps and glutes deliver substantial torque through the pedals on the downstroke, while the hamstrings and calves help maintain rapid turnover on the upstroke at cadences in excess of 150 rpm.

To achieve maximal torque and high peak powers, track sprinters use their upper bodies significantly with isometric contractions (muscle activation that doesn't change the length of the muscle or cause joint movement) of arms, shoulders, and back and core muscles onto the bars, stabilizing through the upper body, and coordinating explosive movements from top to bottom.

For endurance-focussed track events, a combination of sprint, road cycling, and time trialling muscle recruitment is used.

MOUNTAIN BIKING

Mountain biking involves varied terrain and technical handling. Power demands come in quick bursts, including short, steep climbs and accelerations. It engages all major muscle groups, requiring coordination and high levels of proprioception (the body's ability to sense its position, balance, and movement without looking). The core is highly active in managing the bike over rough terrain, with cyclists frequently shifting weight and absorbing impacts. The upper body, including triceps, shoulders, and forearms, work isometrically to maintain control and absorb impacts.

CYCLOCROSS

Cyclocross is a multidirectional, multi-modal athletic endeavour that combines cycling, sprinting, jumping, and running within a single race. It elevates the challenges of mountain biking by incorporating a variety of terrains, including grass, mud, and sand and involves short, intense efforts, frequent dismounts and remounts on and off the bike, and sections where competitors must run with their bikes on their shoulders. This necessitates a distinct recruitment pattern, particularly engaging the hip flexors, glutes, and muscles like the tibialis anterior, which are crucial for running. Additionally, the core and upper body play a significant role in cyclocross, aiding in lifting and carrying the bike and providing stability during repeated accelerations and technical segments. While fast-twitch muscle fibres are predominantly engaged, a strong aerobic base is needed to endure the 45–60 minutes of high-intensity racing.

Muscle activation in different cycling disciplines

DISCIPLINE	DOMINANT MUSCLES	UPPER BODY INVOLVEMENT	KEY FIBRE TYPES	SPECIAL CONSIDERATIONS
Road	Quadriceps, glutes, calves	Medium (sprint)	Type I + IIa	Variability, surges
Time trial	Glutes, quadriceps, core	Low	Type I + IIa	Aero position load
Track	Quadriceps, glutes, hams, calves	High (starts/sprints)	Type IIx	Peak power, high rpm
MTB	All lower and upper core	High	Mixed	Technical handling
Cyclocross	Quadriceps, glutes, hip flexors	High	Type IIa + IIx	Running, lifting bike
Gravel	Glutes, quadriceps, core	Medium	Type I + IIa	Vibration, long duration

CYCLING KINEMATICS

Kinematics is the study of the motion of body segments and joints in cycling without considering the forces behind them. It is used to improve cycling performance, and biomechanics, and reduce injury risk. In road, time trial, and track cycling, kinematic analysis examines joint angles, pedal stroke efficiency, and cyclist positioning to enhance outcomes and musculoskeletal health.

PEDAL STROKE AND JOINT KINEMATICS

Cycling involves the hip, knee, and ankle joints in a closed kinetic chain. A pedal revolution has two major phases: the power phase (0–180 degrees) with hip and knee extension and ankle flexion, generating mechanical power; and the recovery phase (180–360 degrees), which repositions the leg for the next stroke.

Joint angle ranges vary depending on cyclist position, crank length, and saddle height. Optimal joint angles enhance mechanical efficiency and minimize the risk of overuse injuries. For example, excessive knee flexion due to a low saddle height can cause knee pain, while insufficient hip flexion from a saddle positioned too far back can restrict power generation from gluteal muscles.

Innovations
Advancements in wearable sensors and machine learning (see pp.92–93) are making it easier to capture and analyse kinematics in real-world settings. Integration with power meters and physiological data allows for comprehensive cyclist profiling and more individualized training prescriptions.

Knee
Around 75–80 degrees range (from around 110 degrees flexion at top dead centre to around 30 degrees at the bottom).

Hip
Around 40–45 degrees range of motion.

Ankle
15–20 degrees total range (varies between individuals).

PEDAL PHASES
There are two main phases to the cycling action: upward (the recovery phase) and downward (the power phase).

COORDINATION AND TECHNIQUE

Efficient pedalling involves smooth coordination between lower limb segments, minimizing unnecessary movement, and maximizing power transfer.

Limb coordination can be assessed using crank torque profiles (the force pushed into the pedals during each stroke) or motion capture techniques that reveal imbalances in body position or inefficiencies. Two key kinematic variables are:

- **Knee tracking:** Lateral movement of the knee during pedalling, should remain minimal to reduce valgus stress (see pp.32–33) and preserve joint alignment.

- **Pelvic stability:** Excessive rocking of the pelvis often indicates poor core engagement or saddle fit. This can lead to wasted energy and potential lumbar discomfort.

Elite cyclists usually show lower variability in inward and outward knee movement and pelvic rocking. This is due to good neuromuscular control and analysis and optimized bike setup.

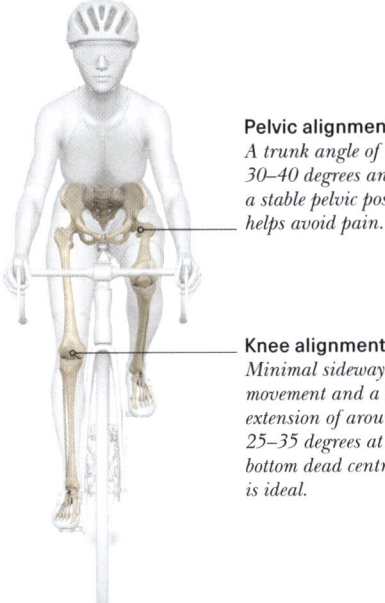

Pelvic alignment
A trunk angle of 30–40 degrees and a stable pelvic position helps avoid pain.

Knee alignment
Minimal sideways movement and a knee extension of around 25–35 degrees at bottom dead centre is ideal.

PERFORMANCE AND INJURY PREVENTION

From a performance perspective, efficient kinematics reduce metabolic cost and improve mechanical output.

Cyclists with more consistent pedal stroke kinematics typically show higher gross efficiency and reduced neuromuscular fatigue. In addition, real-time kinematic feedback is increasingly used in training to reinforce technique changes or to help rehabilitation from injury.

Injuries can also be avoided through kinematic analysis. Repetitive strain injuries such as iliotibial band syndrome, patellar tendinopathy, or lumbar pain are often caused by poor joint alignment or excessive movement variability. Identifying these risk factors early enables intervention through bike fit adjustments, targeted mobility or strength work, or finding the most effective pedal rate (cadence optimization).

BIKE FIT AND POSITIONING

Kinematics are significantly affected by the setup of a bicycle.

Factors such as saddle height, fore-aft saddle position, crank length, and cleat alignment all affect cyclists' joint angles and muscle recruitment patterns.

Positioning the saddle forward reduces hip flexion demands but may increase quadriceps loading, while a rearward saddle shifts the load towards the gluteal and hamstring muscles. Aerodynamic positions, particularly those used in time trials, cause hip closure, which can compromise comfort and power output if not carefully managed. Motion capture and video analysis during bike fitting can help fine tune positions.

AERODYNAMIC POSITIONING

Aerodynamic positioning influences performance in road and time trial cycling, especially at higher speeds. The cyclist contributes approximately 80–90 per cent of total aerodynamic drag, making body position important. Optimizing aerodynamic posture involves balancing reduced frontal area and drag with biomechanical efficiency, joint loading, and maintaining power output over time.

The *primary* **FORCE** opposing motion *on the flat* is **air resistance.**

KINEMATIC DEMANDS

Aerodynamic optimization typically involves altering the cyclist's posture to reduce the projected frontal area and streamline body segments.

Positional changes affect joint angles and range of motion, especially at the hip. In an aggressive time trial position, hip angles can drop to 40 degrees at top dead centre, compared to 55–60 degrees in a traditional road position. While helping to streamline the body, this acute flexion may limit gluteal and proximal hamstring function, particularly in less flexible or poorly adapted cyclists.

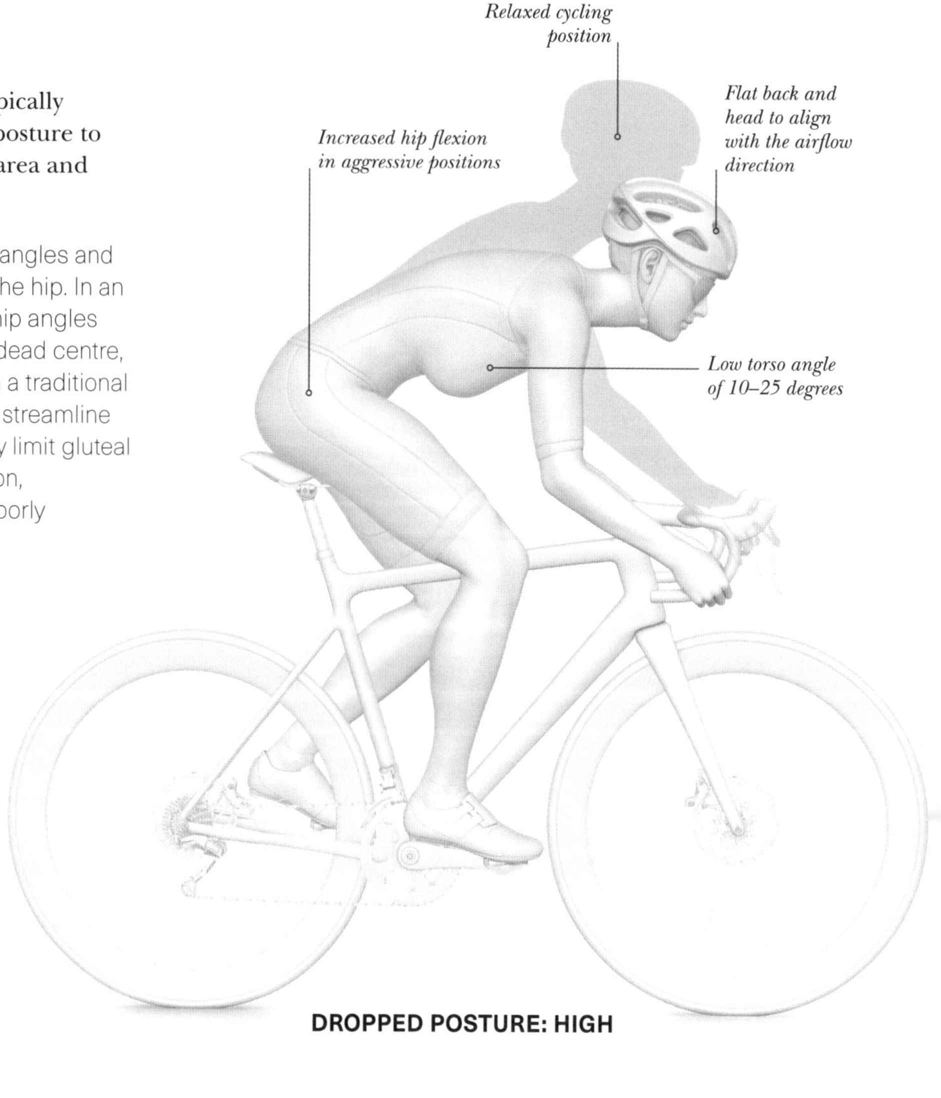

Relaxed cycling position

Flat back and head to align with the airflow direction

Increased hip flexion in aggressive positions

Low torso angle of 10–25 degrees

RELAXED AND DROPPED
Reducing the amount of aerodynamic drag by altering posture is an important part of competitive road cycling.

DROPPED POSTURE: HIGH

BIOMECHANICAL TRADE-OFFS

Reducing frontal area enhances aerodynamics but may decrease biomechanical efficiency.

A more closed hip angle can reduce peak power output by limiting the contribution of large posterior chain muscles and increasing reliance on hip flexors and quadriceps. The goal is to maintain a position that enables effective force application through the pedal stroke while minimizing joint stress. Cyclists should find a position where the aerodynamic gain does not excessively compromise neuromuscular function or energy economy.

> **HIGHLY FLEXED OR ROTATED POSITIONS MAY CAUSE:**
> - **Decreased movement** of the diaphragm, impairing breathing
> - **Increased lumbar** and cervical spine strain
> - **Compromised joint loading** patterns at the knee and ankle

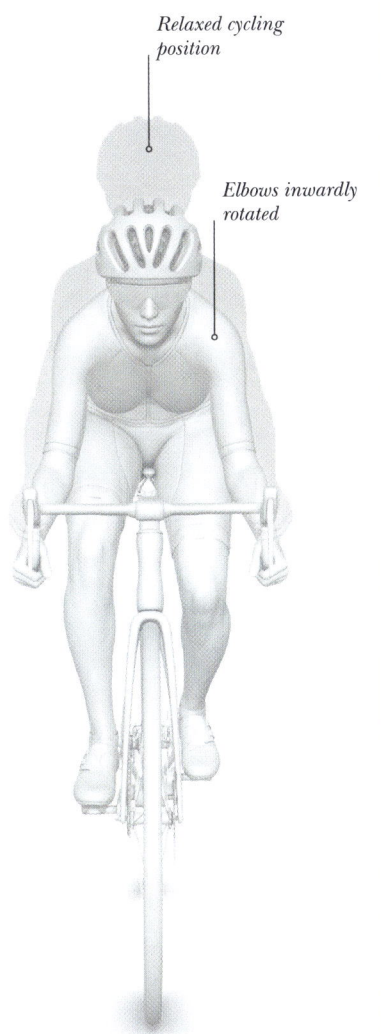

Relaxed cycling position

Elbows inwardly rotated

DROPPED POSTURE: HIGH

COORDINATION AND POSTURAL STABILITY

Kinematically, a stable aerodynamic position requires high coordination and control of multiple segments:

- **Pelvic tilt** must remain consistent to prevent rocking, which can waste energy and affect power transfer.
- **Shoulder and scapular stabilization** become essential to hold narrow elbow pad positions without overloading the upper trapezius or cervical spine.
- **Neck posture** must allow for safe forward vision without excessive extension, which can lead to tension headaches or cervical strain over time.

Cyclists who are able to maintain better positional stability by minimal head movement, steady elbow width, and consistent hip and knee angle, are more likely to sustain both aerodynamic and biomechanical efficiency.

TRAINING AND POSITION SUSTAINABILITY

Aerodynamic positioning is trainable. Cyclists can adapt to more aggressive postures over time with targeted flexibility and core strength training. Bike fitting that integrates motion capture allows real-time assessment.

Aerodynamic positions need to be sustainable for long durations, particularly in time trials or during breakaways. Cyclists with enhanced trunk stability, hamstring flexibility, and muscular shoulder strength can endure extreme positions and maintain joint kinematics and efficient force transfer. Training in an aerodynamic position for long periods of time and integrating the posture into sub-threshold and high-intensity efforts helps to neurologically and physiologically embed the posture.

PEDALLING MECHANICS

Efficient pedalling is crucial for cycling performance, as it impacts power output, efficiency, fatigue resistance, and injury prevention. A comprehensive understanding of the pedal stroke mechanics allows cyclists, coaches, and sport scientists to optimize technique, equipment configuration, and training strategies to achieve maximal performance results.

> *Studies show that skilled cyclists demonstrate smoother activation patterns with better phase timing between muscle groups.*

PHASES OF THE PEDAL STROKE

The cycling pedal stroke is often divided into two main phases – the power phase, where the pedal goes from top to bottom, and the recovery phase, where it moves from the bottom back to the top. The stroke can be further divided into four quadrants to better describe muscle recruitment and torque characteristics.

- **Power phase (0–180 degrees crank angle):** this goes from top dead centre (TDC) to bottom dead centre (BDC), where most of the positive work is done through hip and knee extension and ankle plantarflexion.

- **Recovery phase (180–360 degrees crank angle):** this goes from bottom dead centre (BDC) back to top dead centre (TDC), where the leg returns to the starting position. This part of the stroke predominantly involves hip and knee flexion with ankle dorsiflexion.

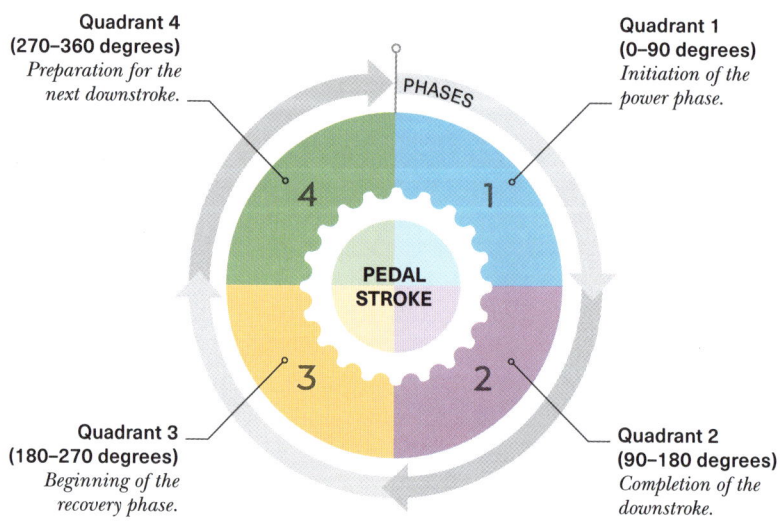

Quadrant 4 (270–360 degrees)
Preparation for the next downstroke.

Quadrant 1 (0–90 degrees)
Initiation of the power phase.

Quadrant 3 (180–270 degrees)
Beginning of the recovery phase.

Quadrant 2 (90–180 degrees)
Completion of the downstroke.

FORCE APPLICATION AND TORQUE PRODUCTION

The effectiveness of pedalling mechanics can be quantified through force vectors applied to the pedal. The two main force components are tangential force and radial force. Efficient cyclists aim to maximize tangential force and minimize radial force, especially when training with power meters or studying pedalling technique.

The effective force ratio (EFR) measures the proportion of applied force that is tangential. Elite cyclists usually have higher EFR than amateurs, especially during steady-state conditions, because they make smooth transitions between power and recovery phases. Less skilled cyclists often show larger radial forces and negative torque during the upstroke because of poor coordination or fatigue.

The hightest force usually occurs between a 70–110 degree angle during the downstroke. Efficient cyclists minimize back-driving (resisting the movement of the pedal) during the upstroke by either active pulling or by lifting their leg up to avoid adding resistance (passive unloading) on the pedal.

Tangential force: *Turns the crank when you push the pedal in a circle.*

70–110 degrees is where the highest force occurs.

Radial force: *The force directed toward the crank axis, which does not contribute to rotational power and is considered wasted.*

Crank: *The metal arm that connects the pedal to the bottom bracket.*

FORCES AT WORK
With training you can increase the tangential force you create and reduce radial force, which is wasted.

MUSCLE RECRUITMENT

Pedalling mechanics rely on well-coordinated activation of multiple muscle groups:

- **Quadrant 1:** The glutes and hip flexors initiate the downstroke, with early activation of the quads.
- **Quadrant 2:** The quads and glutes produce peak power through knee and hip extension, assisted by the calves.
- **Quadrant 3:** The calves and hamstrings guide the transition, with dorsiflexion beginning to lift the pedal.
- **Quadrant 4:** Hip flexors, hamstrings, and dorsiflexors complete the upstroke.

Electromyography (EMG) studies show that skilled cyclists demonstrate smoother activation patterns with better phase timing between muscle groups. Novice cyclists or fatigued states often result in overlapping or poorly timed muscle activation, reducing mechanical efficiency.

PEDAL POWER
During pedalling, the primary power-generating muscles are the quadriceps, hamstrings, and glutes. These muscles work together to produce the force needed to push the pedals down and through the rotation.

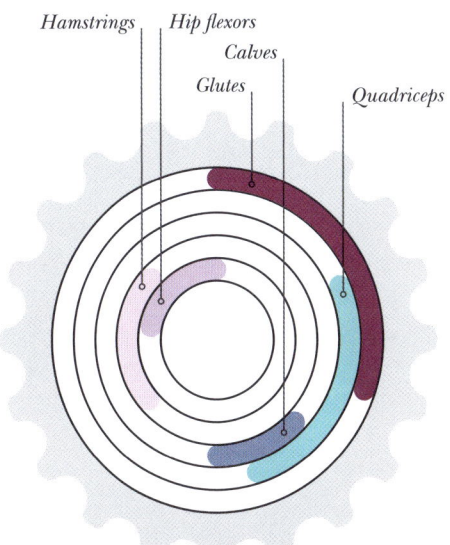

MUSCLE USE DURING PEDAL STROKE

UNDERSTANDING CADENCE

Cadence in cycling refers to the number of pedal revolutions per minute (rpm). It plays a key role in performance and efficiency – higher cadences can reduce muscle fatigue, while lower cadences may rely more on muscular strength. Finding the right cadence depends on the cyclist's fitness, terrain, and cycling goals.

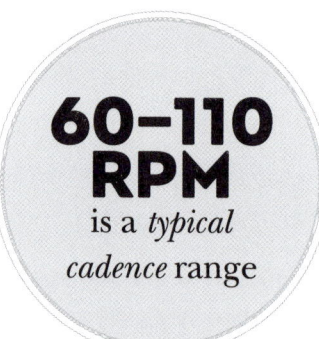

60–110 RPM is a *typical cadence* range

HIGH AND LOW CADENCE

Cadence, or pedal revolutions per minute (rpm), directly influences pedalling mechanics by altering muscle contraction velocity and joint loading patterns. Higher cadence typically:

- **Reduces peak torque** which spreads the force applied to the pedals throughout each pedal stroke.
- **Shifts muscular demand** in favour of endurance-type fibres.
- **Reduces** the load on joints.

Conversely, lower cadence typically:

- **Increases torque demands** (meaning you have to push the pedals harder).
- **Recruits** fast-twitch fibres, which can be beneficial for strength development.
- **Increases** neuromuscular strain.

Optimal cadence varies between cyclists and is influenced by fitness level, muscle fibre composition, and the event (for example, time trial versus climbing or sprinting).

Cadence ranges and context

Cyclists typically operate within a cadence range of 60–110 rpm, depending on discipline, terrain, and physiological characteristics. General classifications include:

▼ **Low cadence:** less than 80 rpm

● **Preferred cadence:** 80–100 rpm (common in road cycling)

▲ **High cadence:** more than 100 rpm (track cycling, sprints, time trials)

Track cyclists and sprinters may exceed 150 rpm in maximal efforts, while climbers may drop below 70 rpm on steep gradients due to gear limitations and the need for higher torque.

PEDAL STROKE AND FATIGUE

Fatigue can affect the way you pedal by reducing effective force application and disrupting muscle coordination. Common signs of fatigue include more radial (sideways) force on the pedals instead of pushing straight down, less consistent torque (force) throughout each pedal stroke, and delayed or prolonged muscle activation patterns. These changes make pedalling less efficient and can increase the risk of injury. Monitoring them through pedal-based power meters or biomechanical assessment tools can provide valuable feedback for pacing, recovery management, and technique refinement.

❝ ❞

Cadence is the silent metronome of cycling – get it right, and everything else falls into rhythm.

EFFICIENCY AND ECONOMY

Mechanical efficiency refers to the ratio of external work output (the power you generate on the bike) to the total metabolic energy your body inputs to produce that work. Pedalling efficiency improves with:

- **Higher IFE (Index of Force Effectiveness)** – a measure of how much of a cyclist's applied force on the pedals is tangential force that contributes to turning the crank and producing power.
- **Consistent and smooth** torque application
- **Reduced co-contraction** of antagonist muscles (those that oppose each other such as hamstrings and quadriceps).

Cyclists with more efficient pedalling mechanics demonstrate lower oxygen consumption at a given workload, better fatigue resistance, and improved endurance performance. Training strategies that emphasize neuromuscular coordination, single-leg drills, and high-cadence intervals can enhance pedalling economy.

GROSS EFFICIENCY AND WORK RATE
At low power outputs, lower cadences are more efficient. As power output increases higher cadences result in better efficiency.

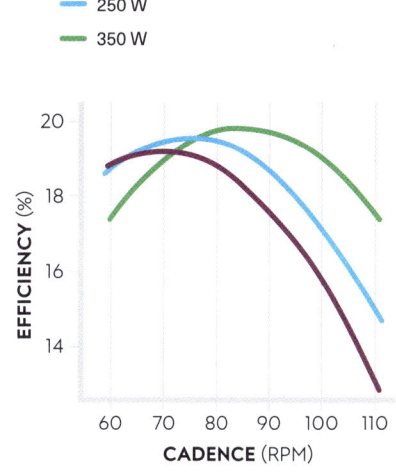

IMPACT OF CRANK LENGTH

Crank length, defined as the distance from the bottom bracket's centre to the centre of the pedal axle, affects both the arc of pedal travel and the cyclist's mechanical leverage. Research shows that gross efficiency and power output remain stable across crank lengths from 145–190mm when cyclists self-select cadence. Cyclists adjust cadence downwards with longer cranks and upwards with shorter ones to optimize power output and minimize physiological cost. The interaction between crank length and cadence reinforces the importance of personalized equipment.

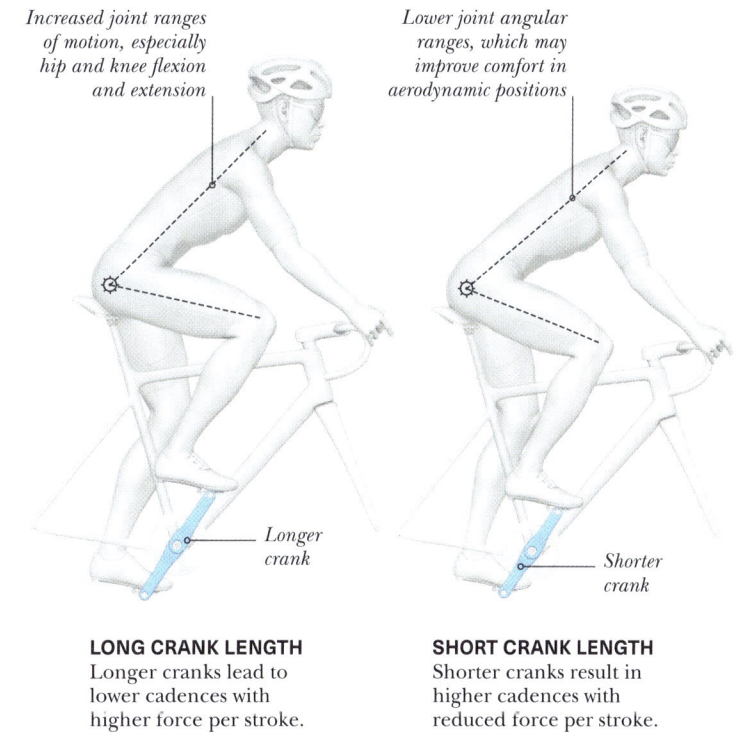

Increased joint ranges of motion, especially hip and knee flexion and extension

Lower joint angular ranges, which may improve comfort in aerodynamic positions

LONG CRANK LENGTH
Longer cranks lead to lower cadences with higher force per stroke.

SHORT CRANK LENGTH
Shorter cranks result in higher cadences with reduced force per stroke.

CADENCE AND BIOMECHANICS

Cadence affects torque (angular force) at the crank. Lower cadences need higher torque (more force) per pedal stroke, straining the quadriceps and gluteal muscles. Higher cadences reduce torque (less force per pedal stroke) but increase contraction speed and neuromuscular activation frequency.

MUSCLES AND CADENCE

Choosing your cycling speed also affects which muscles your body uses and how they work.

Low cadences typically use type II (fast-twitch) muscle fibres due to higher force demands. In contrast, high cadences shift work towards type I (slow twitch) fibres, relying more on oxidative metabolism (see pp.54–55). There's a trade-off because while low cadences may be metabolically more efficient they lead to muscle fatigue due to high force. High cadences may reduce muscle strain but can increase cardiovascular and neuromuscular fatigue. Experienced cyclists tend to have smoother muscle coordination at high cadences, while beginners often use co-contracting muscles at the same time, which is inefficient.

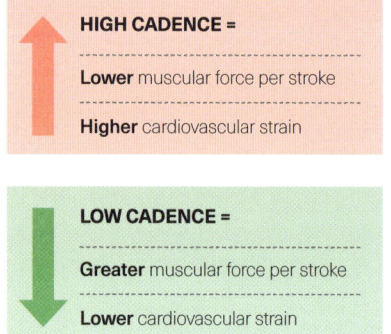

HIGH CADENCE =
Lower muscular force per stroke
Higher cardiovascular strain

LOW CADENCE =
Greater muscular force per stroke
Lower cardiovascular strain

> *The trade-off between force per stroke and stroke frequency is the balance between cardiovascular and muscular load.*

ALIGNMENT

In the context of pedalling biomechanics, valgus and varus refers to the angular relationship between the segments of the lower limb. These issues particularly affect the knee joint, but are also influenced by the hip, ankle, and foot mechanics.

The valgus and varus alignments affect how force is transmitted from the hips through the knees to the pedals and can significantly impact both pedalling efficiency and injury risk.

Deviations of these alignment patterns can be either static (at rest) or dynamic (during cycling) and can impact power to the pedals and pedalling efficiency. This, in turn, increases joint stress at the knee, hip and ankle, ultimately resulting in overuse injuries.

Some cyclists naturally exhibit slight lateral movement of the knee and are able to cycle without pain or discomfort. The key concern is when excessive or asymmetrical movement occurs, as this can indicate a biomechanical issue that needs addressing.

ANATOMY OF CYCLING | Cadence and Biomechanics

Anatomical contributors to valgus and varus in cycling

Several factors influence whether a rider presents with valgus or varus alignment during the pedal stroke:

CONTRIBUTING FACTOR	INFLUENCE ON ALIGNMENT
Hip internal/external rotation	Poor hip control may cause knee valgus collapse
Femoral anteversion or retroversion	Structural rotation of the femur alters knee tracking
Foot pronation/supination	Overpronation tends to drive valgus; supination may promote varus
Q-angle (quadriceps angle)	Larger Q-angle predisposes to valgus alignment
Cleat positioning	Medial or lateral cleat adjustments directly affect knee tracking
Core and glute stability	Weak gluteus medius allows hip drop and internal rotation, contributing to valgus collapse

Cadence and performance strategy

Optimal cadence can be cycling discipline specific:

Climbing is often associated with a lower cadences (60–80 rpm) and may be necessary due to gear ratios and gradient but should be balanced with strength endurance capacity. Modern gearing should provide sufficient ratios for all gradients below 10 per cent.

Time trials are typically performed at moderate to high cadences (85–100 rpm) to help manage fatigue and maintain consistent power output.

Sprinting, track cycling, and BMX racing are associated with very high cadences (110–140+ rpm) in order to leverage speed and momentum, requiring specific neuromuscular training. Depending on muscle fibre composition, peak powers achieved in these disciplines are directly linked to peak cadence, often >140 rpm.

Mountain biking often has the largest cadence ranges due to the varied terrain and technical demands of the sport. Short, steep climbs over rough rocky terrain can result in cadences <50 RPM and faster sections on flatter downhill terrain >110 rpm.

Training across a range of cadences can improve neuromuscular adaptability and prevent overuse injuries associated with fixed cadence habits.

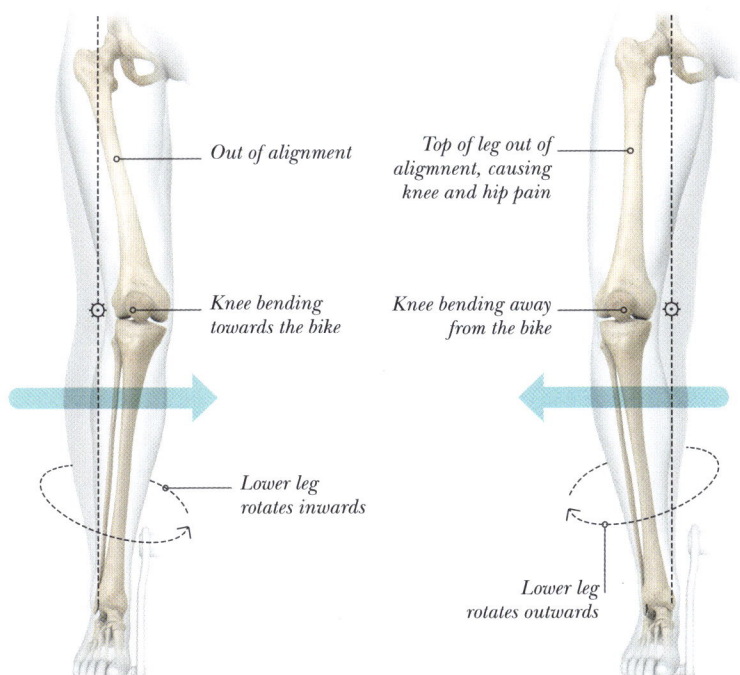

MEDIAL KNEE DRIFT
The knee moves in towards the bike (valgus).

OUTWARD FLARING
The knee moves out, away from the bike's midline relative to the foot (varus).

1 IN 3 cyclists will develop *knee pain* at some point

MONITORING BIOMECHANICS

Monitoring biomechanics in cycling entails a systematic examination of a cyclist's body movements and interactions with the bicycle. Instead of concentrating solely on external metrics such as power or speed, biomechanical monitoring aims to understand how power is generated and the effectiveness of movement patterns in different conditions.

WHY TRACK BIOMECHANICS?

Observing biometrics allows cyclists to improve pedal power and efficiency and detect movement asymmetries. It also helps evaluate fatigue-related changes in pedalling technique, make informed positional adjustments during bike fitting, and facilitates evidence-based rehabilitation following injury.

There are a number of tools and technologies that are used to assess a cyclist's biomechanics, including 3D motion capture systems for the precise assessment of joint kinematics and segment coordination analysis. High-speed video analysis for field-based assessments of knee tracking, pelvic stability, and general posture is also used. Further tools include instrumented pedals or cranks that provide kinetic data on torque, force vectors, and balance. Wearable IMUs (inertial measurement units) allow monitoring of body segment motion and EMG sensors provide neuromuscular assessment.

Field-based tools like pedal-based power meters with force analysis are available to buy for use by non-professional cyclists, enabling kinetic monitoring during regular training and competition.

Key variables in cycling biomechanics monitoring:

1. KINEMATIC VARIABLES (MOVEMENT PATTERNS)
Consistent, symmetrical movement patterns with minimal unnecessary lateral deviation are hallmarks of efficient kinematics.

Joint angles: hip, knee, and ankle angles throughout the crank cycle.

Knee tracking: medial-lateral movement of the knee relative to the pedal spindle.

Pelvic tilt and stability: excessive rocking can indicate poor core control or saddle mismatch.

Trunk angle: related to aerodynamic positioning and lumbar strain.

Foot angle and ankle movement (dorsiflexion/plantarflexion pattern).

2. KINETIC VARIABLES (FORCE PRODUCTION)
Advanced power meters and instrumented crank systems provide real-time data on force vectors, allowing kinetic monitoring outside laboratory settings.

Torque profiles: measurement of crank torque throughout the pedal stroke.

Effective force ratio (EFR): proportion of applied force that contributes directly to crank rotation.

Pedal balance (left/right symmetry): uneven force production can indicate fatigue, injury, or suboptimal fit.

Radial vs. tangential force components: higher tangential force improves efficiency.

3. MUSCLE ACTIVATION PATTERNS (NEUROMUSCULAR ACTIVITY)
Surface electromyography (EMG) can assess the timing, intensity, and coordination of key muscles used.

Timing: captures when a muscle turns on and off during the pedal stroke, giving insight into coordination and recruitment sequencing.

Duration of activation: how long a muscle remains active within each pedal cycle, revealing compensatory strategies or inefficient patterns.

Fatigue indicators: increased signal amplitude over time, delayed activation timing, and shifts in the frequency spectrum, reflecting changes in neuromuscular efficiency as fatigue develops.

INTERPRETING THE INFORMATION

Successful biomechanical monitoring is not just about data collection but about meaningful interpretation.

For a comprehensive veiw of a cyclist's condition, biomechanical data, such as that shown below, should be integrated with power, heart rate, and metrics (such as the rate of perceived exertion or RPE).

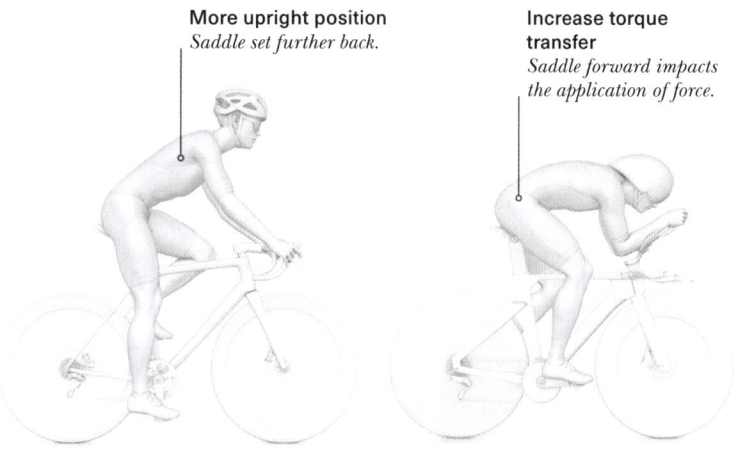

More upright position
Saddle set further back.

Increase torque transfer
Saddle forward impacts the application of force.

ROAD BIKE — TIME TRIAL BIKE

ANTHROPOMETRIC MEASURES	ROAD SETUP	TT SETUP
Height	174cm (5ft 8in)	174cm (5ft 8in)
Inside leg length	83cm (33in)	83cm (33in)
Arm length	59cm (28in)	59cm (28in)
Shoulder width	38cm (15in)	38cm (15in)
Cervical ROM (extension)	78 degrees	78 degrees
Hamstring flexibility	85 degree straight leg raise	85 degree straight leg raise

BIKE FIT PARAMETERS	ROAD SETUP	TT SETUP
Saddle height (bottom bracket to top)	76.2cm (30in)	75.5cm (29.7in)
Saddle setback	6.0cm (2.4in)	1.5cm (0.6in)
Saddle tilt	-2 degrees	-3.5 degrees
Crank length	172.5mmn (6.8in)	165mm (6.5in)
Handlebar drop	12cm (4.7in)	13.2cm (5.2in) to pads
Bar width/pad width	40cm (15.7in)	21cm (8.3in)
Stem/reach	110mm/39cm (4.3/15.4in)	Custom/47cm (18.5in)
Extension tilt	N/A	10 degrees upward

JOINT ANGLES	ROAD SETUP	TT SETUP
Knee angle at bottom dead centre (BDC)	145 degrees	146 degrees
Hip angle at top dead centre (TDC)	50 degrees	39 degrees
Ankle angle at bottom dead centre (BDC)	100 degrees	102 degrees
Torso angle	Around 40 degrees	Around 10 degrees
Pelvic tilt (anterior)	10 degrees	15 degrees
Shoulder angle	100 degrees	95 degrees
Elbow flexion	Around 20 degrees (hands on brakes)	Around 92 degrees (hands on aerobars)

BIKE POSITIONING

Bike fitting or positioning aligns the bike's mechanical demands with the cyclist's anatomy to improve comfort, efficiency, and performance, and prevent injuries. Whether aiming for power output, aerodynamic positions, or pain-free riding, a proper fit is essential for unlocking a cyclist's potential.

The foundation upon which successful cycling performance is built, bike fitting creates a seamless connection between cyclist and machine, allowing power to flow efficiently while minimizing stress on the body. An expert fit respects the individuality of each cyclist, balancing biomechanics, physiology, and riding demands to help them ride longer, faster, and with greater enjoyment.

THE CYCLIST
An effective bike fit starts with looking at the cyclist's discipline (for example, road cycling or mountain biking), experience, injuries, and goals. A good fit considers the cyclist's level, body, flexibility, core stability, previous injuries, leg length discrepancies, and foot structure. The fitter should try to find a balance of efficiency, comfort, and durability, rather than forcing an "ideal" position.

THE PHYSICAL ASSESSMENT
A physical assessment evaluates flexibility and core strength. It helps determine achievable posture and identify possible causes of pain. Restricted ankle dorsiflexion, for example, might explain knee issues or hip discomfort during pedalling.

EVALUATION
Once on the bike, the cyclist's posture and movement are analysed dynamically. This includes

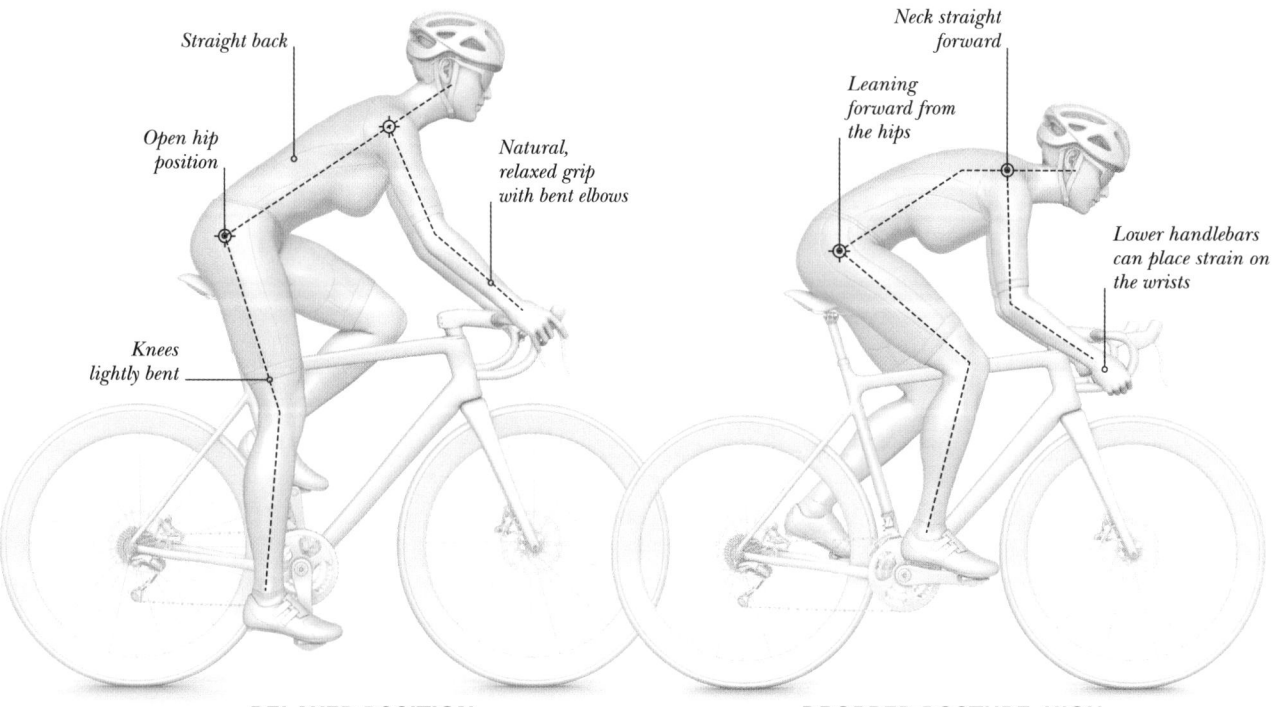

RELAXED POSITION

- Straight back
- Open hip position
- Natural, relaxed grip with bent elbows
- Knees lightly bent

DROPPED POSTURE: HIGH

- Neck straight forward
- Leaning forward from the hips
- Lower handlebars can place strain on the wrists

ANATOMY OF CYCLING | Bike Positioning

Proper bike fit can improve pedalling efficiency by up to 20%

observing joint angles at key points in the pedal stroke, monitoring knee tracking, assessing pelvic stability, and evaluating upper body posture. Technologies such as video analysis or 3D motion capture systems can assist this process, but the fitter's observations and the cyclist's feedback remain crucial to decision-making.

Adjustments are made to the key contact points of the bike: saddle height and fore-aft position, handlebar reach and drop, stem length and angle, and cleat positioning. Crank length may also be considered, particularly for riders with limited hip mobility or specific performance needs. Each adjustment is carefully evaluated to determine its impact on comfort, pedalling smoothness, and power delivery.

Throughout this process, cyclist feedback is essential. The goal is not simply to place the cyclist within theoretical joint angle ranges but to create a position that feels natural, balanced, and sustainable. Positions are often tested across a range of intensities and cadences to ensure that the fit remains effective when put to the test under real-world conditions.

PERFORMANCE AND HEALTH

A key principle of expert bike fitting is that the process is dynamic, not static. Fit positions should evolve alongside the cyclist's fitness, flexibility, and objectives. A position that works well during the early season may require refinement for peak race condition. Likewise, injury history or ongoing discomfort may necessitate ongoing adjustments.

The fit process should also integrate with strength and conditioning, rehabilitation, and technical skills development. For example, a cyclist with recurring lower back pain may benefit from targeted core strengthening alongside the fit.

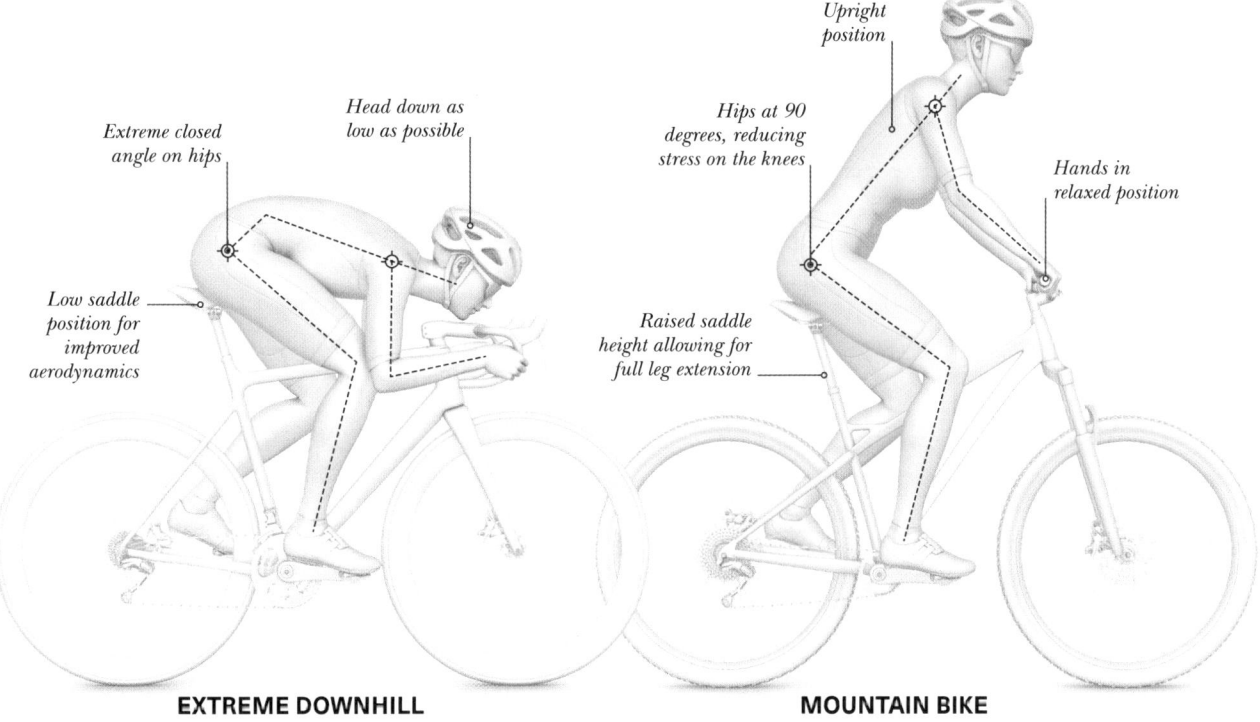

EXTREME DOWNHILL
- Extreme closed angle on hips
- Head down as low as possible
- Low saddle position for improved aerodynamics

MOUNTAIN BIKE
- Upright position
- Hips at 90 degrees, reducing stress on the knees
- Hands in relaxed position
- Raised saddle height allowing for full leg extension

FIT ADJUSTMENTS

The main aim of bike fitting is to optimize the three primary contact points: the saddle, handlebars, and pedals, so that the cyclist can produce power efficiently while minimizing discomfort and injury risk.

However, every adjustment comes with potential challenges and trade-offs, and understanding these interactions is critical to achieving an effective and sustainable fit.

HANDLEBAR REACH AND DROP

Handlebar reach and drop control the upper body posture, trunk angle, and weight distribution between the saddle and the hands. Excessive reach or drop can place undue strain on the cervical spine, shoulders, and wrists, leading to numbness, discomfort, or chronic neck and shoulder tension. Insufficient reach, on the other hand, can lead to cramped posture, reduced chest expansion, and compromised breathing efficiency.

The primary trade-off in reach and drop adjustments is between aerodynamics, comfort, and respiratory function. Time trialists and triathletes often push for aggressive, low positions for aerodynamic advantage, but such positions must be supported by adequate hip mobility, lumbar stability, and shoulder flexibility. Otherwise, performance may be compromised by discomfort, restricted breathing, or unsustainable posture.

FORE-AFT POSITION

Fore-aft saddle position influences hip engagement, balance between gluteal and quadriceps muscle recruitment, and weight distribution across the bike. A forward saddle shifts the cyclist towards the front of the bike, opening the hip angle and favouring quadriceps contribution but potentially overloading the knees. A rearward saddle supports glute activation and can improve climbing leverage but may close the hip excessively in aggressive aerodynamic positions.

The key trade-off here is between power output, muscle recruitment balance, and joint loading. Adjusting saddle position often means balancing gluteal muscle engagement against quadriceps muscle dominance while ensuring pelvic stability and knee alignment.

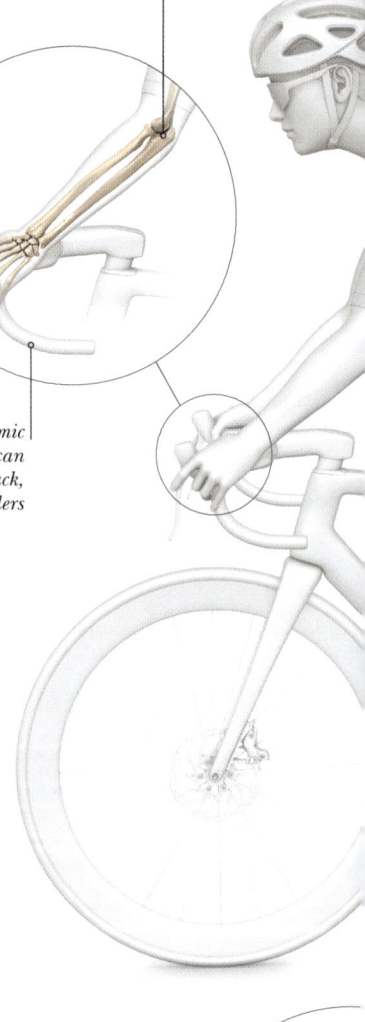

Handlebar reach allows rider to maintain a natural, relaxed grip with bent elbows, for stability and control

Handlebars
The height and reach of the handlebars influence the rider's posture and overall comfort. Amateur cyclists often favour a more upright position that prioritizes comfort over aerodynamics, making long rides more sustainable.

Lower handlebars offer aerodynamic benefits vital for racing, but they can place extra strain on the lower back, neck, and shoulders

Cleats
Cleat positioning is another important consideration, especially for cyclists using clipless pedals. Even small adjustments to cleat angle or position can significantly affect a rider's comfort and efficiency, particularly during long rides or intense efforts.

CYCLING ANATOMY | Fit Adjustments

Saddle

Saddle height is one of the most critical aspects of bike positioning as it directly impacts pedalling mechanics and power output, as well as comfort. A saddle that is too low can compromise power delivery and increase knee discomfort. By contrast, proper adjustment of the position of the saddle allows for balanced pedalling and prevents undue stress on the lower body.

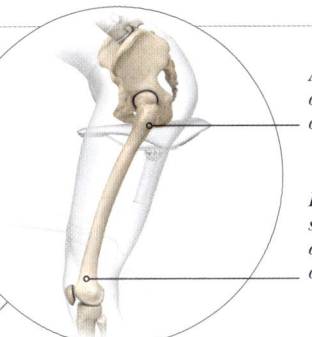

A high saddle can cause overextension and strain on hips and knees

Fore-aft position of the saddle determines weight distribution and alignment of knees relative to pedals

CRANK LENGTH

Longer cranks (see p.31) increase leverage and torque potential but also increase joint range demands, especially hip and knee flexion at top dead centre. This can exacerbate hip closure, restrict diaphragmatic breathing, and cause hip discomfort, especially in aero positions. Shorter cranks, while reducing torque per revolution, allow higher cadence potential, lower joint ranges, and may improve hip clearance and breathing in aggressive positions. The trade-off here is between leverage for torque production and joint mobility for cadence fluency and comfort.

CLEAT POSITION

The foot-pedal interface plays a crucial role in force transmission and knee tracking. Cleat positioning, fore-aft (relative to the metatarsal heads), medial-lateral alignment, and float, must support the cyclist's natural joint mechanics while maintaining stability.

Positioning the cleat too far forward increases leverage but raises calf strain, while moving it back may reduce calf engagement and enhance stability, particularly in time trial setups. Moving cleats from side to side and Q-factor (the distance between the outside edges of the crank arms) directly influence knee alignment. Improper positioning here is a common source of medial or lateral knee pain and can contribute to lack of symmetry in pedal position.

Align cleats so that the ball of the foot is over the pedal axle, to maximize power transfer and minimize discomfort in feet, knees, and hips

Common fit challenges

Several recurring issues tend to surface during bike fitting, each with its own set of solutions and compromises:

- **Knee pain** often relates to saddle height, fore-aft position, or cleat alignment. The challenge is finding the saddle position that balances power delivery with joint loading.

- **Lower back** discomfort frequently involves saddle height, hip mobility limitations, or excessive reach. Addressing core strength and lumbar flexibility often complements positional changes.

- **Saddle discomfort** or numbness can result from inappropriate saddle tilt, poor pelvic support, or excessive pressure on soft tissue areas. In some cases, equipment changes like saddle type or shorts chamois/padding are required, alongside positional adjustments.

- **Hand numbness and wrist pain** are usually signs of weight imbalance between saddle and handlebars, potentially requiring changes to reach, drop, or handlebar type.

39

COMMON INJURIES

Cycling, while considered a low-impact endurance sport, presents a unique set of injury risks driven by its repetitive movement patterns, fixed body positions, and high training volumes. Cycling is characterized by a high prevalence of overuse injuries, which often develop gradually due to mechanical inefficiencies, poor bike position, or training errors.

Crashes and acute injuries do occur, while chronic injuries like knee pain on longer rides or persistent lower back ache commonly derail training consistency and performance. Understanding the mechanisms behind these injuries and finding prevention strategies relies on biomechanics, bike position, tissue load management, and cyclist-specific factors such as mobility, strength, and anatomical structure. Prevention is not simply about avoiding injury but about creating a resilient cyclist who can train and perform consistently across seasons.

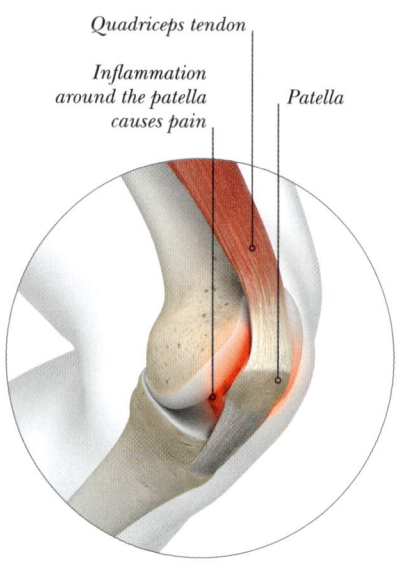

ANTERIOR-MEDIAL VIEW

KNEE PAIN

Among all overuse injuries, anterior knee pain, often linked to patellofemoral pain syndrome, is one of the most frequently encountered issues in cyclists.

The condition often results from the repetitive loading on the patellofemoral joint, where the kneecap is pulled off track or compression leads to irritation of the surrounding tissues. A saddle that is positioned too low or too far forward, poor cleat alignment, or insufficient control of the hip musculature can lead to misaligned knee tracking, amplifying stress at the joint.

Prevention hinges on improving the relationship between the hip, knee, and ankle through proper bike positioning and strength training. Saddle height and position should allow efficient knee extension without hyperflexion, and cleat alignment must respect the natural rotational profile of the rider's foot and leg. Strengthening the gluteus medius and maximus is important, as these muscles provide lateral control of the femur, reducing valgus drift (see p.33) and improving lower limb alignment.

ILIOTIBIAL BAND SYNDROME AND LATERAL KNEE ISSUES

Iliotibial band syndrome (ITBS) is an injury on the side of the knee caused by friction or compression of the iliotibial band. ITBS is not uncommon in cyclists, especially those with poor hip stability, excessive saddle height, or cleat misalignment that encourages lateral knee motion. Prevention comes from improving hip strength and control in the abductors and external rotators, while ensuring that the bike setup minimizes excessive lateral knee deviation.

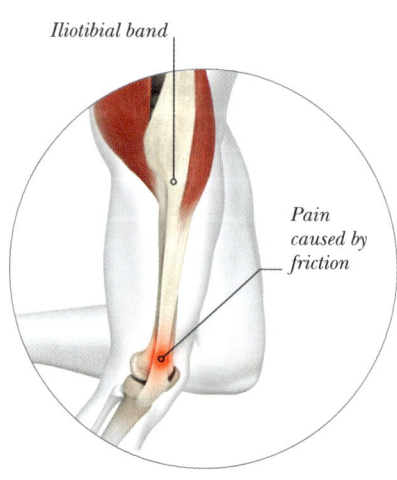

LATERAL VIEW

LOWER BACK PAIN

Lower back pain is one of the most commonly reported musculoskeletal issues in cyclists, largely due to prolonged static postures, flexed lumbar positioning, and insufficient core endurance.

The forward-leaning position required for aerodynamic efficiency places sustained tension on the muscles on the back of the body. These include the glutes, hamstrings, calves, and back muscles. Cyclists with tight hip flexors, weak gluteal muscles, or poor lumbar control are particularly vulnerable, as these deficiencies force the lower back to compensate to main stability.

Preventing back pain in cyclists requires a holistic approach that combines bike setup with off-the-bike conditioning. Handlebar reach and drop must allow the rider to maintain a neutral spine without over-reliance on passive joint structures. Mobility work on the hips and thoracic spine, paired with core stability exercises, helps build the capacity to hold posture without fatigue-induced compensation.

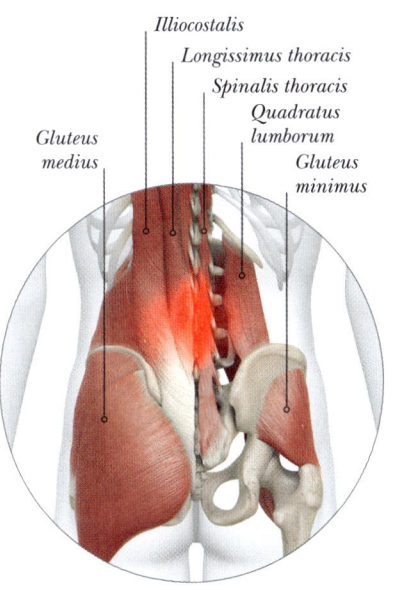

POSTERIOR VIEW

NECK AND SHOULDER PAIN

Neck and shoulder discomfort often occurs with prolonged time spent in positions where the cervical spine remains extended, and the shoulders bear a disproportionate share of body weight.

This type of pain is especially common in time trialists who maintain aggressive aerodynamic postures for extended periods. Overload of the upper trapezius and neck extensors, combined with insufficient scapular stability, leads to fatigue and discomfort.

Effective prevention involves a balance between positional adjustments and strengthening of the supporting musculature. Ensuring the handlebars are not excessively low or far away helps reduce strain on the upper body, while targeted strengthening of the scapular stabilizers, particularly the lower trapezius and serratus anterior, enhances postural endurance.

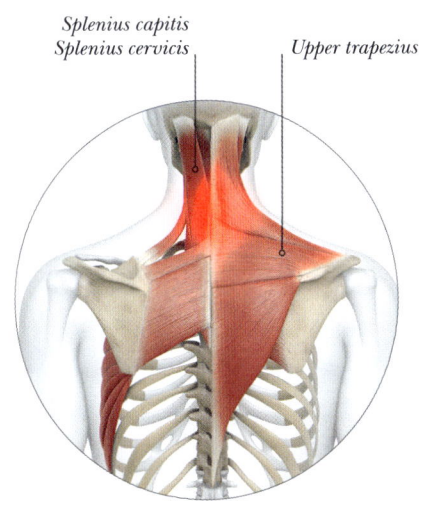

POSTERIOR VIEW

> *" Time spent in aggressive positions should be introduced progressively, giving the body time to adapt to the demands.*

HAND NUMBNESS AND CYCLIST'S PALSY

Compression of the ulnar nerve at the wrist, commonly known as cyclist's palsy, often results in numbness, tingling, or weakness in the fingers, especially the ring and little fingers.

The cause is typically sustained pressure on the hands due to improper weight distribution between saddle and handlebars, poor handlebar setup, or failure to vary hand positions during long rides.

To prevent this, cyclists must ensure that their reach and drop encourage a balanced posture where the core supports much of the body weight, not the hands alone. Padded gloves, high-quality bar tape, and frequent hand position changes are simple yet effective strategies. Scapular control and core endurance also play an indirect but important role, as they reduce the tendency to collapse onto the handlebars.

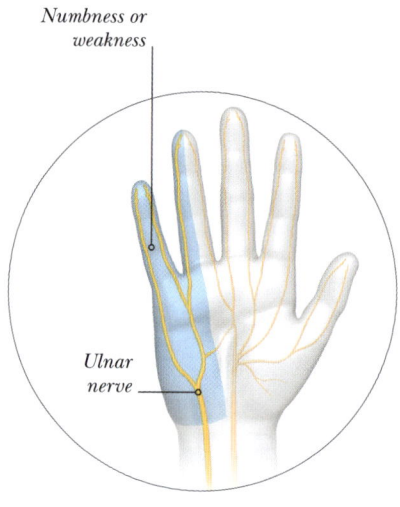

NERVE COMPRESSION

SADDLE SORES AND PERINEAL NUMBNESS

Saddle sores and perineal numbness are significant issues, especially for cyclists spending many hours in the saddle.

These problems arise from excessive pressure on soft tissue areas, poor saddle choice, suboptimal position, and inadequate hygiene. A saddle that is too narrow or too wide, combined with poor pelvic stability, increases the risk of chafing, skin breakdown, and nerve compression.

Prevention involves matching saddle shape and width to the cyclist's pelvic anatomy, ensuring proper saddle tilt (usually slightly nose-down), and allowing the cyclist to move slightly on the saddle without being locked into one rigid position. Standing periodically during longer rides to relieve pressure, combined with appropriate shorts, chamois cream, and good hygiene, further reduces the risk of saddle sores.

 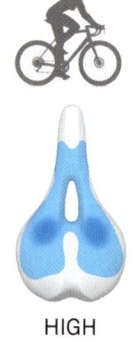   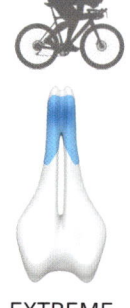

RELAXED　　HIGH　　DROPPED　　EXTREME

SOFT TISSUE AREAS
Depending on the position of the cyclist, blood flow to soft tissue areas can be reduced, causing numbness and sores if held over a long period of time.

ANATOMY OF CYCLING | Common Injuries

ACHILLES TENDINOPATHY AND FOOT ISSUES

Achilles tendinopathy in cycling is typically a product of poor ankle control, calf tightness, or biomechanical inefficiencies such as a saddle set too high or cleats positioned too far forward.

Overextension at the ankle increases tensile load on the Achilles tendon, particularly during high-torque scenarios like climbing. Addressing this requires a careful balance between equipment setup and muscular conditioning. Proper calf flexibility, eccentric strength, and cleat placement that respects the cyclist's natural foot mechanics all contribute to reducing strain.

Foot numbness or "hot spots" under the forefoot often result from local nerve compression due to shoe fit, cleat position, or insufficient foot support. Ensuring shoes accommodate the foot's width, cleats are positioned to reduce excessive pressure, and arch support is adequate are all critical components of prevention.

Training to avoid injury

While biomechanics and equipment setup are important in reducing injury risk, training load errors remain one of the most significant contributors to overuse injuries. Rapid increases in volume, intensity, or frequency often exceed the body's adaptive capacity, leading to tissue overload and breakdown. Effective injury prevention requires structured training with gradual increase in loading, adequate recovery, and monitoring of both objective metrics like power output or heart rate and subjective markers such as perceived exertion, sleep quality, and muscle soreness.

Strength and conditioning play an equally vital role. Cyclists benefit from off-the-bike work that targets the areas cycling alone does not adequately develop, particularly the gluteal muscles, core, scapular stabilizers, and posterior chain (the muscles on the back of the body). These interventions improve movement quality, enhance force transfer, and build tissue resilience, reducing the risk of breakdown under repetitive load.

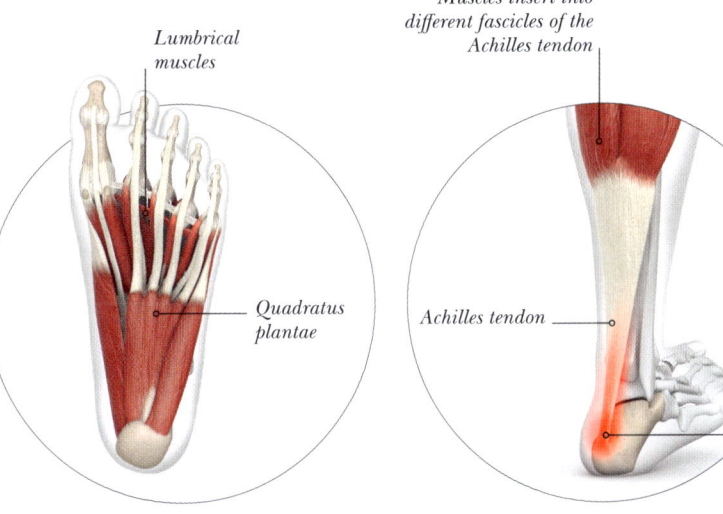

Lumbrical muscles

Quadratus plantae

INFERIOR VIEW

Muscles insert into different fascicles of the Achilles tendon

Achilles tendon

Pain is often felt along the narrow part of the tendon and the heel

POSTERIOR–LATERAL VIEW

> *Rapid increases in volume, intensity, or frequency often lead to tissue overload.*

PHYSIOLOGY OF **CYCLING**

Cycling physiology is about balancing energy supply and demand over various intensities and durations. It involves integrating multiple physiological qualities to address diverse cycling challenges. Whether climbing, bridging to a breakaway, or sprinting, performance reflects how well these systems are trained, adapted, and synchronized.

PHYSIOLOGICAL DEMANDS

Cycling is an endurance sport that places many demands on the body. To perform well, different body systems must work closely together over a range of speeds, terrains, and distances, which makes the sport both challenging and complex. All cyclists benefit from building strong aerobic fitness, good anaerobic power, muscular strength and endurance, and efficient movement.

PERFORMANCE VARIABLES

A helpful framework for understanding the physical components of cycling is provided by three key variables: VO_2 max, anaerobic capacity, and movement economy. Together, these three components explain endurance cycling performance and help guide the structure of training plans aimed at maximizing sustainable power and race efficiency.

VO_2 MAX

The maximum amount of oxygen a person can use during intense physical activity is known as VO_2 max. It's a key indicator of cardiovascular fitness and aerobic endurance and is measured in millilitres of oxygen consumed per kilogram of body weight per minute (ml/kg/min). Cyclists must be able to perform at a high percentage of VO_2 max for long periods without accumulating lactate acid to levels that impair muscle function (see p.57). Elite road cyclists typically have VO_2 max values of between 75–85ml/kg/min for males and 60–75ml/kg/min for females, indicating their ability to transport and use oxygen efficiently. In contrast, an average male has a VO_2 max of 45 and an average female 38–41.

ANAEROBIC CAPACITY

Anaerobic capacity is the amount of energy supplied by non-oxygen dependent pathways when the demand for energy exceeds the rate of aerobic supply (see pp.48–49), such as during accelerations, attacks, or the start of a race. Research shows that muscles' ability to oxidize fuels to produce energy is one of the most important factors in the power a cyclist can sustain over time.

EFFICIENCY

In cycling, efficiency refers to the amount of energy a cyclist uses to produce a given power output or maintain a certain speed. In practical terms, it describes how efficiently a cyclist converts metabolic energy (oxygen and calories) into mechanical power on the bike.

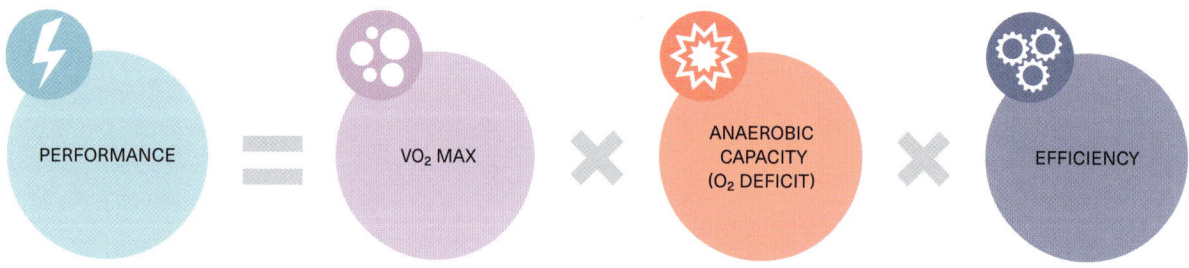

PERFORMANCE = VO_2 MAX × ANAEROBIC CAPACITY (O_2 DEFICIT) × EFFICIENCY

TRANSITIONS, ENDURANCE, AND RECOVERY

As race demands are in constant flux, cyclists transition between aerobic and anaerobic systems (see pp.48–49). Sudden accelerations, attacks, climbs, sprints, and breakaways need recurrent high-intensity efforts that exceed the lactate threshold and are primarily fuelled by the glycolytic energy system.

Recovery is just as important as the high-intensity efforts and relies on efficient lactate acid clearance and replenishment of the energy compound phosphocreatine (Pcr). Inadequate recovery can lead to a rapid decline in a cyclist's performance during races that involve repeated surges, such as criteriums or the final stages of a road race.

BUILDING ENDURANCE

Muscular endurance is fundamental to a cyclist's ability to sustain power output over extended distances. The repetitive nature of cycling necessitates physiological adaptations such as increased mitochondrial density, enhanced capillarization of muscle fibres, and improved fat oxidation (see pp.60–61). These adaptations not only delay glycogen depletion but also help efficient pedalling across numerous pedal strokes. Neuromuscular coordination is also crucial; elite cyclists optimize torque (the force used to push the pedals) application throughout the pedal stroke, minimizing energy loss and improving efficiency.

THERMOREGULATION AND HYDRATION

Maintaining thermoregulation and hydration management (see pp.74–75) is critical to supporting the body. Extended exposure to elevated temperatures without enough cooling or fluid intake can degrade cardiovascular function, hasten fatigue onset, and diminish performance. Equally significant is psychological endurance – the capacity to tolerate discomfort, maintain concentration amid fatigue, and make strategic decisions under physical stress.

Energy use in different cycling disciplines

DISCIPLINE	PRIMARY ENERGY SYSTEMS	KEY DEMANDS
Road cycling	Aerobic and anaerobic	Endurance, repeated efforts, recovery, efficiency
Track sprint	Phosphagen and glycolytic	Explosive power, peak force, fast-twitch recruitment
Track pursuit	Aerobic and anaerobic	Sustained high power, aerodynamics, pacing
Mountain biking (XCO)	Aerobic and anaerobic	Variable intensity, technical handling, strength endurance
Cyclocross	Aerobic and anaerobic	Short surges, lactate tolerance, technical skills
Time trial	Aerobic	Threshold power, sustained effort, aerodynamics
BMX	Aerobic and anaerobic	Sprint power, explosiveness, technical jumping
Gravel and ultra-endurance	Aerobic	Fuelling efficiency, metabolic flexibility, resilience

> *Each cycling discipline uses different levels of energy from the anaerobic and aerobic systems.*

ENERGY SYSTEMS

The foundation of cycling performance lies in the human body's ability to produce and use energy efficiently. To meet the demands of cycling – from short, explosive sprints to long, steady endurance rides – it relies on three distinct energy systems that work together to produce the fuel your muscles need.

HOW ENERGY IS RELEASED

The molecule adenosine triphosphate (ATP) stores, transports, and releases the energy used for muscle contractions. The body has three ways of accessing ATP, or three energy systems. The system it draws on depends on the duration and intensity of the exercise.

> The *oxidative system* has the **highest ATP yield**, but is the *slowest to activate*.

Phosphagen system

The phosphagen or anaerobic alactic (no lactate production) energy system instigates the breakdown of creatine phosphate, a high-energy molecule found in muscles, to provide immediate energy (ATP) for a duration of around 10 seconds. This system is used during short, explosive activities such as sprinting.

Glycolytic system

The glycolytic or anaerobic lactic system causes the breakdown of glucose from muscle glycogen or circulating blood glucose to produce ATP rapidly without the need for oxygen. This system can produce energy for up to two minutes. However, it is less efficient, and lactate acid is a by-product, which contributes to fatigue (see p.57).

Oxidative system

Commonly known as the aerobic system, the oxidative system provides a more efficient but slower production of ATP. This system is the main metabolic pathway for energy production from carbohydrate, fat, and protein (to a lesser extent) for prolonged endurance activities. The majority of ATP production occurs in the mitochondria of cells.

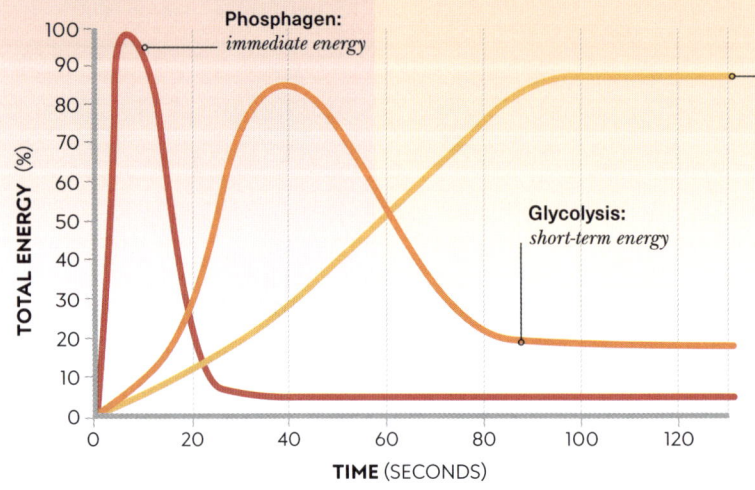

ENERGY SYSTEMS

These energy systems work simultaneously and are integrated. The phosphagen system meets immediate needs, glycolysis takes over as exercise extends, and the oxidative system dominates under prolonged lower-intensity conditions.

PHYSIOLOGY OF CYCLING | Energy Systems

WHICH SYSTEM?

The aerobic system is the most critical energy pathway for cycling performance, as nearly all events in the sport last longer than one minute. It is highly fatigue-resistant and suited for sustaining energy expenditure at low to moderate intensities. However, cycling is subject to frequent and significant fluctuations in intensity. These are influenced by the sport's unique dynamics, such as periods of freewheeling and the drafting effect (see pp.100–101), which allow for recovery at low or zero intensity. Consequently, cycling performance is characterized by more pronounced intensity variability than other endurance sports like swimming or running, where the effort is typically more consistent. This is why all the energy systems play a critical role in performance. For instance, road sprinters often possess a highly developed phosphagen system, enabling them to deliver powerful 10-second sprints, even after hours of racing. Similarly, mountain bikers rely on their glycolytic capacity to meet the demands of cross-country events, which require short, explosive efforts to navigate technical climbs and overcome obstacles.

MITOCHONDRIA AND EXERCISE

Mitochondria are small, membrane-bound organelles found in nearly all human cells and are often referred to as the cell's "powerhouses". Their primary role is to produce energy in the form of adenosine triphosphate (ATP) through a process called oxidative phosphorylation, which occurs across the inner mitochondrial membrane. To do this, mitochondria use oxygen and the breakdown products of carbohydrates, fats, and proteins – making them essential for sustained, aerobic energy production.

Mitochondria also clear, recycle, and use lactate, playing a key role in regulating the cellular redox balance – the equilibrium between oxidation and reduction reactions. This redox control is vital for normal metabolism, cell signalling, and protection against oxidative damage.

ATP and phosphocreatine shuttles help link mitochondrial ATP production to myofibrillar ATP consumption, where energy is used for muscle contraction.

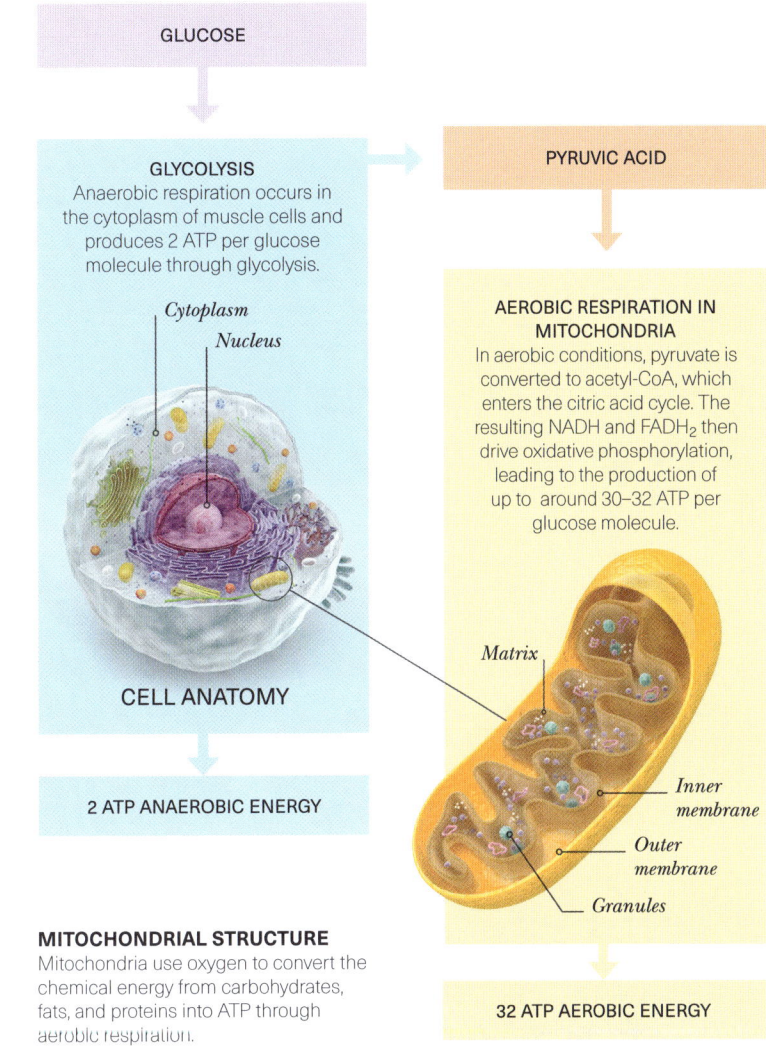

GLUCOSE

GLYCOLYSIS
Anaerobic respiration occurs in the cytoplasm of muscle cells and produces 2 ATP per glucose molecule through glycolysis.

Cytoplasm
Nucleus

CELL ANATOMY

2 ATP ANAEROBIC ENERGY

PYRUVIC ACID

AEROBIC RESPIRATION IN MITOCHONDRIA
In aerobic conditions, pyruvate is converted to acetyl-CoA, which enters the citric acid cycle. The resulting NADH and $FADH_2$ then drive oxidative phosphorylation, leading to the production of up to around 30–32 ATP per glucose molecule.

Matrix
Inner membrane
Outer membrane
Granules

32 ATP AEROBIC ENERGY

MITOCHONDRIAL STRUCTURE
Mitochondria use oxygen to convert the chemical energy from carbohydrates, fats, and proteins into ATP through aerobic respiration.

ENERGY DURATION

Cycling performance centres on the ability to produce power over varying durations, from a 5-second sprint to a 5-hour race. Underpinning this ability is the body's capacity to generate ATP through the three primary energy systems.

UNDERSTANDING THE POWER–DURATION CURVE

The three energy systems don't function in isolation but work together to sustain muscular work. Their interaction shapes what is known as the power–duration curve, a foundational concept in endurance sports physiology and performance modelling. The power–duration curve is a graphical representation of the maximal average power output an individual cyclist can sustain over various durations.

SHORT DURATION: 0–30 SECONDS

In efforts lasting up to about 30 seconds, the phosphagen (ATP-PCr) system is the dominant energy source. Cyclists can produce extremely high power outputs – over 1200 W in elite sprinters – but only for a short duration. This limitation is due to the rapid depletion of phosphocreatine stores in muscle. This system is critical for maximal, explosive efforts such as:

- Track cycling sprints
- Road race accelerations
- Final 200m of a sprint finish

On the curve, this segment represents the highest point on the y-axis and the steepest part of the decline.

MID DURATION: 30 SECONDS–2 MINUTES

As phosphocreatine depletes, anaerobic glycolysis becomes the main energy source. Glucose is broken down without oxygen, producing ATP along with lactate and hydrogen ions. The buildup of hydrogen ions lowers muscle pH, impairs enzymes, and contributes to fatigue.

- Repeated race surges
- Short, steep climbs
- 1-minute efforts in criteriums

This section of the power curve shows a continued drop in power but still reflects high anaerobic contribution.

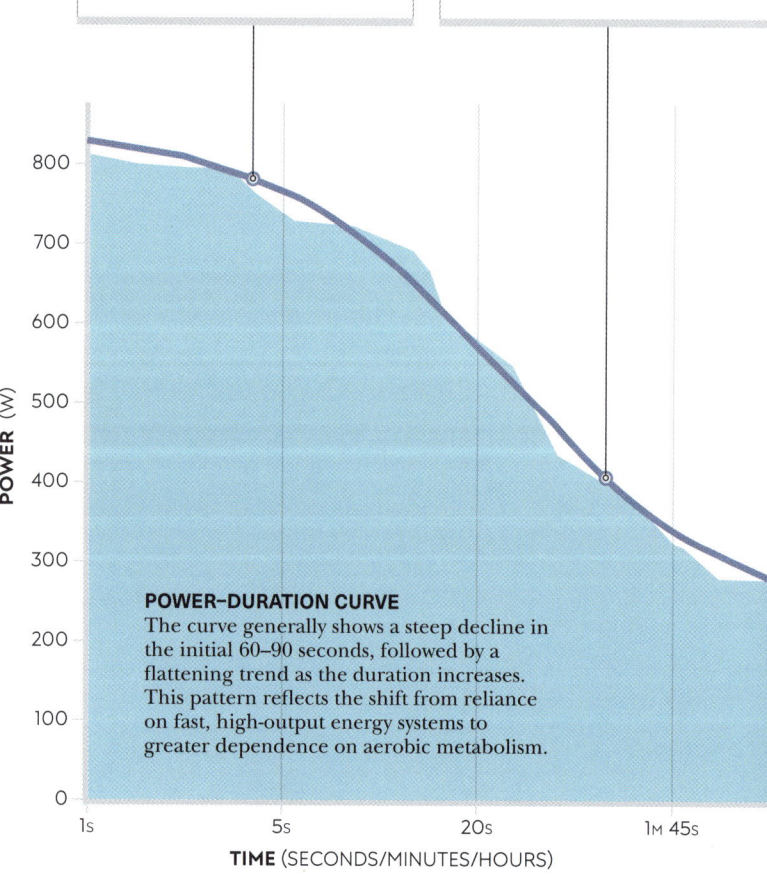

KEY
— Power–duration curve
▨ Mean maximal power curve

POWER–DURATION CURVE
The curve generally shows a steep decline in the initial 60–90 seconds, followed by a flattening trend as the duration increases. This pattern reflects the shift from reliance on fast, high-output energy systems to greater dependence on aerobic metabolism.

LONGER DURATION: 2 MINUTES–60 MINUTES PLUS

For efforts lasting several minutes to hours, the oxidative system is the main energy source. Carbohydrates and fats are metabolized in mitochondria using oxygen to produce ATP efficiently, but more slowly. Performance depends on VO_2 max, lactate threshold, and muscular efficiency. Key durations and physiological markers include:

- 5-minute power: VO_2 max-dominant
- 1-hour power: reflects aerobic endurance and fatigue resistance

These regions of the curve flatten progressively, reflecting a shift towards sustainable, lower-intensity aerobic energy production.

CRITICAL POWER (CP) AND W'

The Critical Power (CP) model provides a practical framework to measure a cyclist's performance capacity across the energy systems.

This model divides power output into two components:

- **Critical Power (CP):** The highest power a cyclist can sustain over a prolonged period (about 60 minutes) without a continual rise in fatigue markers. It represents the limit of aerobic energy supply and is closely related to functional threshold power (FTP, see p.57). Riding at or below CP draws primarily on aerobic metabolism.
- **W' (W prime):** Often referred to as anaerobic capacity, W stands for "work" and W' is the finite amount of work that can be performed above CP. W' is composed of energy from the phosphagen and glycolytic systems. It is depleted during high-intensity bursts and recharged during CP.

For example, a cyclist attacking from the peloton uses W' to surge above CP. If the attack fails and they continue riding at threshold, recovery of W' is slow or absent. A cyclist with a high W' can absorb and respond to repeated surges; one with a high CP can sustain long, hard efforts with less fatigue.

MEASURE OF ANAEROBIC CAPACITY

W' can be used to measure a cyclist's anaerobic capacity at high intensities over the lactate threshold (LT, see p.57).

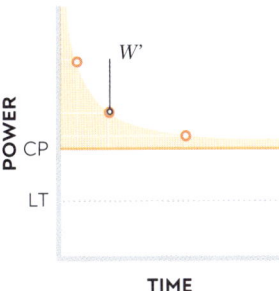

TRAINING IMPLICATIONS

Understanding how energy systems map on the power–duration curve allows cyclists and coaches to target specific adaptations:

- **Sprint training** develops neuromuscular and phosphagen system capacity.
- **Anaerobic intervals** (for example, 30 seconds–2 minutes) enhance glycolytic power.
- **Threshold intervals** (for example 8–20 minutes) raise FTP and lactate clearance (see p.56).
- **Endurance rides** promote mitochondrial biogenesis, fat oxidation, and glycogen sparing (see p.60).

Changes in curve shape over time can indicate training effect.

- A rise in short-duration power suggests improved anaerobic capacity.
- A flatter tail or higher plateau at long durations reflects improved aerobic conditioning.

RAPID TRANSITIONS

Cyclists rarely ride at steady intensities. A well-trained cyclist will have a broad, durable power–duration curve, enabling high outputs across time domains. This allows them to race effectively, recover quickly from efforts, and repeat decisive moves.

THE CARDIORESPIRATORY SYSTEM

Cycling imposes complex demands on the cardiorespiratory system, which ensures the delivery of oxygen to active muscles, the removal of carbon dioxide and metabolic by-products, as well as the regulation of body temperature.

At the core of this system is the oxygen transport cascade – the process by which oxygen is transferred from the atmosphere to the mitochondria of active muscle fibres. The sequence begins with pulmonary ventilation, during which the frequency and depth of breathing increases in response to exercise.

Elite endurance cyclists may exceed **200ML** of blood per beat during **intense** efforts.

INCREASE IN OXYGEN
Following pulmonary ventilation, more oxygen enters the lungs. It diffuses across the lung's alveolar-capillary membrane into the bloodstream, where it binds to haemoglobin. The cardiovascular system then enhances cardiac output to deliver this oxygen-enriched blood to peripheral tissues. Ultimately, at the muscular level, oxygen diffuses from capillaries into muscle cells and is used in the mitochondria for aerobic metabolism.

BREATHING
During cycling, the frequency (ventilation rate) and depth of breathing (tidal volume) increases dramatically, allowing more oxygen to enter the lungs. While a normal resting rate of volume of air moved into and out of the lungs per minute is 60–100, in elite athletes, it may exceed 150–200 litres per minute. Certain aerodynamic positions or high-intensity efforts can reduce respiratory muscle function. When this happens, highly trained endurance athletes may experience a phenomenon called exercise-induced arterial hypoxemia in which blood oxygen saturation declines due to restrictions in diffusion or a mismatch in the amount of air and blood reaching the alveoli.

CARDIOVASCULAR RESPONSE
As exercise intensity rises, heart rate and stroke volumes (the amount of blood pumped out of the heart per beat) increase and can reach 35–40 litres per minute in elite cyclists. This levels out before maximum intensity is reached – further increases in cardiac output are driven mainly by rising heart rate. Trained cyclists typically have high stroke volumes and low resting heart rates (bradycardia).

IMPROVING OXYGEN UPTAKE
Oxygen extraction at the muscle level is crucial as cycling intensity increases. Endurance training improves oxygen uptake through increased mitochondrial density, muscle capillarization, and myoglobin content in skeletal muscle, making oxygen use more efficient and delaying fatigue.

The fluctuating demands on the cardiorespiratory system during racing require rapid cardiovascular and pulmonary adjustments. Elite cyclists are able to adapt to these needs over time.

Assessing cardio fitness

Measuring cardiorespiratory fitness in cyclists involves assessing the peak oxygen uptake (VO_2 max, see p.46), changes in breathing rate and volume (see pp.56–57), and the rate at which the body consumes oxygen during exercise. VO_2 max indicates the upper limit of the cardiovascular system and is closely linked with performance in prolonged climbs and time trials. However, shifts in breathing below maximum intensity, such as the first and second ventilatory thresholds (see pp.56–57), often provide more practical insights into sustainable performance. These thresholds denote the points where breathing increases disproportionately to workload, usually in response to lactate accumulation and acid-base disturbances (see pp.56–57).

PHYSIOLOGY OF CYCLING | The Cardiorespiratory System

HEART AND CIRCULATION
Arteries (red) carry oxygenated blood away from the heart, while veins (blue) carry de-oxygenated blood towards the heart. This is reversed in the pulmonary loop of the circulatory system, which connects the heart and lungs.

HEAD AND UPPER BODY

Veins
Return de-oxygenated blood from the head and upper body to the heart

Arteries
Deliver oxygenated blood to the upper body

RIGHT LUNG

HEART

LEFT LUNG

Pulmonary artery
Delivers de-oxygenated blood to the lungs for expulsion of carbon dioxide

Pulmonary vein
Carries oxygenated blood from the lungs to the heart for circulation

LIVER

Arteries
Deliver oxygenated blood to the lower body

Veins
Return de-oxygenated blood from the legs to the heart

GASTRO-INTESTINAL TRACT

Capillaries
Oxygen diffuses into tissues in exchange for carbon dioxide

LOWER BODY

Artery wall
Thick muscular wall changes diameter to regulate blood flow

Vein valve
One-way valves prevent backflow

Vein
Veins carry de-oxygenated blood from working muscles to the heart and lungs for the removal of carbon dioxide.

Capillary
Capillary networks connect arteries and veins with tissue cells. This is where the exchange of oxygen for waste products takes place.

Artery
Arteries carry oxygen-rich blood from the heart and lungs to working muscles.

Training adaptations
Training causes physical adaptations that result in improvements in the energy systems that power movement.

Aerobic training
This improves the body's efficiency in aerobic cell respiration, leading to enhanced aerobic endurance and VO_2 max. Adaptations result in:

- **lactate accumulation** occurring at higher exercise intensities
- faster **lactate clearance**
- increased **stroke volume**
- increased **blood volume**
- greater **red blood cell volume**, improving oxygen transport
- Increased **capillarization of muscles** (see p.49)
- increased **mitochondrial density** (see p.49), allowing for improved aerobic cell respiration
- elevated **oxidative enzyme** activity, improving mitochondrial efficiency
- improved **efficiency of existing capillaries**
- improved **blood redistribution**
- increased size of **slow-twitch muscle fibres** (see p.54)
- Better blood flow redistribution to working muscles.

Anaerobic training
This increases your body's ability to tolerate and clear blood lactate and also raises your lactate threshold (see p.57). Adaptations result in:

- increased **muscular strength**
- improved **mechanical efficiency**
- increased **muscle oxidative capacity**
- increased **muscle buffering capacity** (enabling muscles to withstand the build-up of acidity produced during high-intensity exercise
- improved **lactate-clearance capacity**.

MUSCLE FIBRE TYPES

Cycling performance is influenced not only by cardiovascular fitness and metabolic efficiency, but also by the skeletal muscles responsible for generating force. Muscle fibre types, which vary in contractile speed, force production, fatigue resistance, and metabolic pathways, form the foundation of muscular performance.

The two main types of muscle fibres are Type I and Type II. This classification reflects a continuum, rather than distinct categories, with further hybrid fibres (such as Type IIx) also present, particularly in recreationally trained or transitioning athletes.

SLOW-TWITCH OXIDATIVE MUSCLE FIBRES

Type I fibres have a slow contraction speed, produce low force output, and are highly resistant to fatigue. These fibres are densely populated with mitochondria, display a high capillary density, and contain a substantial amount of myoglobin, which increases their ability to use oxygen for ATP production through aerobic metabolism. They are used for endurance activities and play a predominant role in sustaining low- to moderate-intensity efforts.

In cycling, Type I fibres are extensively used during long climbs, moderate-intensity rides, and endurance training. Their capacity to maintain force with minimal fatigue makes them essential for stage racers, time trialists, and domestiques who engage in prolonged sub-threshold efforts. Additionally, these fibres exhibit higher efficiency in oxidizing fats, which contributes to glycogen saving and improved metabolic resilience during extended competition times.

FAST-TWITCH OXIDATIVE-GLYCOLYTIC MUSCLE FIBRES

Type IIa fibres represent a hybrid between endurance and power. They have faster contraction speeds than Type I, moderate fatigue resistance, and can use both aerobic and anaerobic metabolism. They have a lower density of mitochondria and capillaries than Type I fibres but also maintain significant glycogen and phosphocreatine stores for high-intensity output.

These fibres are crucial when both speed and endurance are

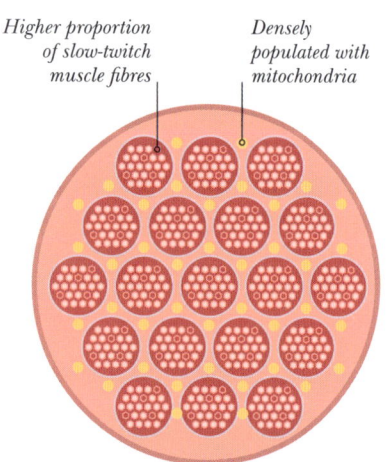

Higher proportion of slow-twitch muscle fibres — *Densely populated with mitochondria*

Muscle with characteristics of Type I: slow-twitch oxidative
Suited for endurance

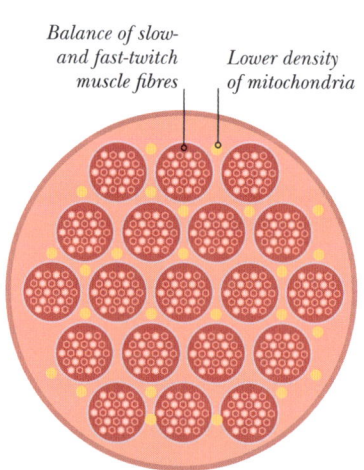

Balance of slow- and fast-twitch muscle fibres — *Lower density of mitochondria*

Muscle with characteristics of Type IIa: fast-twitch oxidative-glycolytic
Suited for medium-distance events

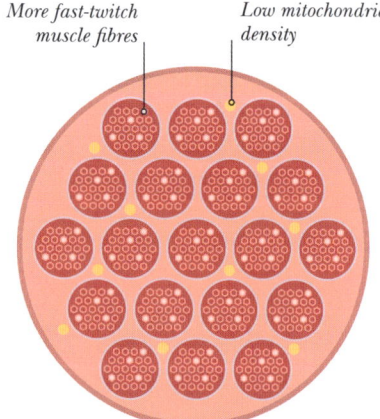

More fast-twitch muscle fibres — *Low mitochondrial density*

Muscle with characteristics of Type IIx: fast-twitch glycolytic
Suited for sprinting

PHYSIOLOGY OF CYCLING | Muscle Fibre Types

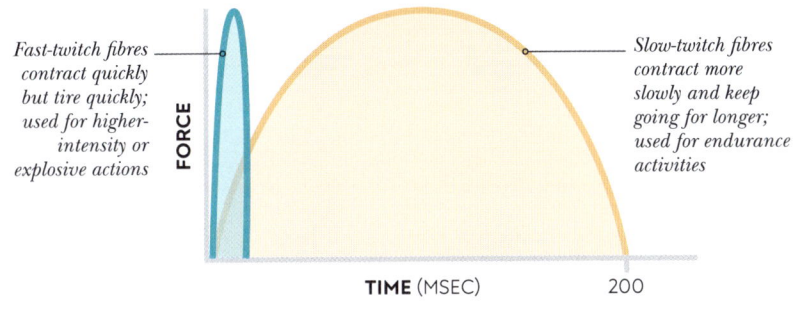

Slow- and fast-twitch muscle fibres

Your nervous system automatically recruits the appropriate muscle fibre types based on the demands of the exercise. Most skeletal muscles contain a mix of slow-twitch (Type I) and fast-twitch (Type II) fibres, enabling a wide range of movements across different intensities and durations.

Fast-twitch fibres contract quickly but tire quickly; used for higher-intensity or explosive actions

Slow-twitch fibres contract more slowly and keep going for longer; used for endurance activities

HOW SLOW- AND FAST-TWITCH MUSCLES COMPARE

---

needed, such as sustained efforts, accelerations out of corners during criterium races, or repeated bursts during breakaways. In trained cyclists, Type IIa fibres can develop aerobic characteristics and enhance fatigue resistance through endurance and interval training. These fibres are engaged during prolonged time trials or strenuous climbing efforts when Type I fibres begin to tire or sustained force production is required under duress.

FAST-TWITCH GLYCOLYTIC

Type IIx fibres are the most powerful and fastest-contracting, but also the most fatiguable. They have low mitochondrial density, rely primarily on anaerobic glycolysis and the ATP-PCr system, and are designed for explosive, short-duration efforts. They are suited for rapid, maximal force production, such as sprinting or attacking on a steep incline. While essential for peak power outputs, Type IIx fibres are minimally active during endurance riding. However, in track sprinters and road sprinters, they play a critical role in jump acceleration, final sprints, and short, all-out bursts. With endurance training, Type IIx fibres can convert to Type IIa, demonstrating the body's prioritization of fatigue resistance over maximal output when exposed to repeated submaximal demands. These fibres also require consistent high-intensity or resistance-based work to maintain their functional role.

TRAINING

Type IIa fibres are responsive to training, capable of increasing mitochondrial content, oxidative enzyme activity, and fatigue resistance. Through training, cyclists can improve the metabolic efficiency of their fast-twitch fibres while maintaining their ability to generate power. Sprint interval training, neuromuscular torque development, and strength training help maintain the recruitment capacity of larger muscles.

MUSCLE FIBRE USE

During lower-intensity exercise, type I fibres are recruited first, with type II units activated as demand increases. High-intensity intervals, sprint efforts, and resistance training stimulate high-threshold fibres, while low-intensity volume enhances oxidative capacity in both slow- and fast-twitch fibres. A cyclist climbing at threshold for 30 minutes may rely on Type I and IIa fibres, but if a decisive acceleration occurs in the final kilometre, Type IIx fibres must be recruited.

> *Fast-twitch muscle fibres are **ideal** for short bursts of power*

KEY
- Mitochondria
- Fast-twitch muscle fibres
- Slow-twitch muscle fibres

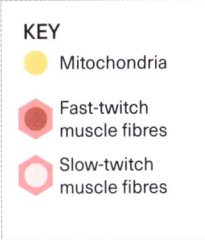

FAST AND SLOW
Bundles of muscles called fascicles contain varying proportions of slow-twitch and fast-twitch fibres, which are suited for different types of activity.

VENTILATORY THRESHOLDS

Cycling performance is linked to the body's ability to produce energy aerobically while managing fatigue-related byproducts. Key concepts in this process include ventilatory thresholds, blood lactate dynamics, and functional threshold power (FTP).

PHYSIOLOGICAL MARKERS

Ventilatory thresholds mark key points where breathing increases more rapidly during intense exercise to meet rising oxygen demand and remove extra carbon dioxide from metabolism and acid buffering.

50–70% of VO_2 max is the point at which **most cyclists** experience VT1.

As exercise intensity increases, the body requires more oxygen to produce energy and generates more carbon dioxide – both from aerobic metabolism and from buffering acids during anaerobic efforts. Ventilatory markers offer insight into a cyclist's endurance, metabolic efficiency, and ability to sustain high power over time. Understanding these markers is essential for guiding training intensity, evaluating performance, and planning race pacing strategies in cycling.

Ventilatory Threshold 1
Often referred to as the aerobic threshold, VT1 is the point at which ventilation (the mechanical act of breathing) increases disproportionately relative to oxygen uptake. This threshold coincides with the onset of blood lactate accumulation (LT1). Below this intensity, energy is primarily supplied aerobically, with fat oxidation being prominent. VT1 is important for long endurance rides as it sets the upper limit for sustainable low-intensity work that encourages aerobic adaptation without significant fatigue.

Ventilatory Threshold 2
Often referred to as the anaerobic threshold, this is a higher intensity marker than VT1 and happens when ventilation increases rapidly and disproportionately in relation to carbon dioxide production. This threshold closely corresponds with Lactate Threshold 2 (LT2, see right). Beyond this point, metabolic acidosis and fatigue increase markedly, limiting the duration for which exercise can be sustained.

VT1 AND VT2
Ventilatory thresholds serve as non-invasive indicators of metabolic events and are used in both laboratory tests and field evaluations. Recognizing VT1 and VT2 helps in establishing training zones and enables coaches to design sessions aimed at specific physiological objectives, like improving fat metabolism or boosting lactate clearance capacity.

ZONE 1 — LOW INTENSITY
ZONE 2 — MODERATE INTENSITY
ZONE 3 — HIGH INTENSITY

LACTATE THRESHOLD

Lactate is continually produced even at rest, but at low intensities production and clearance are balanced. As intensity increases, Type II muscle fibres are recruited, and lactate production rises.

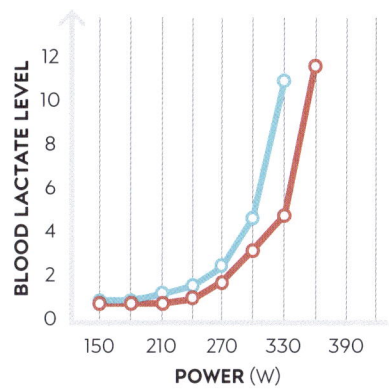

The lactate threshold (LT) represents the highest intensity at which lactate remains stable in the blood; beyond this point, it begins to accumulate. LT1 is when blood lactate starts to slowly accumulate but can be cleared and LT2 is when it accumulates faster than can be cleared, leading to fatigue and reduced muscle function. Well-trained cyclists have higher lactate thresholds, meaning they can work at a higher percentage of their VO_2 max before lactate accumulation forces a reduction in intensity.

Effective lactate clearance is an indicator of aerobic efficiency and can be improved through high-intensity interval training and tempo efforts near threshold.

FUNCTIONAL THRESHOLD POWER (FTP)

FTP is defined as the highest average power output a cyclist can sustain for one hour without experiencing fatigue-induced decline. Though not a direct physiological measure, FTP correlates closely with VT2 and LT2, especially in trained cyclists.

Functional threshold power (FTP) is often estimated by measuring 95 per cent of your average power output during a 20-minute test. The resulting value is multiplied by 0.95 to to get an idea of sustainable power over one hour. Training based on functional threshold power (FTP) is important in cycling as intensity zones are typically defined as percentages of FTP. Regular training near your threshold can raise FTP by increasing both aerobic enzyme activity and lactate tolerance. At this level you'd be working hard but not at maximal intensity.

> *Changes in FTP over time are a useful indicator of improved metabolic fitness.*

MAXIMAL LACTATE STEADY STATE (MLSS)

MLSS refers to the highest exercise intensity at which blood lactate concentration remains stable over time, usually during a 30-minute steady-state effort.

It represents the balance point between lactate production and clearance. MLSS is closely associated with LT2 and VT2, serving as a standard for determining a cyclist's sustainable performance capacity. Unlike FTP or a 20-minute power test, identifying MLSS is done by measuring lactate levels several times a day or week. This allows cyclists and trainers to pinpoint when lactate begins to accumulate progressively, indicating a shift to anaerobic metabolism.

THE ALTITUDE EFFECT

Altitude exposure affects cycling performance due to changes in oxygen availability. As altitude increases, atmospheric pressure decreases, lowering the air pressure. This results in reduced arterial oxygen saturation and challenges the body's ability to supply oxygen to working muscles, affecting endurance performance. However, these physiological stressors can be managed to drive adaptations that enhance both sea-level and altitude performance.

ACUTE EFFECTS OF ALTITUDE

At elevations above approximately 1,500m (4921ft), cyclists experience a significant reduction in VO_2 max, with a decrease of around 6–8 per cent per 1,000m (3281ft) of altitude gained.

This reduction is primarily attributable to diminished oxygen delivery, which restricts aerobic energy production. Consequently, power output at lactate threshold, maximal aerobic power, and high-intensity efforts are adversely affected. At the same time, heart rate increases at a given submaximal workload due to the greater demand placed on the cardiovascular system to make up for reduced oxygen availability.

Neuromuscular performance can also be impaired, particularly during efforts requiring repeated high-intensity outputs. Although sprint capacity is primarily anaerobic, it is indirectly influenced by decreased recovery between efforts owing to lower rates of oxygen replenishment in the muscles. Cyclists may also feel higher levels of fatigue and exertion for the same mechanical workload. Finally, sleep disturbances and appetite suppression are common in the initial days of altitude exposure, further impacting training quality and recovery.

ACCLIMATIZATION AND ADAPTATION

When exposed to high altitude (over 2,000m/6,562ft) for extended periods, the human body starts to adapt to increase oxygen transport and use. One of the most well-documented responses is an increase in erythropoietin (EPO) production, which stimulates the production of red blood cells (RBC)

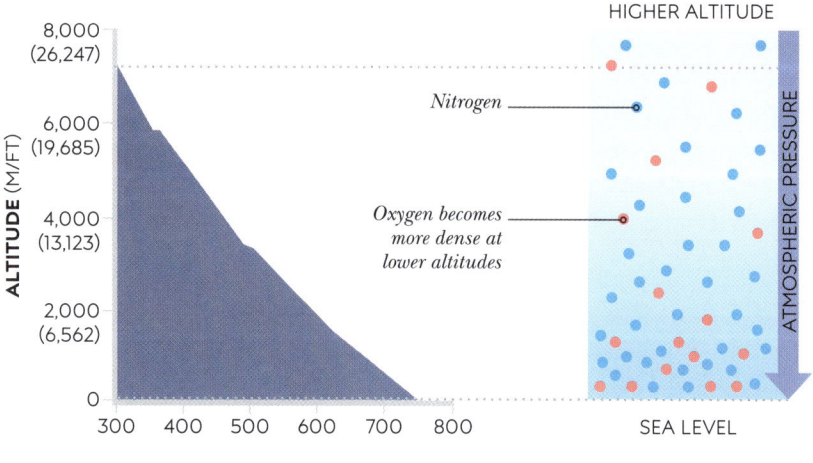

THE EFFECT OF ALTITUDE
Pressure decreases as altitude rises. This leads to the oxygen molecules moving further apart, making it harder to breathe.

and increases haemoglobin mass. This enhances the blood's oxygen-carrying capacity, which helps to improve athletic performance upon return to sea level.

Additional adaptations can include increased capillary density, enhanced mitochondrial efficiency, and, in some instances, improved muscle buffering capacity (how your muscles manage acidity during hard exercise, helping you delay fatigue and keep performing). These responses can vary among individuals, with factors such as genetic predisposition, iron availability, training status, and the duration and intensity of altitude exposure influencing the extent of these adaptations.

There are various strategies to use altitude for performance improvement, each with advantages and disadvantages:

1. Live High, Train High (LHTH)
The traditional model of altitude training involves residing and exercising at moderate altitudes, of 2,000–2,500m (6,562–8,202ft). This method encourages changes in the blood although reduced oxygen availability may impede training intensity. While this approach is suitable for base training periods, it proves less effective when high-intensity efforts are required.

2. Live High, Train Low (LHTL)
Recognized as an effective approach, this method combines the advantages of haematological adaptations to residence at high-altitude with the capability to train at low altitude (or simulated sea level), allowing high-quality sessions. Typically requiring a minimum of 3 weeks of exposure with 12–16 hours daily at altitude, this approach is logistically challenging but has been shown to enhance sea-level endurance performance.

3. Intermittent hypoxic exposure
This involves brief exposures, ranging from minutes to hours, to low oxygen conditions (hypoxic), typically using simulated altitude with tents or masks. Although less effective in stimulating erythropoiesis compared to prolonged exposure, Intermittent Hypoxic Exposure (IHE) and Intermittent Hypoxic Training (IHT) may improve ventilatory and metabolic responses, especially for cyclists unable to participate in extended altitude camps.

4. Race preparation at altitude
For races at high altitudes, it is crucial to engage in pre-acclimatization. Lack of acclimatization substantially diminishes performance, with some cyclists experiencing reductions in threshold power exceeding 10 per cent. It is advisable to undergo a gradual exposure period of 10–14 days before competition to enable partial physiological adaptation.

Practical considerations
Aside from the training needed to cycle at altitude, other factors must be taken into account:

- **Individual response monitoring:** Due to variation in response, athletes should undergo haematological assessments and monitor training metrics (such as heart rate and perceived exertion). Altitude may also increase stress on the immune system, so monitoring for illness is essential.

- **Timing of performance gains:** The peak benefit from an altitude camp typically occurs 2–3 weeks after returning to sea level, when red blood cell mass remains elevated but fatigue from the camp has reduced. This period can be targeted for key races.

- **Logistics and recovery:** Altitude training requires careful management of nutrition, hydration, sleep, and recovery. The additional physiological stress can increase fatigue and injury risk if not properly planned.

- **Altitude simulation:** Hypoxic tents or rooms offer flexible access to altitude strategies, especially for athletes limited by travel or budget. These require consistent use and should follow effective LHTL protocols. Teams may also use portable altitude systems during travel or in-season periods to maintain adaptation or induce mild hypoxic stress between camps.

AT 2,000M (6,562FT) THE **AVAILABLE OXYGEN** PRESSURE IS ABOUT **25%** LOWER THAN AT **SEA LEVEL**.

FUELLING OUR ENERGY SYSTEMS

Fuel sources for both the aerobic and anaerobic energy systems include carbohydrates, fats, and proteins. Carbohydrates are the primary fuel for moderate to high-intensity exercise in both systems. At lower intensities, fats are used more predominantly to support energy production.

DIET AND HYDRATION

Macronutrients – carbohydrates, proteins, and fats – form the foundation of a cyclist's diet, each playing distinct physiological roles.

Carbohydrates provide glucose for immediate energy and replenishment of muscle and liver glycogen stores. Proteins support the repair and remodelling of muscle tissue and contribute to immune and hormonal function. Fats, in addition to serving as a dense, sustained energy source, are essential for the hormone synthesis, absorption of fat-soluble vitamins, and maintenance of cell integrity.

Optimizing the balance of these macronutrients is essential for supporting a cyclist's energy availability, recovery, and overall performance. Dietary fibre is often considered a "fourth" macronutrient due to its significant health benefits, despite being indigestible. Fibre plays a critical role in maintaining gastrointestinal health by supporting the gut microbiome, which aids digestion, immune function, metabolic health, and mental wellbeing via the gut–brain axis.

Carbohydrates should constitute the largest proportion of a cyclist's energy intake, with an emphasis on nutrient-dense, minimally processed sources such as whole grains, legumes, vegetables, and fruits. These provide sustained energy release and additional micronutrients and phytonutrients.

Protein should account for 15–20 per cent of daily energy intake, with attention to both quality and diversity. High-quality sources include lean meats, dairy, and eggs, as well as plant-based options like legumes, soy, nuts, and seeds. Adequate protein is critical for muscle repair, adaptation to training, and immune support. Dietary fats, particularly monounsaturated and polyunsaturated fats, are vital for hormone production, absorption of fat-soluble vitamins (A, D, E, and K) and long-term energy availability. Healthy fat sources include olive oil, oily fish, avocados, nuts, and seeds. Saturated fats should be consumed in moderation, and trans fats avoided. Prioritizing whole, minimally processed foods across all macronutrient groups helps sustain energy levels, supports training adaptations, and promotes long-term health in cyclists.

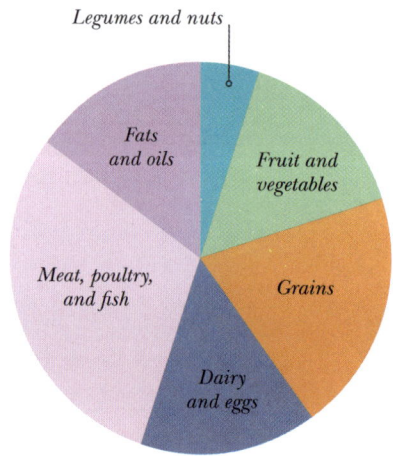

ANIMAL-BASED DIET

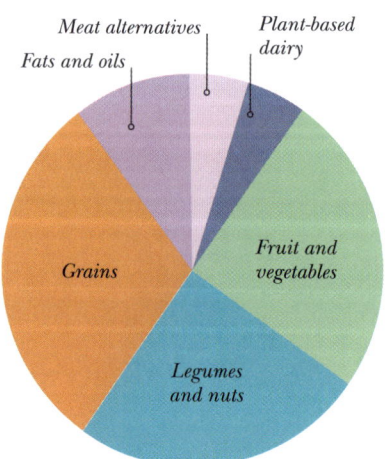

PLANT-BASED DIET

VARYING FOOD INTAKE
A balance of food types whether you're eating an animal- or plant-based diet is important to overall health.

PRIMARY FUEL

Carbohydrates should be considered the primary fuel source for cycling performance across all disciplines. Both the aerobic and anaerobic energy systems rely on a continuous supply of carbohydrates, either as glycogen stored in muscles and the liver or as glucose circulating in the bloodstream, to produce ATP, the body's primary energy currency.

The fundamental role of carbohydrates in fuelling our bodies explains why sports nutrition companies have invested heavily in research to optimize the type and amount of carbohydrate that maximizes energy production. Strategies such as carbohydrate loading, precise fuelling during exercise, and efficient recovery protocols are central to enhancing performance and sustaining energy levels during competition.

Additionally, nutrition both before and after competition is focussed on replenishing glycogen stores in muscles and the liver to ensure optimal readiness for subsequent training or racing.

Advances in our scientific understanding of these strategies, including the timing, composition, and dosing of carbohydrate intake, have been a key factor in many of the recent improvements in cycling performance at the highest levels of the sport. These may continue to be made in the future as we discover more about how best to fuel cyclists.

FAT IN THE DIET

Fat yields more energy per gram than carbohydrates, but requires more oxygen to metabolize.

As a result, fats serve as the primary energy source during lower-intensity exercise when oxygen delivery to working muscles is not a limiting factor. At higher intensities, where oxygen availability is reduced due to the cardiorespiratory system working harder, carbohydrates become the dominant fuel source because they can be used by the body more efficiently with less oxygen.

THE ROLE OF PROTEIN

Protein is rarely used as a fuel source since carbohydrates and fats are more efficient for energy production during exercise.

Protein may contribute to energy production during:

- **Prolonged exercise:** when energy stores are depleted, and carbohydrate intake is insufficient.
- **Deliberate fuel restriction:** for instance, during low-carbohydrate diets or prolonged fasting.

These situations highlight the importance of appropriate nutritional strategies. Any intentional restriction of carbohydrates or reliance on alternative fuel sources should be conducted with proper nutritional guidance and a clear performance goal.

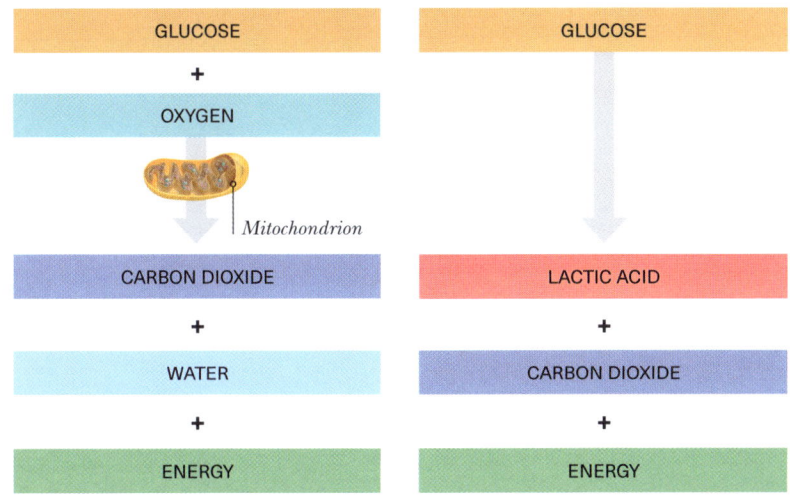

AEROBIC RESPIRATION
More energy is generated aerobically: one molecule of glycogen can yield 30–32 of ATP. Aerobic respiration occurs in the mitochondria of cells.

ANAEROBIC RESPIRATION
A mere 2 ATP molecules are produced anaerobically, along with the lactic acid that causes muscle fatigue. This process occurs in cell cytoplasm.

TIMING AND AVAILABILITY OF FUEL SOURCES

Fuelling for cycling performance requires a precise understanding of the timing, availability, and composition of energy substrates (carbohydrates, fats, and proteins). The relative contribution of each one depends on exercise intensity, duration, the cyclist's training status, and nutrition strategy.

Strategic manipulation of fuel timing and composition can improve performance, recovery, and adaptation. You need to coordinate what you eat before you exercise (glycogen loading and meal timing), during exercise (strategic carbohydrate intake), and afterwards (recovery nutrition focused on glycogen replenishment and muscle repair).

FUEL FOR MODERATE TO HIGH INTENSITY EXERCISE

Carbohydrate is the most important macronutrient for moderate to high intensity cycling. Stored in the body as muscle and liver glycogen, it is also available from blood glucose and ingested sources. Glycogen availability is a key limiting factor in endurance performance, as depletion leads to fatigue and impaired output.

*Cyclists have enough **glycogen** for about **90–120** minutes of hard exercise*

Cyclists should begin a ride or training session with fully replenished glycogen stores. A high-carbohydrate meal (2–3g/kg of body weight) consumed 3–4 hours before the start of training or competition supports this. Closer to the ride (within 1 hour), a small carbohydrate-rich snack (1g/kg) is useful to help maintain blood glucose and top up liver glycogen. This is especially important for early morning sessions when it's not possible to have a large meal earlier.

Consuming carbohydrate during exercise helps maintain glucose levels in the blood, which is one way of preventing the depletion of muscle glycogen. It also supports central nervous system function, delays fatigue, and is essential for sessions lasting longer than 60–75 minutes. The recommended intake is 30–60g (1–2oz) of carbohydrate per hour for rides of up to 2.5 hours. For longer events or high-intensity efforts exceeding 2.5–3 hours, intake should be increased to 90g (3oz) per hour or more. This can be consumed in the form of sports drinks, gels, chews, or bars, depending on preference and event logistics. Small, regularly-timed intake (every 15–20 minutes) of these carbs are more effective than infrequent large amounts. Practice in training is essential to fine-tune individual tolerance to transportable carbohydrates and optimize gut adaptation.

Sustaining energy

During long endurance events like a Tour de France stage, elite cyclists typically consume 90–120g (3–4oz) of carbohydrate per hour, with some reaching as high as 150g (5oz) per hour. These high intake rates support sustained high energy expenditure, delay fatigue, and aid recovery between stages. Such intakes are only possible with the use of a range of transportable carbohydrates that use separate absorption pathways in the gut. Cyclists train their digestive systems to tolerate these amounts.

120g carbohydrate **PER HOUR**

PHYSIOLOGY OF CYCLING | Timing and Availability of Fuel Sources

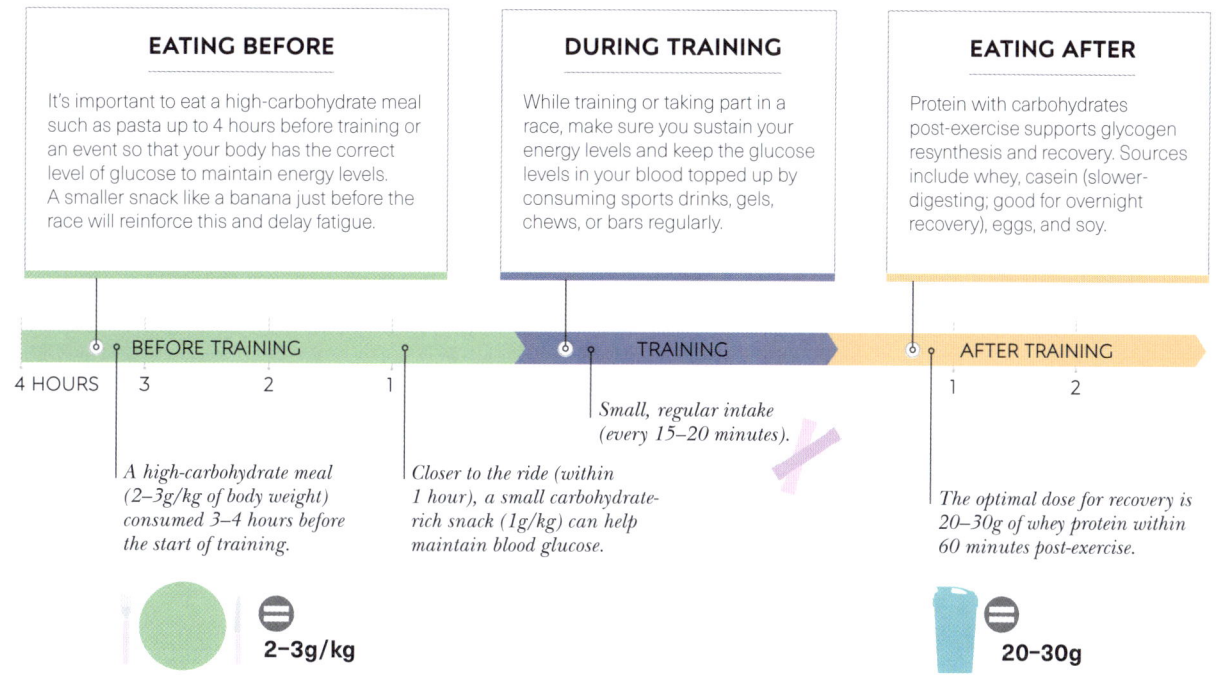

EATING BEFORE

It's important to eat a high-carbohydrate meal such as pasta up to 4 hours before training or an event so that your body has the correct level of glucose to maintain energy levels. A smaller snack like a banana just before the race will reinforce this and delay fatigue.

DURING TRAINING

While training or taking part in a race, make sure you sustain your energy levels and keep the glucose levels in your blood topped up by consuming sports drinks, gels, chews, or bars regularly.

EATING AFTER

Protein with carbohydrates post-exercise supports glycogen resynthesis and recovery. Sources include whey, casein (slower-digesting; good for overnight recovery), eggs, and soy.

BEFORE TRAINING — TRAINING — AFTER TRAINING

4 HOURS 3 2 1 | Small, regular intake (every 15–20 minutes). | 1 2

A high-carbohydrate meal (2–3g/kg of body weight) consumed 3–4 hours before the start of training.

Closer to the ride (within 1 hour), a small carbohydrate-rich snack (1g/kg) can help maintain blood glucose.

The optimal dose for recovery is 20–30g of whey protein within 60 minutes post-exercise.

2–3g/kg

20–30g

FAT AS FUEL

Fat oxidation (fat burning), is the process by which the body breaks down fatty acids to produce energy. This contributes significantly to ATP production during low- to moderate-intensity cycling.

Even in lean athletes there are stores of fat, mainly in muscle tissue and body fat, that can fuel many hours of submaximal exercise. The ability to efficiently use fat spares glycogen and extends endurance capacity.

Fat is not typically ingested during exercise because it takes longer to digest and has a limited contribution to immediate energy needs. However, fat adaptation through low-carbohydrate, high-fat (LCHF) diets or strategic "train low" sessions (training with low glycogen availability) may enhance mitochondrial adaptations and increase fat burning capacity. They can also impair high-intensity performance due to reduced carbohydrate availability and shouldn't be used long term.

Pre-ride meals for longer, lower-intensity sessions may include moderate fat content, particularly when carbohydrate demands are lower. For example, a mixed meal 3–4 hours before a long endurance ride may include nut butter, eggs, or full-fat dairy and carbohydrate sources like oats or fruit.

PROTEIN FOR RECOVERY

Protein is not a primary fuel for exercise but contributes to energy production during prolonged endurance events when glycogen stores are low.

Branched-chain amino acids (BCAAs) may be used for energy late in exhaustive rides. However, the more relevant role of protein is in muscle repair, recovery, and adaptation. A 20–30g (0.7–1oz) dose of whey protein within 60 minutes post-exercise includes enough leucine to maximally stimulate muscle repair and growth. For heavier cyclists or after prolonged sessions, up to 40g (1½oz) may be beneficial.

FUELLING STRATEGIES FOR DIFFERENT EVENTS

The best fuelling strategies for cycling depend on the type, duration, and intensity of the event. While the underlying principles of energy availability, hydration, and macronutrient timing remain consistent, the specific demands of events like criteriums, time trials, stage races, and ultra-endurance rides require tailored approaches.

Planning fuelling strategies in advance of cycling events will help you get the best performance.

CRITERIUMS, CYCLOCROSS, AND MOUNTAIN BIKE RACES

These cycling events comprise repeated bursts of speed, tactical attacks, and numerous technical turns. Due to the highly glycolytic nature of these events, even though they are relatively short, maintaining high carbohydrate availability is crucial for the repeated high-intensity efforts. Preparation includes starting with fully loaded glycogen stores. During the race, in-race fuelling is often limited by course dynamics and intensity. For races under 75 minutes, carbohydrate ingestion may not be necessary, though a small bottle of carbohydrate-electrolyte drink can help maintain hydration and blood glucose. If the event exceeds an hour, a small amount of carbohydrate during the race may improve performance, especially in hot conditions or when the race includes prolonged surges. Post-race recovery should include moderate portions of carbohydrate within 30 minutes to replenish glycogen and support muscle repair.

TIME TRIALS

Time trials (TTs) are efforts at or near threshold intensity, requiring steady pacing and minimal disruption. Pre-race carbohydrate loading the day before, along with a meal 3–4 hours before ensures optimal glycogen stores. A sports drink or snack 30–60 minutes before the start helps maintain blood glucose and hydration.

For shorter TTs (under 30 minutes), in-race fuelling is unnecessary. For efforts longer than 45 minutes, sipping a carbohydrate drink can maintain cognitive focus and delay fatigue, especially in hot or humid conditions. As with other events, rapid post-race nutrition accelerates recovery, particularly when back-to-back races or training follow.

FUEL NEEDS

- CYCLISTS IN CRITERIUMS REQUIRE **HIGH-CARB** MEALS
- TIME TRIALS – CYCLISTS **CARB LOAD** BEFORE THE EVENT
- ROAD RACE CYCLISTS NEED **SUSTAINED CARBOHYDRATE INTAKE**
- STAGE RACE CYCLISTS NEED **FREQUENT CARBOHYDRATE FUELLING**
- ULTRA-ENDURANCE RIDERS HAVE MIXED MEALS INCLUDING **FAT AND PROTEIN**
- MTB RIDERS NEED **HIGH-CARBOHYDRATE** MEALS AND SNACKS

Fuelling strategies

EVENT TYPE	PRE-RACE	DURING	POST-RACE RECOVERY
Criteriums, cyclocross, MTB (45–90 min)	High-carbohydrate meal (2–3g/kg) 3–4 hours before; snack (1g/kg) within 1 hour	Usually minimal; carbohydrate drink (around 30–40g/1–1.4oz) if over 1 hour or hot conditions	20–30g (0.7–1oz) protein and 1–1.2g/kg carbohydrate within 60 minutes
Time trials (20–60 min)	Carb load day before; meal 3–4 hours before; light snack or drink pre-start	None for under 30 minutes; around 20–30g (0.7–1oz) carbohydrate if over 45 min	Same as criteriums; rapid recovery if more events follow
Road races (2–5 hrs)	Carbohydrate-rich meal (3–4g/kg) 3–4 hours before; snack/gel 15–30 minutes before	60–90g (2–3oz) carbohydrates per hour using drinks, gels, bars; up to 150g per hour for pros	Rapid CHO + protein intake; hydrate and prepare for next stage if needed
Stage races (multi-day)	Daily intake 8–12g/kg carbohydrates; high-carb meals and snacks throughout day	90–120g (3–4oz) carbohydrates per hour; consistent intake every 15–20 minutes; electrolytes crucial	Recovery shakes (carbohydrates:protein) within 60 minutes; high-carb meals/snacks
Ultra-endurance (6+ hrs)	Mixed meals; carbohydrate-rich with some fat/protein depending on intensity	60–120g (2–4oz) carbohydrates per hour with solids and sports products; fat/protein included	Focus on ongoing intake; carbs, protein, fluids, electrolytes for recovery

ROAD RACES

Fuelling for road races is complex due to their duration and unpredictability. Carbohydrate availability is paramount. Cyclists should load up the carbohydrates before the race and then sustain intake during the race to preserve glycogen and maintain blood glucose. For races longer than 2 hours, cyclists should aim for 60–90g (2–3oz) of carbohydrate per hour. This is best achieved using multiple transportable carbohydrates (glucose and fructose in a 1:0.8 ratio) via a mix of drinks, gels, bars, and real food like rice cakes or energy chews.

Fuelling should be proactive and evenly spaced, every 15–20 minutes when possible. Hydration and sodium intake (500–1000mg per hour) must be adjusted for sweat rate and environmental conditions. In hot races, maintaining fluid and electrolyte intake is just as important as carbohydrate availability.

STAGE RACES

Stage races such as the Tour de France involve a comprehensive fuelling strategy that addresses daily performance and overall recovery. Carbohydrate intake is emphasized during the stages, immediately after, and throughout the evening. Recovery drinks with carbohydrate and protein, followed by frequent high-carbohydrate meals and snacks are consumed to restore glycogen stores ahead of the next stage.

Nutrition strategies may vary during the race, with intake adjusted for stage demands (for example, lower for easier stages, higher for mountain stages). Sleep, hydration, and gastrointestinal health are supported through supplements and meal planning.

ULTRA-ENDURANCE EVENTS

Ultra-endurance events demand the highest level of fuelling planning. The energy demands far exceed glycogen stores, making other fuel types essential. Unlike shorter events, fat plays a meaningful role, especially at lower intensities. Mixed meals including carbohydrate, fat, and small amounts of protein can be beneficial. Solid foods (such as sandwiches, bars, and rice cakes) are more commonly used alongside conventional sports products. Nutrition strategies should be practised and refined in training.

PERIODIZATION OF NUTRITION

In high-performance cycling, nutrition evolves in response to a cyclist's competitive season. Nutrition periodization is planned dietary intake to meet specific demands, optimize race-day readiness, and achieve target weight and body composition without compromising performance or health.

> **Endurance** cyclists may consume *>120g* of *carbohydrates per hour*

MANAGING BODY COMPOSITION

Rather than focussing only on energy expenditure, periodized nutrition is used as a strategic tool to enhance training outcomes, support racing, and manage fluctuations in body mass and composition throughout the season.

Depending on the demands of their chosen discipline, professional cyclists often face the challenge of optimizing power-to-weight ratio while maintaining energy intake to train and race effectively.

Nutrition plays a central role in managing this balance. Key periods, such as pre-season, during race blocks or taper weeks, may involve a focussed effort to optimize body composition or reach a target race weight. During these times, energy intake may be slightly restricted, or carbohydrate intake selectively reduced around training sessions to create an energy deficit.

However, this must be monitored carefully to avoid impairing metabolic health, recovery, or training adaptation. Cyclists may incorporate strategies like "low energy availability" on selected days, where calorie intake is slightly below expenditure, while ensuring sufficient protein intake (at least 1.8–2.2g/kg/day) to preserve lean mass. This can be achieved relative to the total energy expenditure and strategically targeting at times that do not compromise training quality, such as during post-training fuelling or on training days with low to moderate load.

Body composition changes are best made during low-pressure periods of the season. During intense race phases or critical preparation periods performance and recovery must take priority.

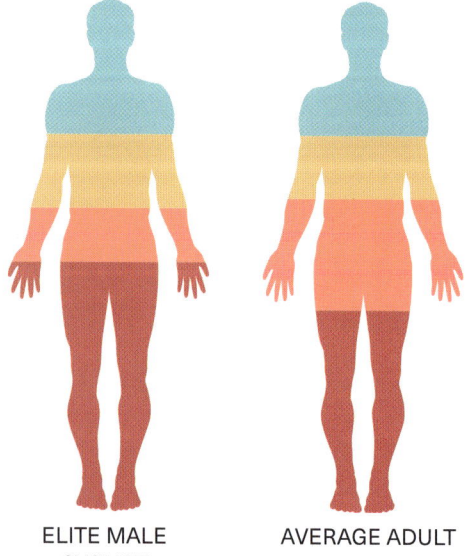

KEY
- Muscle mass
- Fat mass
- Skeletal mass
- Other (organs, water, etc.)

ELITE MALE CYCLIST

AVERAGE ADULT MALE

HUMAN BODY COMPOSITION
Body composition of an elite male cyclist vs. an average male, highlighting greater lean mass and lower fat in the cyclist.

HIGH-CARBOHYDRATE AVAILABILITY

High-carbohydrate availability should be prioritized in periods that demand peak performance, such as during racing phases, high-intensity training, and major events. Energy intake is raised to match or exceed energy expenditure. The objective is to maximize muscle glycogen storage; support, sustained high power output, and improve cognitive function and motor skill under fatigue.

NUTRITION AND TRAINING

Synchronizing nutrition with your training plan will increase its effectiveness. For instance, during base training emphasizing aerobic development, "train low" strategies (reduced carbohydrate availability in certain workouts) can boost metabolic adaptations like fat oxidation and mitochondrial function. As training intensifies and becomes more race-specific, carbohydrate intake increases to fuel higher workloads and maintain training quality. During the taper phase (when training intensity is reduced before competition to allow the body to recover), nutrition focusses on reducing gut content while maximizing glycogen stores without overconsuming energy.

PLANNING FOR EVENTS

Different race types and roles in a team influence nutritional periodization. Nutrition planning becomes individualized, especially

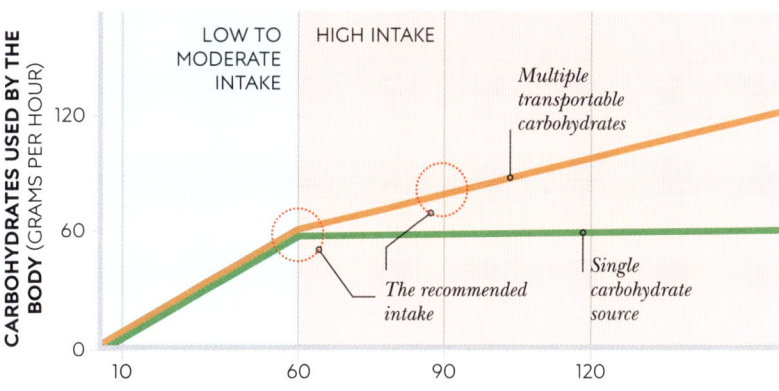

RECOMMENDED INTAKES

Increased absorption of carbohydrates takes place with multiple transportable carbohydrates, such as those available in sports drinks, gels, and bars.

in the weeks leading into a key race. This taper period may involve high glycogen intake, hydration management, and low-fibre diets to reduce gastrointestinal weight. Small changes, such as decreasing fibre and salt intake two days out, can make a difference without affecting energy stores.

Energy deficiency

Extended periods of low energy availability are problematic and can lead to relative energy deficiency in sport (REDS), a condition that impairs hormonal function, bone health, immunity, and overall performance.

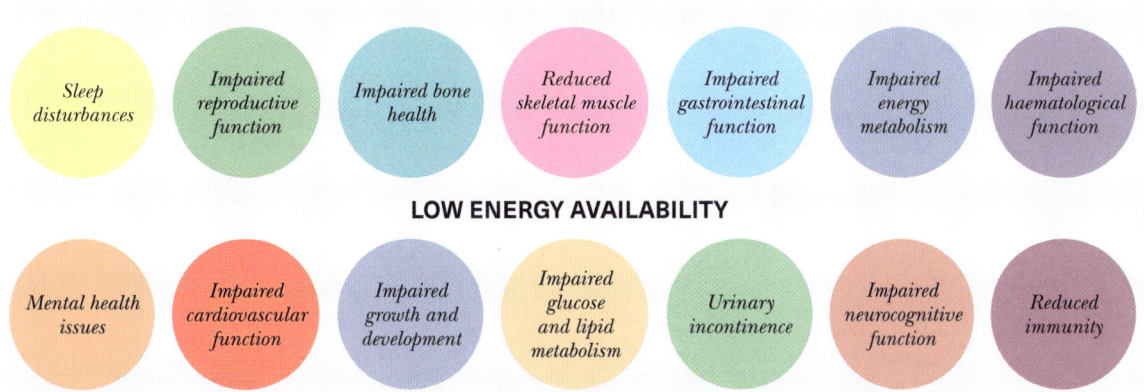

NUTRITION SUPPLEMENTS

Effective nutrition strategies, including the use of supplements, can support training adaptations, enhance competition performance, and reduce recovery time. The Australian Institute of Sport (AIS) categorizes sports supplements into four evidence-based groups – A, B, C, and D – to guide responsible use in high-performance sport.

Group A
These supplements have strong scientific evidence supporting their use and are considered safe when used appropriately.

SPORTS FOODS
- **Carbohydrate products (gels, drinks, bars):** Essential in training and racing during moderate to high-intensity efforts lasting more than 60 minutes.
- **Protein supplements (whey, casein):** Used to support muscle repair and adaptation post-exercise.
- **Electrolyte drinks:** Aid hydration, especially in hot conditions or during high sweat losses.

MEDICAL SUPPLEMENTS
- **Iron:** Important for endurance cyclists, particularly females, due to the risk of iron deficiency. Supplementation should be accompanied by blood screening for iron status where possible.
- **Vitamin D:** May be relevant for indoor-based cyclists or those training in low sunlight environments.

PERFORMANCE SUPPLEMENTS
- **Caffeine:** Enhances alertness and reduces perceived exertion. Typically used in doses of 3–6mg/kg taken around 60 minutes before exercise or in smaller doses (1–3mg/kg) throughout a race.
- **Creatine monohydrate:** While traditionally linked with strength sports, it may benefit sprint performance and recovery in track or criterium cyclists.
- **Sodium bicarbonate:** Effective for high-intensity efforts (for example, time trials, track cycling events, BMX races, criteriums), though gastrointestinal side effects may limit practical use.
- **Beta-alanine:** Buffers muscle acidity; best suited for anaerobic efforts or disciplines involving short, repeated bursts like BMX or track sprint.
- **Nitrates (beetroot juice):** Can enhance exercise efficiency and endurance by enhancing blood flow and reducing oxygen cost. Effective in time trials and submaximal efforts.

Group B
This group includes supplements with promising early findings but requiring further validation.

- **Tart cherry juice:** Shows potential in reducing inflammation and improving sleep and recovery post-race.
- **Collagen/gelatin with vitamin C:** Being explored for tendon and joint health, possibly helpful for injury prevention or management in cyclists with overuse issues.
- **Pre- and/or probiotics:** May support immune function and reduce gastrointestinal distress during training blocks or travel.

These supplements should only be tried on an individual basis under professional guidance, particularly when tailored to specific goals like recovery, travel-related illness prevention, or injury rehabilitation.

Group C
This includes supplements with no meaningful scientific backing for performance enhancement in athletes.

- Examples often promoted in cycling communities include **BCAAs, glutamine, and alkaline water** – none of which show clear benefits in well-fed, trained cyclists.
- Use of these products represents a potential waste of resources and can distract from evidence-based practices.

Group D
This group includes substances are either banned or carry a high risk of contamination and positive doping tests.

- Examples include **DMAA, methylhexanamine, and select prohormones or testosterone boosters** marketed as "natural".
- Cyclists should strictly avoid Group D products. Supplement selection should always be third-party batch tested (for example, Informed Sport, NSF Certified for Sport).

SAFETY FIRST

When cyclists consider taking supplements, there are several practical considerations to keep in mind to ensure safety, effectiveness, and compliance with sport regulations. These include:

- **Purpose and goals:** Consider why you are taking a supplement. Are you looking for improved endurance, strength, or speed?

- **Personalization:** Supplement use should be tailored to the cyclist's event demands, training phase, physiological needs, and medical status.

- **Trial in training:** Supplements, particularly performance-enhancing aids like caffeine or sodium bicarbonate, must be evaluated for effective dosage and tolerance during training, not on race day.

- **Nutrition first:** Supplements are intended to complement, not replace, a well-structured diet. Emphasis should remain on whole foods and consistent eating patterns.

Caffeine taken **60 MINS** *before a race enhances alertness*

CREATINE

Creatine monohydrate is a well-established sports supplement with substantial evidence supporting its effectiveness. It is widely used in professional cycling.

Creatine plays a key role in the phosphagen energy system (see p.48), as creatine phosphate serves as the primary fuel source for this pathway. Supplementation with creatine over time increases muscle creatine stores, enhancing the capacity of this energy system to support single and repeated short, high-intensity efforts such as sprint track cycling and road cycling.

BLOOD BUFFERING

Supplements that buffer acidosis help reduce the buildup of hydrogen ions (H+) in the blood and muscles, which can cause fatigue during high-intensity exercise.

These supplements work by increasing the body's buffering capacity or removing excess acid. **Sodium bicarbonate** can be used to enhance the blood's buffering capacity by increasing its pH. This is important for high-intensity exercise, where lactic acid accumulates in the blood, impairing energy production, muscle contraction, and overall performance. By increasing the blood's ability to buffer acidosis, sodium bicarbonate helps improve tolerance to lactate accumulation. It can be taken a few hours before exercise although appropriate forms and dosage are needed to minimize gastrointestinal discomfort.

Beta-alanine helps buffer acidosis and is used to enhance performance during high-intensity efforts. Unlike sodium bicarbonate, it works at the muscle level, increasing intramuscular carnosine levels, which act as a buffer against hydrogen ions produced during anaerobic metabolism. Beta-alanine needs to be taken over several weeks to take effect.

SODIUM BICARBONATE
0.2–0.4g/kg
90–120 MINS BEFORE <10 MIN HIGH-INTENSITY EFFORTS

BETA-ALANINE
65mg/kg/day
4–24 WEEKS

CAFFEINE
3–6mg/kg
1 HOUR BEFORE

CREATINE
20g a day in 4 doses OR 3–5g single dose
IN BASE TRAINING PERIODS

NITRATES
500–600mg/day
1–7 DAYS PRIOR AND 1–5 DAYS AFTER A RACE

COMMON NUTRITION PITFALLS

While the importance of nutrition is widely acknowledged, many cyclists, from amateur riders to seasoned professionals, fall into recurring pitfalls that compromise training quality, recovery, and performance. Understanding these mistakes is the first step towards effective nutritional planning and execution.

Up to **50%** *of cyclists suffer* gastrointestinal *issues* **during racing**.

Sound nutrition significantly enhances performances in cycling. Avoiding the common pitfalls shown here requires education, planning, and habit formation.

Nutrition should be as deliberate and individualized as training plans. Coaches, cyclists, and their support team must work collaboratively to embed evidence-based nutrition into daily routines. The most common problems are concerned with over- or under-eating, bad timing, and lack of planning around fuelling the body and exercise, race, or training times. All of these can have negative consequences and affect performance, decision-making, and fatigue levels. Planning and education are the best ways to avoid these pitfalls and make sure you get the most out of the food you consume before, during, and after cycling events.

1. Under-fuelling

One of the most common issues is chronic under-fuelling, particularly among young riders or those aiming for low body mass. Extended periods of low energy availability, whether intentional or unintentional, can lead to relative energy deficiency in sport (REDs, see p.67). Symptoms can include poor recovery, impaired training adaptation, frequent illness, hormonal disruption, and decreased bone density. In the short term, it reduces the quality of training and impairs decision-making during races.

Solution: Establish a minimum energy intake threshold, ensure meals and snacks are consistent throughout the day, and prioritize fuelling both before and during rides lasting over 60 minutes.

2. Poor timing of nutrient intake

Even when overall intake is sufficient, mistimed fuelling can reduce training effectiveness. Skipping breakfast before a hard morning session, delaying post-ride recovery intake, or failing to fuel long rides until "empty" are common examples. This can impair glycogen replenishment, blunt training adaptations, and increase muscle breakdown.

Solution: Implement a "fuel for the work required" model, eating more on high-load days and ensuring carbohydrate intake aligns with key sessions. Aim for a carb-rich meal 2–3 hours pre-ride and recovery nutrition within 60 minutes post-session.

3. Inadequate in-ride fuelling

Many cyclists underestimate how much fuel they need during rides and races, particularly during moderate to high-intensity efforts lasting 90 minutes or more. Under-fuelling increases perceived exertion, reduces power output, and impairs cognitive function, impacting pacing and decision-making.

Solution: Train the gut to tolerate more than 90g (3oz) of carbohydrate per hour, depending on duration and intensity. Fuel consistently, start early in the ride, and avoid long gaps.

PHYSIOLOGY OF CYCLING | Common Nutrition Pitfalls

4. Fear of carbohydrates

Carbohydrate phobia exists in endurance sports with trends like low-carb high-fat (LCHF) diets. While some periodized carbohydrate intake strategies may benefit metabolic flexibility, chronic carb restriction impairs high-intensity performance and recovery. This is particularly damaging in race scenarios where glycolytic capacity is essential.

Solution: Educate cyclists on the role of carbohydrates in performance. Tailor carbohydrate intake to session demands rather than removing them entirely. Use fast-acting carbs during training and races, and complex carbs in meals to replenish stores.

5. Neglecting hydration and electrolytes

Hydration errors can lead to dehydration, hyponatremia (low levels of sodium in the blood), or impaired thermoregulation. Cyclists often lack a structured hydration strategy, especially in heat or during long stages.

Solution: Determine individual sweat rate and electrolyte loss, especially in hot conditions. Replace fluids and sodium accordingly, aiming for 400–800ml (0.7–1.4pt) of fluid per hour with added electrolytes during rides longer than 90 minutes.

6. Inconsistent recovery practices

Recovery nutrition is often neglected, especially after easier sessions or travel. Inadequate protein and carbohydrate intake post-session slows muscle repair, glycogen replenishment, and adaptation.

Solution: Target 20–40g (0.7–1.4oz) of protein and 1–1.2g/kg of carbohydrate within 60 minutes of finishing key training sessions or races. For multiple-day events, such as stage races or training camps, recovery nutrition becomes critical and must be habitual.

7. Overuse or misuse of supplements

Many cyclists rely on supplements to fill perceived gaps or gain an edge, often at the expense of food-first principles. This includes inappropriate timing, unverified claims, or use of untested products which pose health and anti-doping risks.

Solution: Emphasize a food-first approach and use evidence-based supplements (for example, caffeine, carbs, and creatine) when appropriate. Choose batch-tested products (certified by Informed Sport or NSF) and seek advice from a qualified sports dietitian or nutritionist.

8. Lack of nutrition planning for travel and racing

Poor food access during travel, unfamiliar environments, and inconsistent meal timing often disrupt nutrition during races. Cyclists without a proactive plan may experience energy deficits, GI distress, or food anxiety.

Solution: Plan travel meals and snacks in advance. Identify reliable food sources at hotels or race locations. Bring staple items (such as oats, protein powder, and recovery shakes) and communicate with hotel staff early to ensure needs are met.

> *If you don't fuel before cycling – especially before a moderate to hard session – your body quickly runs into energy problems.*

THERMOREGULATION

Thermoregulation is the process by which the human body maintains its core temperature, despite fluctuations in environmental conditions and internal heat production. Hyperthermia (over heating) is the fundamental limiter of endurance performance irrespective of duration and intensity, and so efficient thermoregulation is at the core of optimal performance.

THE HYPOTHALAMUS

During exercise, most of the energy produced by skeletal muscle is converted into heat. This must be dissipated to maintain homeostasis (stable body temperature).

When heat production exceeds heat loss, core body temperature begins to rise. In well-trained athletes, core temperature may approach or exceed 39.5–40.0°C (103–104°F) during high-intensity efforts in hot environments, which increases physiological strain and contributes to central and peripheral fatigue. The hypothalamus is the central regulator of thermoregulation in the brain. It receives input from thermoreceptors located in the skin, muscles, spinal cord, and brain. These detect changes in both skin and core temperatures. In response to thermal stress, the hypothalamus activates effector mechanisms, most notably sweating and cutaneous vasodilation (see right), aimed at promoting heat loss.

Internal temperature changes
When you cycle, the activity raises internal temperature. If cycling in cold conditions, body temperature may reduce if you are inadequately protected from the cold. Internal body temperature must be maintained between 37°C (98.6°F) and 37.8°C (100°F).

Suitable clothing
Wearing the right clothing and layers that can be added or removed helps regulate body temperature.

THE BRAIN'S RESPONSE
The hypothalamus controls the body's response to temperature. Training can help us acclimatize to extreme temperatures.

Heat loss during cycling

METHOD OF HEAT LOSS	DESCRIPTION	EFFECTIVENESS
Evaporation	Conversion of sweat on the skin to water vapour, removing heat from the body.	Primary mechanism during cycling, especially in warm conditions.
Convection	Transfer of heat to air or water moving across the skin.	Enhanced by cyclist's motion and wind speed; important at moderate to high cycling speeds.
Radiation	Emission of infrared heat from the skin to cooler surrounding surfaces.	Moderate contribution during early exercise and in cooler environments; minimal in hot conditions.
Conduction	Direct heat transfer to a cooler surface in contact with the body.	Minimal in cycling due to limited contact with surfaces.

PHYSIOLOGY OF CYCLING | Thermoregulation

> When the **hypothalamus** is unable to set off the appropriate corrective responses, core temperature continues to rise or fall beyond the desired window.

RECEPTORS DETECT THE CHANGE

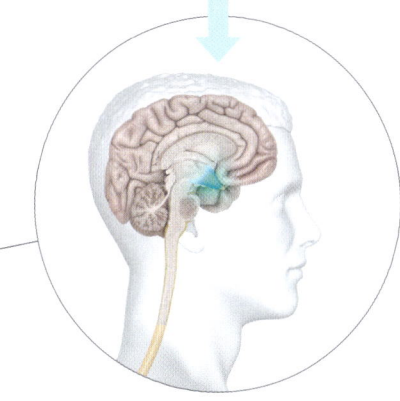

Hypothalamus
This important endocrine gland is situated in the brain and is the body's thermoregulatory centre. It stimulates the appropriate response in order to return the body's internal temperature to normal.

CORRECTIVE RESPONSES ACTIVATED

INTERNAL TEMPERATURE RETURNS TO OPTIMUM LEVEL

CORRECTIVE RESPONSES DEACTIVATED

HEAT REDUCTION

Sweating and vasodilation are the two key ways in which the body is able to reduce its core temperature when raised through exercise like cycling.

Sweat glands are activated by the hypothalamus, secreting fluid onto the skin surface. Evaporation of this sweat is the key cooling mechanism in cycling. The rate of sweating can exceed 2 litres (4 pints) per hour in elite cyclists under heat stress. However, if ambient humidity is high, the air becomes saturated with water vapour, reducing sweat evaporation efficiency. This leads to accumulation of sweat on the skin surface that does not cool the body effectively, causing a disproportionate rise in core temperature and a resulting drop in performance level.

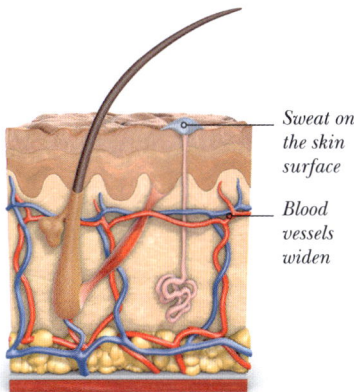

Sweat on the skin surface
Blood vessels widen

VASODILATION
In response to heat stress, blood vessels in the skin dilate, increasing blood flow to the periphery. This helps the transfer of internal heat to the skin surface, where it can be dissipated.

ACCLIMATIZING TO HEAT
Acclimatization is used to improve thermoregulatory efficiency in situations of extreme heat. Over a period of 7–14 days of repeated heat exposure, the body adapts by increasing plasma volume, reducing heart rate at a given workload, initiating sweating earlier in specific regions, and producing more dilute sweat. These adaptations enhance the body's ability to dissipate heat and maintain cardiovascular stability during exercise in the heat.

Many professional cyclists also now incorporate heat acclimation camps, sauna protocols, or environmental chamber training to stimulate these adaptations prior to hot-weather competitions.

COLD CONDITIONS
In cold environments, thermoregulation in cycling shifts to heat conservation, with the body minimizing heat loss through peripheral vasoconstriction and shivering thermogenesis (the production of heat through involuntary muscle contractions). Cyclists are vulnerable due to heat loss from wind exposure, causing muscle cooling, which reduces power output and coordination. Cold-induced diuresis (an increase in urination) and suppressed thirst can lead to dehydration. Cold air may also provoke bronchoconstriction. Effective thermoregulation relies on layered clothing and additional layers and warm-ups to preserve core temperature.

HYDRATION

Hydration plays a central role in thermoregulation, directly influencing a cyclist's ability to dissipate heat, maintain cardiovascular function, and sustain high-intensity performance. The ability to sweat effectively depends on the availability of body water.

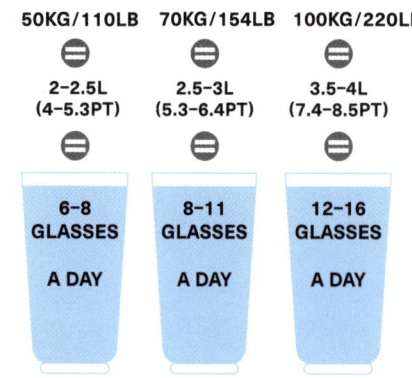

HOW MUCH WATER TO DRINK A DAY
The amount of water needed varies between individuals depending on weight and external conditions.

SWEAT RATES

Cyclists can lose between 0.5–2.0 litres (1–4pt) of fluid per hour through sweat, depending on intensity, duration, environmental conditions, and individual sweat rates. In hot climates or during high-intensity efforts, these losses can be even greater.

If fluid intake does not match sweat loss, dehydration ensues. A 4 per cent reduction in body mass due to fluid loss has been shown to impair thermoregulatory capacity. As dehydration progresses, plasma volume decreases, which has a cascade of detrimental effects. Reduced plasma volume impairs the ability to deliver blood to the skin for heat dissipation and to the muscles for oxygen delivery. The cardiovascular system will compensate by increasing heart rate, which increases perceived exertion and reduces power output.

In addition, dehydration raises core temperature more rapidly. This is because less fluid is available for sweating, and blood flow to the skin is compromised. Consequently, the threshold for the onset of fatigue is lowered. In longer events, failure to hydrate adequately can lead to early fatigue, impaired decision-making, and in severe cases, heat exhaustion or heat stroke. It is not uncommon for elite cyclists to lose 3–4 per cent of their body weight during long events, but such deficits come at a clear performance cost if not managed properly through proactive hydration strategies.

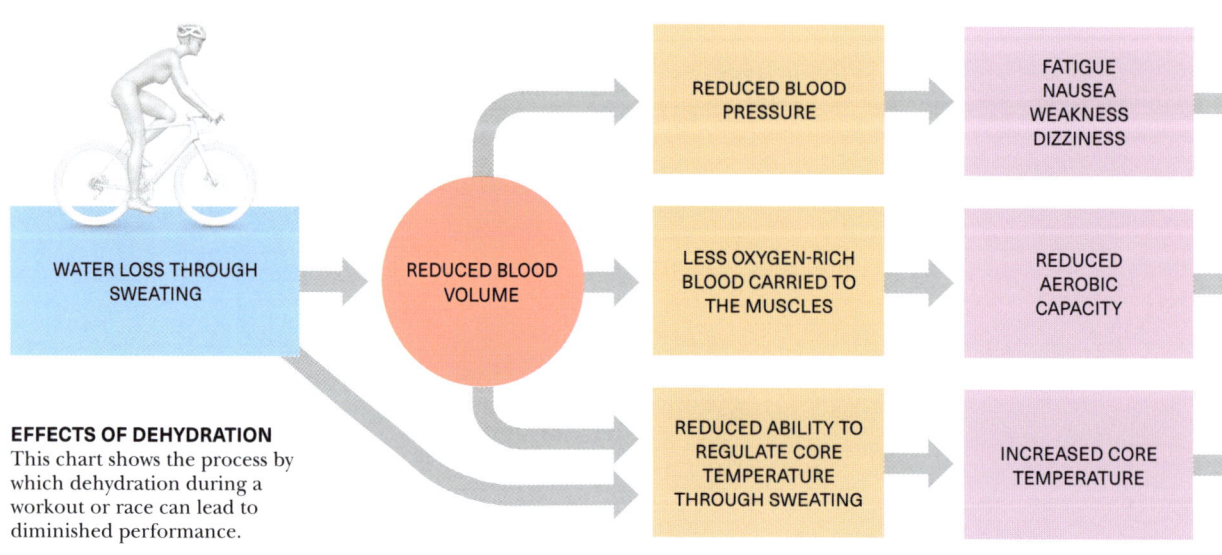

EFFECTS OF DEHYDRATION
This chart shows the process by which dehydration during a workout or race can lead to diminished performance.

ELECTROLYTE BALANCE

The relationship between hydration and electrolyte balance is critical. Sweat contains water, significant amounts of sodium and smaller quantities of potassium, magnesium, and chloride.

Replacing lost fluids without adequate electrolytes can lead to hyponatremia (low levels of sodium in the blood), especially in long events with high fluid turnover. Conversely, failing to replace both fluid and electrolytes leads to compounded dehydration and electrolyte imbalances, which impair nerve conduction, muscle contraction, and thermoregulatory responses.

To support optimal thermoregulation, cyclists should begin exercise in a hydrated state. Pre-race hydration strategies may include drinking 5–7mL/kg of fluid approximately four hours before exercise, and an additional small amount (about 3–5mL/kg) 2 hours before if no urine has been passed. During exercise, fluid replacement should aim to limit body mass loss to under 2 per cent, which translates to drinking between 400–1000mL per hour, adjusted for individual sweat rate, environmental conditions, and gastrointestinal tolerance.

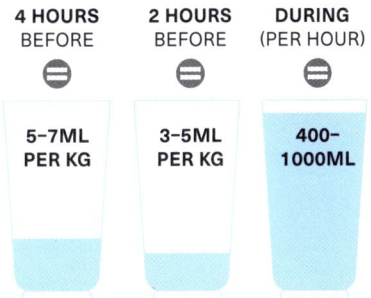

4 HOURS BEFORE	2 HOURS BEFORE	DURING (PER HOUR)
5–7ML PER KG	3–5ML PER KG	400–1000ML

PERSONALIZED STRATEGIES

The concept of dehydration in sport has generated ongoing debate, particularly around its definition, measurement, and impact on performance.

Traditionally, dehydration is defined as a loss of body mass exceeding 2 per cent, with studies suggesting this threshold impairs endurance, thermoregulation, and cognitive function. More recent perspectives argue that moderate body mass losses do not always equate to performance decline, especially in well-trained athletes who may tolerate fluid deficits without noticeable impairments. This has led to an increase in more personalized strategies, with individualized sweat rates, thirst cues, and environmental context being considered.

RACE CONTEXT

Real-world hydration practices in cycling are also influenced by race dynamics and terrain.

Cyclists may delay fluid intake during high-intensity race efforts or technical descents, increasing the risk of dehydration. Support vehicles, feed zones, and nutrition strategies must all align to enable consistent hydration, especially in hot races. Post-exercise rehydration is equally important to restore plasma volume, support recovery, and prepare for subsequent days of racing. This typically involves ingesting 1.25–1.5 times the volume of fluid lost during exercise, along with adequate sodium intake to promote fluid retention, but should be tailored to individual needs.

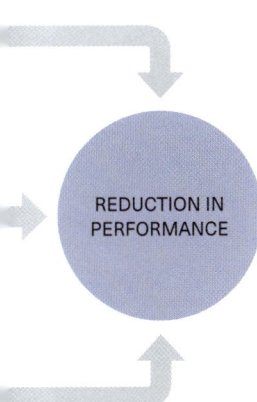

REDUCTION IN PERFORMANCE

Sodium levels

Overhydration can be as dangerous as dehydration. During exercise, we lose sodium through sweating (known as electrolyte depletion). Drinking excessively during exercise dilutes the already depleted sodium levels in your blood. This can lead to sleep disruption, and the potentially life-threatening condition Exercise Associated Hyponatraemia (EAH). Symptoms include headaches, fatigue, nausea or vomiting, muscle spasms, and seizures. Sports drinks contain electrolytes and therefore do not deplete the sodium levels in your blood in the same way that drinking water does. However, even drinking sports drinks to excess can dilute sodium levels.

FATIGUE AND RECOVERY

Fatigue is broadly defined as the inability to maintain force or power for a given exercise intensity and is accompanied by an increased perception of effort.

It is a multifactorial phenomenon that arises from both central (neurological) and peripheral (muscular and metabolic) mechanisms. In the context of endurance sports like cycling, fatigue is not only a physiological challenge but also a complex interplay of metabolic, neuromuscular, psychological, and environmental factors that limit sustained performance.

> **Nutrition** and **sleep** are *crucial* in the *management of fatigue*

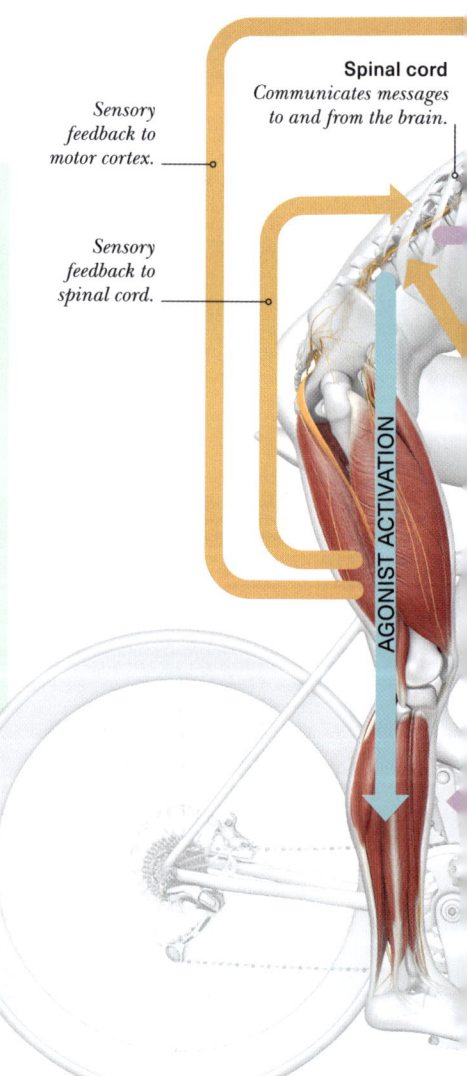

Spinal cord
Communicates messages to and from the brain.

Sensory feedback to motor cortex.

Sensory feedback to spinal cord.

AGONIST ACTIVATION

Acute fatigue
In cycling, fatigue can be classified into acute and chronic forms. Acute fatigue develops during a single session and is typically characterized by reduced power output, increased perceived exertion, and compromised neuromuscular coordination. This type of fatigue is driven by factors such as substrate depletion (especially muscle and liver glycogen), thermoregulatory strain, dehydration, and accumulation of metabolic byproducts like lactic acid (hydrogen ions) and inorganic phosphate. During prolonged or high-intensity cycling, glycogen depletion is a primary contributor to fatigue, as muscle fibres shift to less efficient energy pathways, leading to reduced force production and slower contractile speed.

Central fatigue
Central fatigue reflects a decrease in voluntary activation of muscles due to altered function within the central nervous system. This may involve reduced motor unit recruitment or altered neurotransmitter activity in the brain, often linked to thermal stress, hypoglycaemia, or psychological strain. Environmental stressors such as heat, cold, altitude, or wind can accelerate both acute and chronic fatigue by increasing the physiological cost of a given workload. In racing scenarios, tactical demands, such as surges, attacks, and climbs, compound fatigue by forcing cyclists above their sustainable thresholds, increasing reliance on anaerobic systems and hastening the onset of exhaustion.

Cumulative fatigue
In multi-day races or heavy training blocks, cumulative fatigue becomes a dominant concern. This chronic fatigue results from repeated overload without sufficient recovery and is often accompanied by mood disturbances, reduced motivation, impaired sleep, and declining performance despite continued training, a state sometimes referred to as overreaching or, in more severe cases, overtraining syndrome (OTS).

MANAGEMENT

To manage fatigue, cyclists must employ a range of physiological, nutritional, and psychological strategies. Pacing is fundamental as distributing effort according to terrain, race demands, and energy availability reduces the risk of premature fatigue. Power meters and heart rate monitors provide objective feedback to help cyclists regulate intensity, avoiding excessive time in the "red zone" unless it is tactically necessary. Nutrition is equally vital. Consuming adequate energy in the form of carbohydrates during exercise delays glycogen depletion, maintains blood glucose levels, and reduces central fatigue. Pre-race carbohydrate loading and mid-race fuelling strategies are proven methods for delaying fatigue in cycling events lasting over 90 minutes.

PSYCHOLOGICAL RESILIENCE

Mental strategies such as self-talk, focus cues, and visualization can help cyclists cope with discomfort and maintain motivation under fatigue. Psychological resilience and mental fatigue resistance are increasingly recognized as important factors in elite cyclists, especially during long breakaways, adverse conditions, or final-hour race decisions.

RECOVERY

Effective fatigue management depends on a robust approach to recovery. Recovery is the physiological and psychological process of restoring homeostasis and rebuilding function after exercise-induced stress. Combined with strategic pacing, nutrition, hydration, and mental skills, optimal recovery strategies can aid cyclists and coaches to better manage fatigue, optimize performance, and promote physiological adaptations.

FLUID INTAKE

Hydration management and electrolyte balance also influence fatigue (see pp.74–75), particularly in hot conditions where dehydration impairs cardiovascular function and thermoregulation. Regular fluid intake tailored to individual sweat rates and environmental conditions helps preserve plasma volume and cooling capacity, delaying both central and peripheral fatigue.

Brain
The motor cortex signals the muscles to move, and the sensory cortex receives information from the muscles.

Feedback to spinal cord.

Muscles
The agonist muscle opposes the antagonist.

ANTAGONIST ACTIVATION

WEAKER SIGNALS
Central fatigue originates in the brain or spinal cord and causes fewer or weaker signals to be sent from the brain to the muscles. It can be improved through endurance training.

> " "
> *Low muscle glycogen is one of the strongest contributors to fatigue in endurance cycling.*

RECOVERY STRATEGIES

Recovery is the physiological and psychological process of restoring the body's systems to baseline function after exercise. In cycling, where training loads are high, and race schedules can involve multiple consecutive days of intense effort, effective recovery is critical to performance and adaptation.

> *Recovery is just as important as training in cycling. It allows your muscles to repair, adapt, and grow stronger.*

CYCLING RECOVERY

Recovery strategies should include the neuromuscular, cardiovascular, metabolic, hormonal, and cognitive systems. The goal is not only to repair and restore but also to create conditions for supercompensation – the performance-enhancing adaptations that occur after adequate rest following training stress.

Cycling places unique demands on recovery. Unlike many sports, it can involve high training volumes with variable intensity profiles, long-duration events, and cumulative fatigue during multi-day stage races. Because it is non-weight-bearing, cyclists may tolerate higher loads without the same mechanical trauma as running, but this often leads to underappreciation of the need for structured recovery. Neuromuscular and systemic fatigue can still occur.

NUTRITION

What you eat plays a central role in accelerating recovery. The primary post-exercise objectives are to replenish glycogen, repair muscle tissue, and restore fluid and electrolyte balance. Timing is critical – cyclists should aim to rehydrate and refuel within the first 60 minutes post-exercise to maximize glycogen resynthesis and muscle protein synthesis. During multi-day races or double training session days, this window becomes even more important. In longer recovery periods, daily energy and macronutrient needs are the focus.

HYDRATION

Fluid losses via sweat must be replaced to restore plasma volume and optimize cardiovascular recovery. Sodium intake should accompany fluid to support retention and rehydration. Cyclists should monitor hydration status via simple markers such as urine colour or body mass changes before and after sessions. Inadequate rehydration affects thermoregulation, cardiovascular function, and cognitive performance, especially if training or racing starts before full recovery.

LOW-INTENSITY RIDES

Active recovery in the form of low-intensity rides lasting 30–60 minutes is a widely used strategy in cycling. These sessions promote blood flow, help metabolite

---

Meditation

Practising meditation can help cyclists by aiding relaxation and stress relief. This, in turn, helps to ensure good-quality sleep, so your body can repair efficiently.

In addition, meditation encourages you to practise mental focus, which can boost willpower and self-discipline when you must motivate yourself to keep up your training regimen. It may also help increase your mental resilience to frustration, pain, stress, and tough training days, and bolster you during racing challenges.

SLEEP

Sleep is the foundation of recovery. During deep sleep, the body undergoes key anabolic processes, including growth hormone release, muscle repair, glycogen resynthesis, and immune system regeneration. For cyclists, who often train early or race late into the day, protecting sleep quantity and quality is essential. Strategies to maintain "sleep hygiene" such as consistent sleep-wake timing, minimizing screen exposure before bed, and managing post-race caffeine intake support better recovery outcomes.

Sleep hygiene

Proper sleep hygiene can enhance sleep quality and quantity. Try the following habits and practices:

- Keep the bedroom dark, quiet, and cool at 19–21ºC (66–70ºF)
- Ensure your bed and pillow are comfortable
- Avoid backlit screens in the hour before bedtime
- Avoid caffeine later in the day
- Go to bed and wake up at the same time every day
- Create a nightly routine that starts 30 minutes before bedtime to prepare your body for sleep
- Use relaxation or breathing techniques (see Meditation, opposite) if you are anxious or have difficulty falling asleep

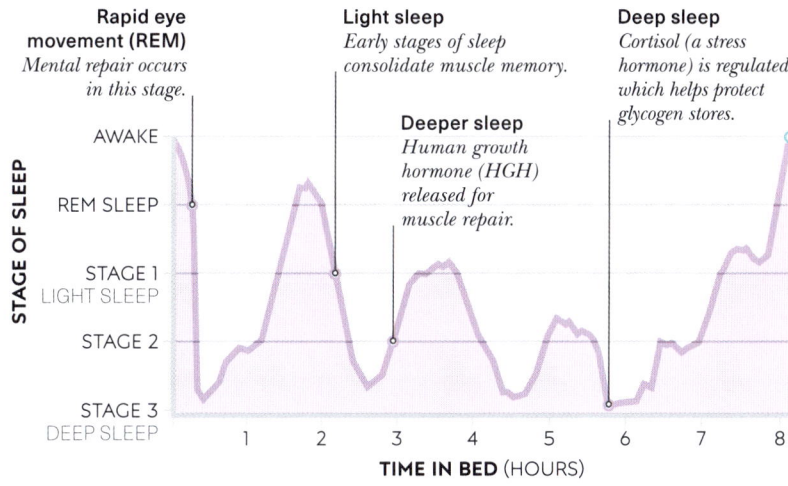

SLEEP FOR RECOVERY
There are distinct stages of sleep that we pass through several times a night. They are all essential for recovery.

clearance, and maintain movement patterns without additional training stress. Active recovery is particularly valuable after races or high-intensity sessions to reduce residual muscle stiffness and aid psychological reset. True rest days, with no or minimal activity, however, are still essential in the overall recovery plan, especially during periods of high cumulative fatigue.

OTHER STRATEGIES

Compression garments, cold water immersion, and cryotherapy (including cold water immersion, cold exposure, and cold application) are used to enhance recovery, especially during multi-day stage races or heavy training blocks. Compression may assist the flow of blood back to the heart and reduce subjective muscle soreness, while cold exposure is believed to reduce inflammation and perceived fatigue. While research shows mixed results regarding performance benefits, these strategies may offer individual or psychological advantages when used appropriately. Importantly, cold water immersion should be used cautiously during key adaptation phases, as it may blunt signalling pathways critical for training gains.

Massage therapy and other physical therapy techniques remain staples of professional cycling. While their physiological effects are debated, benefits can include improved relaxation, reduced perceived soreness, and enhanced mood – factors that contribute meaningfully to recovery.

MONITORING RECOVERY

Tools such as wellness questionnaires, sleep tracking, mood ratings, and perceived fatigue are often used to keep track of recovery and help individualize treatment. However, integrating tools like heart rate variability (HRV), resting heart rate, and power-to-heart rate ratios can provide valuable insight into autonomic recovery and workload tolerance.

CYCLING
TECHNOLOGY

The technological transformation of cycling over the past century has redefined the sport. Over this time we have moved from an era of basic steel frames and rudimentary components to highly sophisticated machines interwoven with digital data streams and meticulously refined aerodynamics. This evolution has not only amplified cycling capabilities but has also fundamentally reshaped training methodologies, racing tactics, and the very analysis of performance.

THE BICYCLE: AN ENGINEERING MARVEL

The bicycle stands as one of the most efficient inventions in human history, an enduring marvel of engineering that has transformed transportation, sport, and society. Despite its apparent simplicity, the bicycle is a finely tuned machine that combines mechanical ingenuity, material science, and human-centred design to achieve remarkable efficiency and performance.

PHYSICAL PRINCIPLES

At its core, the bicycle operates on basic physical principles: conservation of energy, mechanical advantage, and dynamic balance.

A cyclist's pedalling force is transferred through a chain-and-gear system to the rear wheel, producing forward motion with minimal energy loss. Modern drivetrains are highly refined, with precisely machined components that ensure smooth power transmission across a wide range of speeds and terrains. The introduction of derailleurs and indexed shifting systems revolutionized gear changes, allowing cyclists to adapt to gradients and conditions with speed and accuracy.

Riding a bicycle can be up to **5 TIMES** more efficient than walking

1818
Invented by the German Baron Karl von Drais, the Drasine had no pedals.

1839
A Scottish version of the French velocipede with rear wheel cranks and pedals.

1860
A French velocipede with iron banded wheels, known as the "bone shaker".

1870
The ordinary bicycle had a large front wheel that rolled easily over unpaved roads.

1885
The Rover Safety bicycle was a rear-wheel-drive design with the same size wheels.

1960
Road racing bikes of the 1960s had drop handlebars and lightweight frames.

2020
The aerodynamic, lightweight design of a modern time trial racing bike.

SUSTAINABLE DESIGN
In a world of ever-advancing technology, the bicycle remains a model of sustainable design and mechanical excellence. Whether used for commuting, recreation, or elite competition, it shows how engineering can amplify human ability with elegance and style.

MECHANICAL EFFICIENCY

One of the most striking features of the bicycle is its mechanical efficiency. Studies have shown that cycling is more energy-efficient than walking, running, or even swimming. A cyclist can travel four to five times further than a pedestrian using the same amount of energy. This is due in large part to the wheel, the archetypal invention that reduces friction and enhances locomotion. Bicycle wheels, with their lightweight rims and tensioned spokes, strike a delicate balance between strength, aerodynamics, and rolling resistance.

MATERIALS

Frame design and materials have also evolved dramatically. Early bicycles used wood and steel, but modern high-performance bikes leverage advanced materials like aluminium alloys, carbon fibre, and titanium. Each material brings unique engineering properties. Carbon fibre allows engineers to create aerodynamic shapes while improving stiffness-to-weight ratios. It can be precisely layered and moulded, giving designers control over how a frame flexes or resists forces in specific areas, a feature critical for racing performance.

SUSPENSION AND BRAKES

Suspension systems, primarily in mountain biking, demonstrate further engineering sophistication. Front and rear suspension absorb impacts from rough terrain, improving control, comfort, and traction. These systems use finely tuned damping and spring mechanisms, often adjustable to match cyclists' preferences or terrain types. Similarly, braking systems have evolved from simple callipers to powerful hydraulic disc brakes that offer superior modulation, especially in wet or technical conditions.

AERODYNAMICS

Aerodynamics plays a crucial role in modern cycling, particularly at higher speeds where air resistance becomes the primary force opposing motion. Engineers use wind tunnel testing and computational fluid dynamics (CFD) to refine every surface, from frame tube shapes to helmet profiles and cyclist position. Time trial bikes are particularly optimized to reduce drag, sometimes using hidden brakes and truncated air foil tube designs.

DATA ANALYTICS

The integration of electronics and data analytics has also transformed the cycling landscape. Electronic shifting systems provide crisp, reliable gear changes with the press of a button. Power meters and GPS head units deliver real-time feedback on performance metrics like cadence, heart rate, power output, and aerodynamic drag. These tools not only help cyclists train with precision but also provide engineers and coaches with data to fine-tune bike fit, component selection, and pacing strategies.

EFFICIENCY OF TRAVEL MODES
Cycling is the most efficient way of travelling as it uses fewer kcals per kilometre than other methods including walking.

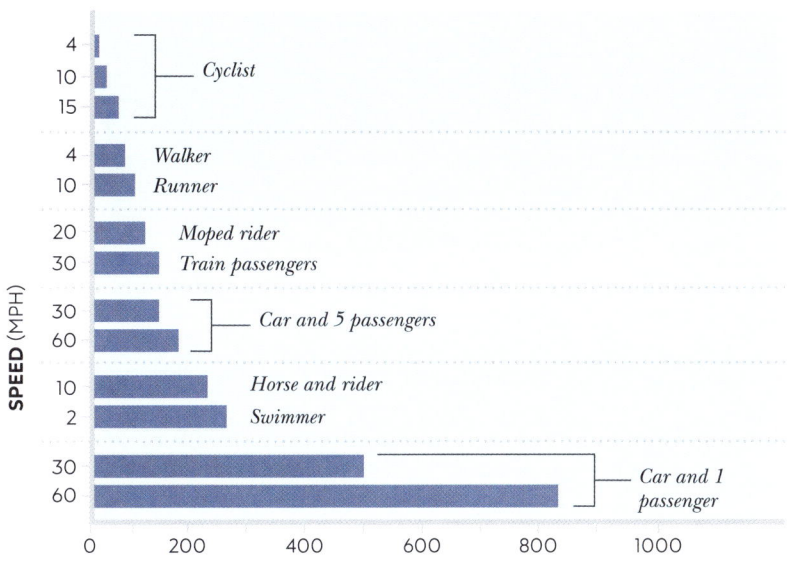

DESIGN ADVANCES

In professional cycling, technological innovation has dramatically reshaped performance capabilities over the past few decades. Advances in frame design, wheel technology, and other components have been driven by a combination of materials science, aerodynamics, biomechanics, and increasingly, data-driven modelling techniques.

FRAME DESIGN

Advances in frame design stem from improvements in materials, aerodynamic profiling, and integrated design philosophies.

MATERIALS

The shift from steel to aluminium to carbon fibre composites has been transformative. Carbon fibre allows manufacturers to tailor stiffness, compliance, and weight characteristics precisely, enabling bikes to be both lighter and more aerodynamically efficient without sacrificing strength. The way the layers of carbon are oriented (layup schedules) are modelled using a process called finite element analysis (FEA). This is done by calculating how each piece of the bike behaves during different cycling conditions and then combining the results to predict how the bike will behave under stress. Companies invest millions into computational simulations before producing a single prototype.

AERODYNAMICS

Modern bicycle frames prioritize aerodynamic drag reduction as a critical performance factor. Aero road bikes feature dropped seat stays, truncated air foil tube shapes, and tightly integrated front ends. These designs balance aerodynamics with weight and stiffness for all-round performance. Computational Fluid Dynamics (CFD) modelling and extensive wind tunnel testing are standard practices used to refine these shapes.

Modern time trial (TT) and track bikes represent the pinnacle of aerodynamic engineering in cycling. They are meticulously designed to minimize air resistance, enabling cyclists to achieve maximum speed with optimal efficiency. Key aerodynamic features include wide-set fork legs and seat stays that channel airflow around the cyclist's legs, acting as aerodynamic "splitter plates". Innovations such as split seat posts and scalloped fork blades further enhance airflow and reduce turbulence.

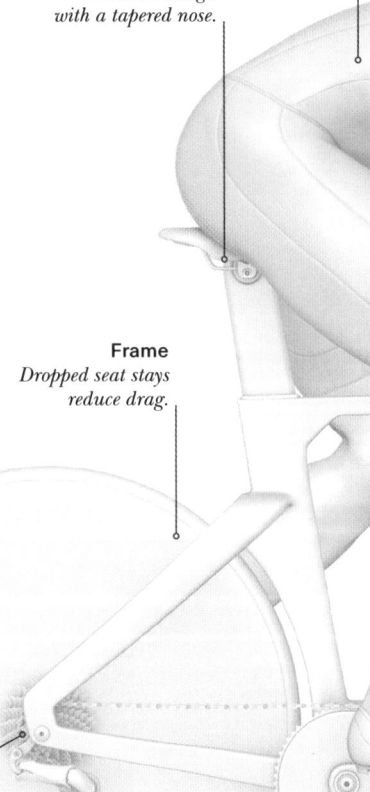

Aerodynamic clothing
Cycling jersey and bib shorts designed to be tight fitting to minimiaze wind resistance.

Saddle
A narrow design with a tapered nose.

Frame
Dropped seat stays reduce drag.

Disc wheels
Designed to reduce aerodynamic drag.

Ceramic bearing systems
Help improve drivetrain efficiency.

CYCLING TECHNOLOGY | Design Advances

INTEGRATION
Beyond tubing shapes, modern frames integrate cables, seat clamps, and even hydration systems within the frame itself to reduce drag. 3D printing and rapid prototyping have enabled designers to quickly repeat and customize solutions, even for specific races or a cyclist's anatomy.

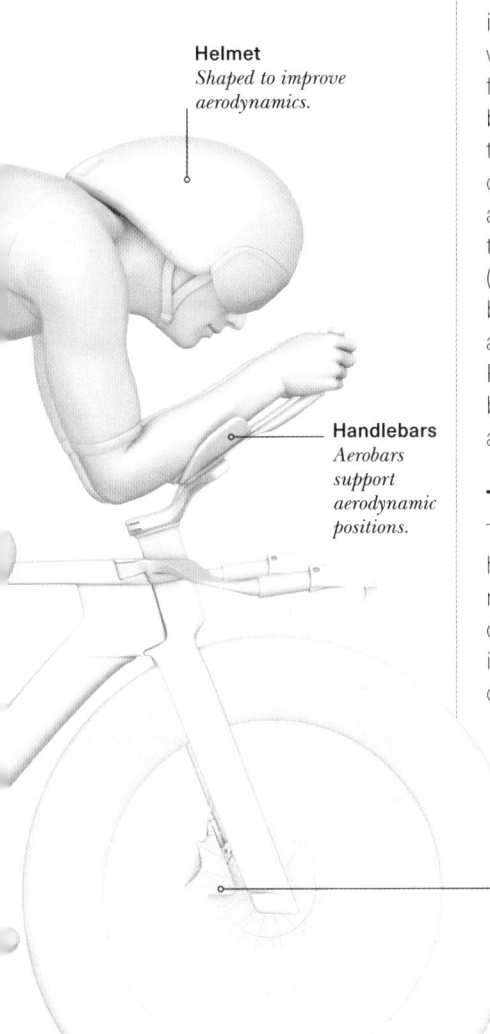

Helmet
Shaped to improve aerodynamics.

Handlebars
Aerobars support aerodynamic positions.

Disc brakes
Allow for better modulation and control in wet conditions.

WHEELS
Wheel development has been as revolutionary as frame design, focussing on aerodynamics, rolling resistance, and material engineering.

AERODYNAMICS
Deeper rim profiles (50–80mm) are now common in road racing, reducing turbulence and smoothing airflow. The trend towards wider rims (19–21mm internal widths) paired with wider tyres (25–32mm) has further optimized aerodynamics by creating a more seamless tyre-to-rim transition. Companies conduct elaborate CFD studies and real-world yaw angle testing to fine-tune wheel performance (yaw angle refers to the angle between the direction of travel and the direction of the wind). Rims are bulging in shape with blunt noses to perform better across variable crosswinds.

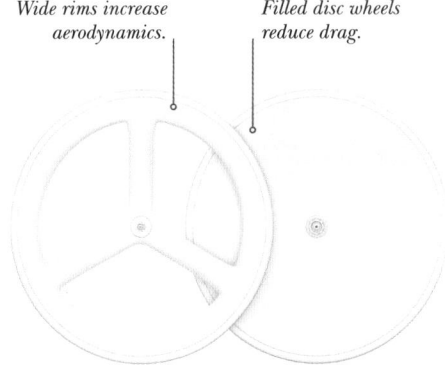

Wide rims increase aerodynamics.

Filled disc wheels reduce drag.

IMPROVING DESIGNS
For time trials and track cycling events tri-spoke and filled disc wheels are used to maximize aerodynamic drag reduction.

TUBELESS TYRES
Tyre systems without tubes have drastically reduced rolling resistance by allowing lower operating pressures without increasing puncture risk. Rim construction techniques, particularly the use of high-temperature-resistant resins and advanced carbon layups (layering), have enhanced wheel durability, braking performance (in rim brake systems), and compliance without sacrificing stiffness. Autoclave carbon moulding combines heat and high pressure in an autoclave (a large pressure vessel) to cure carbon fibre parts that have been impregnated with resin. Robotic layup (layering materials robotically) and resin transfer moulding (RTM) techniques ensure consistency and reduce human error. These manufacturing methods, originally perfected in aerospace industries, now allow for lighter, stronger, and more precisely made wheels.

SELF-SEALING PUNCTURES

An additional advantage associated with tubeless tyre systems is their ability to self-seal small punctures using liquid sealants. These sealants, typically latex-based with added particulates, quickly seal holes as they occur, allowing cyclists to maintain lower tyre pressures for improved rolling resistance, comfort and grip. In select mountain bike and road event cyclists use foam tyre inserts to enhance ride quality by allowing lower tyre pressures for improved traction and comfort, while simultaneously protecting rims from impacts and reducing the risk of pinch flats. These inserts also provide added stability during hard cornering and can enable limited run-flat capability in the event of a puncture.

> The *minimum weight* of a *road racing bike* under the UCI is
> **6.8KG**

OTHER COMPONENTS

Aside from the wheels and frame there continue to be advances in other areas that add to the efficiency and speed of both competitive and non-professional cycling.

DRIVETRAIN EFFICIENCY

Advances in drivetrain components have centred around minimizing friction and maximizing reliability. Ceramic bearing systems, optimized chain coatings, and precise chain line optimization techniques have contributed to measurable savings in a cyclist's power output. Wireless electronic shifting systems allow for faster, more reliable gear changes with cleaner cockpit setups that improve aerodynamics.

HANDLEBARS

Fully integrated handlebars and stems, often developed alongside frames, streamline the front-end profile of the bike. Adjustable single shell (monocoque) handlebars have become common, allowing for aerodynamic gains without compromising on cyclist fit.

BRAKING SYSTEMS

The shift to disc brakes across the professional peloton was initially controversial but is now almost universal. Disc brakes provide superior modulation and braking power, particularly in wet conditions, and have allowed designers to optimize frame and fork designs without needing to accommodate traditional calliper mounts and braking surfaces.

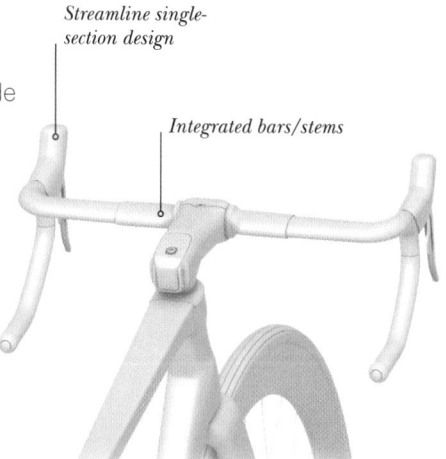

Streamline single-section design

Integrated bars/stems

Power meters and sensors

Almost every professional cyclist's bike today is equipped with a power meter that has advanced significantly in terms of accuracy, weight, and integration. Companies have pushed sensor miniaturization, allowing power measurement at pedals, cranks, or chain ring spider locations.

PEDALS   CRANK ARM

SPIDER   HUB

SENSOR PLACEMENT POSITIONS

CYCLING TECHNOLOGY | Design Advances

UNION CYCLISTE INTERNATIONALE (UCI) REGULATIONS

The UCI plays a vital role in constraining and guiding innovation. Their technical regulations ensure a balance between technological advancement and fair competition but sometimes create friction with engineers and manufacturers.

Key areas where UCI regulations influence bike design include:

DIMENSIONAL LIMITATIONS

Specific dimensional constraints for bicycle components ensure safety and fairness in competition. Handlebars, stems, and frame tubes must have cross-sectional dimensions between 1–8cm. Each component must fit within designated rectangular templates. These templates define the permissible dimensions and positions for various frame elements, including the top tube, down tube, seat tube, seat stays, chain stays, and forks. Additional frame components between the head tube and handlebar stem must also conform to specified dimensions.

MINIMUM WEIGHT LIMIT

The 6.8kg (15lb) minimum weight rule was originally designed to ensure safety as lightweight carbon technology was still maturing. While modern materials easily allow bikes to dip below this weight, manufacturers often add ballast (such as heavier bottom brackets or integrated tools) to meet the requirement.

FRAME SHAPE AND EQUIMENT

Any new frame must pass a UCI approval process, including physical inspection and wind tunnel validation. This bureaucratic hurdle means that certain innovations (for example, extremely aggressive geometries or radical shapes) may not make it to market if they are not approved.

BAN ON MOTORIZED ASSISTANCE

Regulations around hidden motors (after scandals in past races) have led to stricter checks and even the adoption of technologies like magnetic resonance imaging (MRI) to inspect bikes at races.

While the UCI's regulations have limited some aspects of "blue-sky" engineering, they have also encouraged manufacturers to focus on real-world performance gains rather than just laboratory excellence. The result is a modern racing bicycle that is incredibly sophisticated, aerodynamically slippery, and durable, while still appearing visually like a traditional road bike to the casual observer.

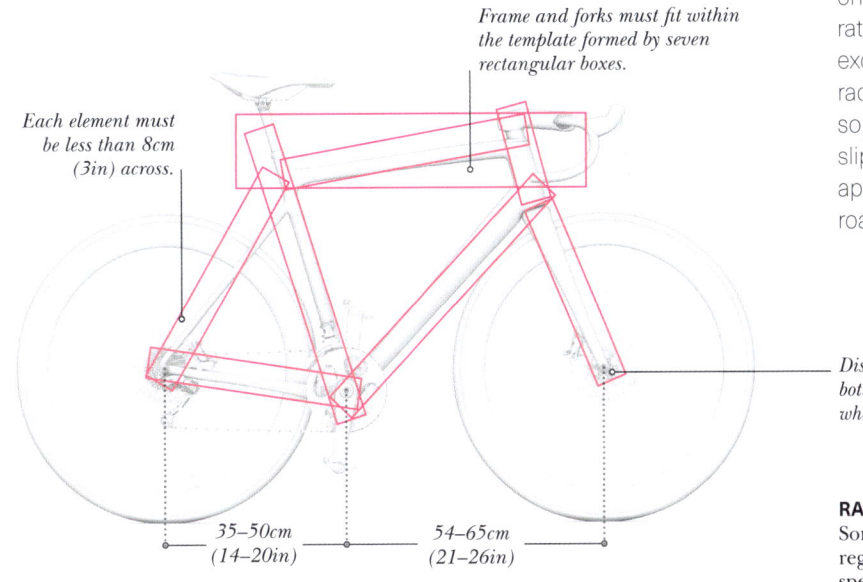

Frame and forks must fit within the template formed by seven rectangular boxes.

Each element must be less than 8cm (3in) across.

Distance allowed between bottom bracket and front wheels is 540–650mm.

35–50cm (14–20in)

54–65cm (21–26in)

ALLOWABLE DISTANCES

RACING BIKE REGULATIONS
Some of the current UCI technical regulations defining the design and specification of competition racing bikes. These are regularly updated.

EFFICIENT SETUP

Making the most of your bicycle for maximal efficiency includes critical component selection, mechanical maintenance, and aerodynamic refinement. These all contribute to enhanced performance and reduced energy expenditure.

Effective gearing can **improve** *mechanical efficiencies* by up to **99%**

COMPONENT SELECTION

The selection of different parts that make up your bicycle profoundly influence its mechanical efficiency.

CRANK LENGTH
Crank length affects pedalling dynamics; shorter cranks can facilitate a more aerodynamic posture and reduce joint strain, particularly beneficial in disciplines like time trials and track cycling.

Additionally, the Q-factor plays a role in pedalling efficiency (see below). A narrower Q-factor aligns more closely with the natural stance of the human body, potentially improving biomechanical efficiency and reducing the risk of knee injuries. Recent trends among professional road cyclists towards shorter crank lengths are thought to have led to further improvements in efficiency, although scientific evidence to support this remains limited.

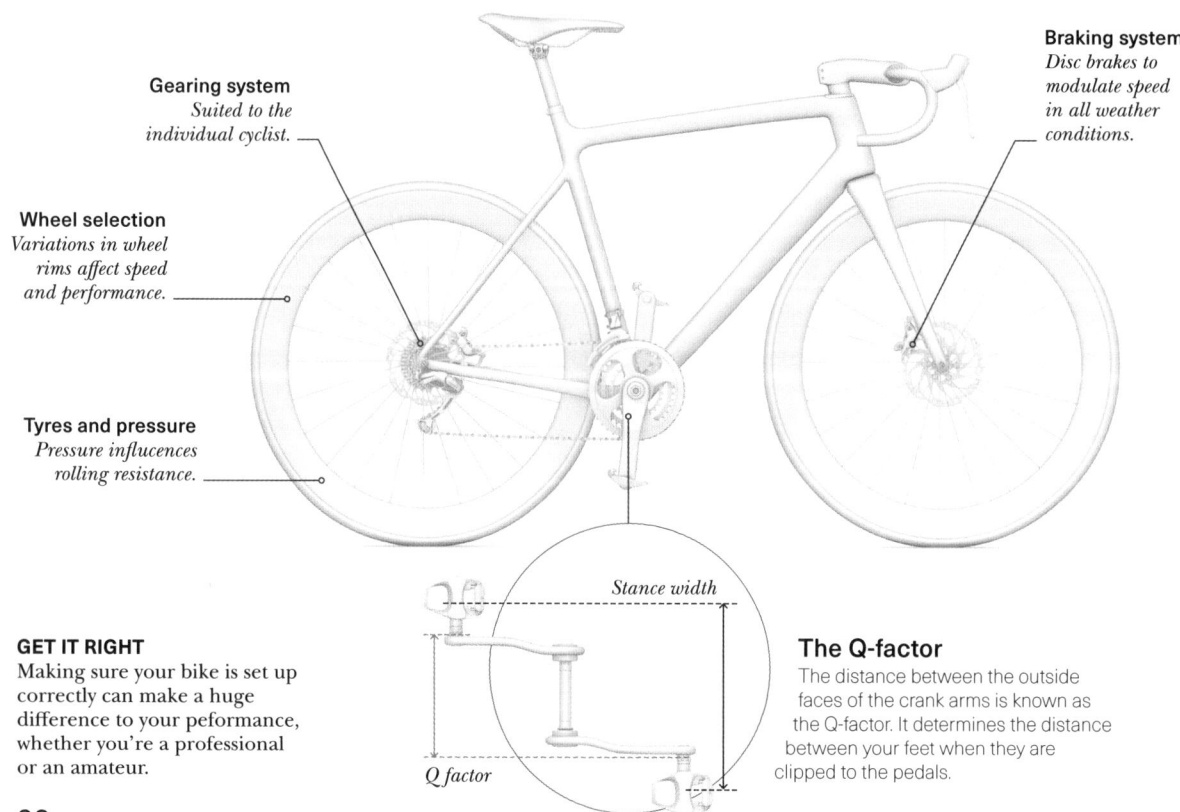

Braking system
Disc brakes to modulate speed in all weather conditions.

Gearing system
Suited to the individual cyclist.

Wheel selection
Variations in wheel rims affect speed and performance.

Tyres and pressure
Pressure influences rolling resistance.

Stance width
Q factor

The Q-factor
The distance between the outside faces of the crank arms is known as the Q-factor. It determines the distance between your feet when they are clipped to the pedals.

GET IT RIGHT
Making sure your bike is set up correctly can make a huge difference to your peformance, whether you're a professional or an amateur.

GEARING SYSTEMS

Gearing systems should be tailored to the cyclist's strength and terrain, ensuring optimal cadence and power transfer. Derailleur systems, when properly maintained, can achieve mechanical efficiencies of between 93–99 per cent, depending on factors such as sprocket size and chain alignment. Larger sprockets and a straight chain line contribute to higher efficiency, while cross-chaining and smaller sprockets can reduce it.

WHEEL SELECTION

Wheel selection plays a pivotal role in efficient setup. Deeper-section rims offer aerodynamic advantages at higher speeds, while shallower rims may provide better handling in crosswinds.

Reducing the rotational mass of wheels, particularly at the rims, enhances acceleration and climbing performance. Carbon fibre rims are lighter than aluminium versions, and so allow for deeper profiles without excessive weight penalties being incurred.

TYRES AND PRESSURE

Tyre width and pressure must be adjusted based on road conditions and cyclist weight to balance rolling resistance and comfort. Wider tyres, when properly inflated, can offer lower rolling resistance and improved comfort compared to narrower ones. Optimal tyre pressure minimizes rolling resistance while maintaining adequate grip and comfort.

ELECTRONIC SHIFTING SYSTEMS

Advances in electronic gear-shifting systems have introduced precise and rapid gear changes, reduced the likelihood of mis-shifts, and enhanced overall drivetrain efficiency. These systems, by eliminating mechanical cables, also reduce maintenance requirements.

Getting the right setup will improve your enjoyment of cycling as well as enhancing your performance.

CRANKS ACROSS THE DISCIPLINES

The setup that provides optimum benefit varies whether you're road racing, mountain biking or using another kind of bike.

Crank length and Q-factor

DISCIPLINE	TYPICAL CRANK LENGTH	Q-FACTOR NOTES
Road racing	160–172.5mm	Narrow preferred for aero and efficiency
Time trial	150–165mm	Very short cranks common to improve aerodynamic position and hip angle
Gravel	165–170mm	Somewhat wider Q for stability
Cyclocross	165–170mm	Slightly shorter for better cadence and pedal clearance
Mountain biking	165–170mm	Moderate Q-factor for stability and technical handling
Track cycling	170–175mm	Longer cranks for maximum leverage and torque
Ultra-distance	165–170mm	Shorter for reduced joint stress over time

Gearing systems

DISCIPLINE	DRIVETRAIN	CHAINRINGS	CASSETTE	NOTES
Road racing (rolling/hilly)	2x12	52/36T or 50/34T	11–30T or 11–32T	Semi-compact for versatility; compact for steeper climbs
Road racing (flat)	2x12	54 or 54/39T	11–25T or 11–28T	Big gears for high-speed racing; tight gearing for cadence control
Gravel racing	1x12 or 2x12	40T (1x) or 48/31T (2x)	10–44T (1x) or 10–36T (2x)	1x for simplicity; 2x for tighter gear jumps
Mountain biking	1x12	32T–36T	10–50T or 10–52T	Wide range for steep climbs and technical terrain
Cyclocross	1x11 or 1x12	40T–44T	11–28T to 11–36T	1x preferred for reliability in mud; moderate gear range
Time trial	2x12 (or 1x12)	56/42T or 54/40T	11–25T or 11–28T	Big rings for flat, fast efforts; small gaps between cassette gears

Tyres and pressure

DISCIPLINE	RECOMMENDED TYRE TYPE	WIDTH RANGE	RECOMMENDED PRESSURE	NOTES
Road racing (smooth tarmac)	Slick road tubeless or clincher	25–28mm	70–90 psi (4.8–6.2 bar)	28mm increasingly common even for pro racing; lower pressure for comfort and grip
Road racing (rough tarmac/ cobbles)	Slick or light file tread tubeless or clincher	28–32mm	60–75 psi (4.1–5.2 bar)	Wider tyres and lower pressures improve control and rolling resistance
Time trial	High-TPI slick racing tubeless or clincher (with latex inner tubes)	25–28mm	80–100 psi (5.5–6.9 bar)	Higher pressure reduces rolling resistance, but too high can be slower over rough surfaces
Gravel racing (fast, hardpack)	Light file tread or semi-slick tubeless	34–44mm	30–45 psi (2.1–3.1 bar)	Balance between speed and grip; tubeless preferred
Gravel racing (mixed/loose)	Knobbier all-round gravel tubeless	38–46mm	25–40 psi (1.7–2.8 bar)	Wider, lower-pressure tyres for rougher conditions and better control
Cyclocross (dry conditions)	File tread CX tubular or tubeless	32–33mm (UCI limit)	22–30 psi (1.5–2.1 bar)	Lower pressures improve grip
Cyclocross (mud)	Aggressive knobby CX tubular or tubeless	32–33mm	18–25 psi (1.2–1.7 bar)	Very low pressure for maximum grip in mud
Mountain biking	Lightweight XC race tubeless	2.2–2.4" (55–61mm)	18–25 psi (1.2–1.7 bar)	Tubeless almost universal; lower pressures for traction

MECHANICAL MAINTENANCE

Regular maintenance is essential to preserve a bicycle's performance. Some key elements that should be considered when improving bike efficiency include:

DRIVETRAIN

Maintaining a clean drivetrain significantly boosts efficiency. Consistent chain care, including regular cleaning and lubrication, is crucial; always wipe off excess oil to avoid attracting dirt. For optimal performance, periodically degrease the chain, cassette, chainrings, and derailleur pulleys before applying fresh lubricant suitable for your riding conditions. Regularly check chain stretch using a chain checker to prevent premature wear on the cassette and chainrings and replace the chain when it reaches the recommended wear limit.

WHEELS AND TYRES

For optimal performance, wheels must remain true. Spin each wheel to check for lateral and vertical movement; significant wobble indicates the need for truing or even rebuilding. Regularly inspect tyres for cuts, punctures, casing damage, and tread wear. Maintain appropriate tyre pressure based on the terrain and cyclist weight, and replace tyres exhibiting considerable tread wear, sidewall fraying, or dry cracking.

GENERAL INSPECTION

Before every ride, a quick check of critical components (stem, handlebars, seatpost, crankarms, wheels) should be done to ensure optimal performance. Listening to the bike while riding can reveal problems early and usually signal that something needs attention.

BRAKES

For rim brakes, inspect the pads for uniform wear and contamination, and ensure the braking surface is clean and without significant grooves. For disc brakes, verify that the pads have adequate material remaining and that the rotors are clean and not warped. If hydraulic brakes feel spongy, the system may need bleeding. When replacing hydraulic brake pads ensure they are correctly bedded in and that the disc rotors are centred and aligned.

CABLES, HOUSING, AND HOSES

For effective shifting and braking, clean and undamaged cables and housing are essential. In mechanical systems, ensure cables move freely without sticking and that housing is free of cracks or corrosion. Replace frayed cables and worn housing when necessary. For hydraulic brakes, inspect hoses for any signs of leaks, cracks, or bulges, particularly at the connections and bends.

BEARINGS

Bearings (headset, bottom bracket, hubs, and pedals) should operate smoothly with no lateral movement. If there is excessive play, the bearings need servicing or replacing. Ceramic bearings are more durable and their lower friction can improve efficiency.

SUSPENSION

Suspension components require regular inspection. Adhering to the recommended service intervals for oil changes and seal replacements is vital for maintaining performance and preventing expensive repairs. Always verify air pressure or damper settings before each ride.

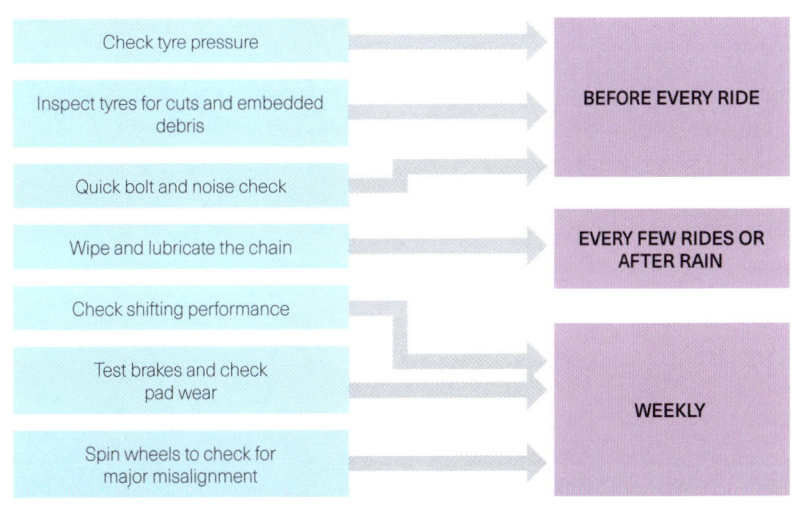

SENSOR TECHNOLOGY AND DATA ANALYTICS

Both amateur and professional cycling has seen rapid transformation driven by advances in sensor technology and data analytics. From training optimization to race-day decision-making and equipment innovation, the sport is now deeply intertwined with data-centric tools.

POWER METERS

Power meters offer objective workload measurements, allowing cyclists to train at specific intensities aligned with their physiological capabilities and performance goals.

These are the cornerstone of modern performance tracking in cycling. Found in cranks, pedals, or wheel hubs, power meters measure the torque and cadence to calculate wattage (work output). They enable quantification of physical load in both training and racing and are used in training monitoring and prescription. Additionally, wattage can be used to determine energy expenditure and facilitate individualized nutrition strategies.

Heart rate data supplements this by indicating internal load and recovery patterns. Continuous glucose monitors, muscle oxygen monitors, and other biochemical assessment tools now provide metabolic insights, facilitating more personalized intensity management and fuelling strategies. Data from power meters and wearable sensors allows cyclists and coaches can make more informed decisions about training prescription. This integrated perspective leads to improved adaptation, a lower risk of overtraining, and enhanced long-term development.

Core temperature sensors
Devices monitor internal temperature via skin-contact wearables, helping manage heat stress which is critical for hot-weather racing and heat adaptation protocols.

Inertial measurement units (IMUs)
Used in high-end smart trainers and some wearables, IMUs measure acceleration, angular velocity, and orientation. They support advanced biomechanics analysis and indoor ride simulation.

Tyre pressure sensor
Helps manage rolling resistance and puncture risk, especially in gravel and time trial disciplines. Some include the ability adjust tyre pressure on the move to adapt to different surfaces.

CYCLING TECHNOLOGY | Sensor Technology and Data Analytics

TRACKING DATA
Sensors on the rider and bike monitor power, heart rate, motion, and physiology to provide real-time performance data.

Heart rate monitors
Worn on the chest or wrist, these use optical or electrical signals to track cardiovascular effort. While less direct than power output, heart rate helps track fatigue, recovery, and stress.

GPS devices
Bike computers use GPS and barometric altimeters to log route data, elevation, speed, and segment times. These are often synched to cloud training monitoring platforms for analysis.

Aerodynamic sensors
Measure real-time drag (CdA) in real-world riding conditions. Devices integrate wind speed measurements with GPS, accelerometers, and barometric data to estimate aerodynamic efficiency without wind tunnels.

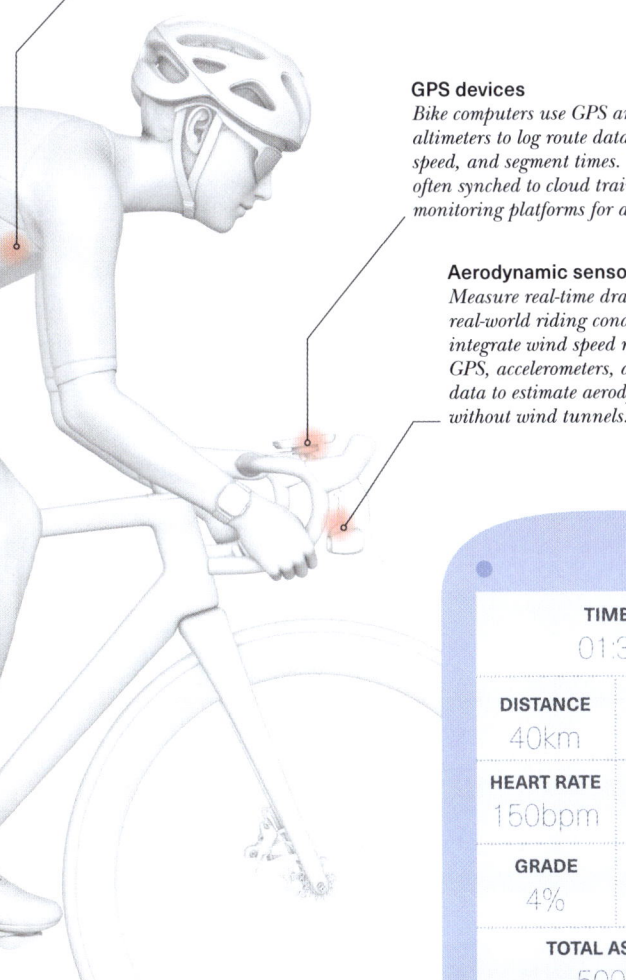

TIMER	
01:30	
DISTANCE	**SPEED**
40km	30kph
HEART RATE	**CADENCE**
150bpm	95rpm
GRADE	**POWER**
4%	250W
TOTAL ASCENT	
500m	

Other devices
Some further devices that are used to convert physical activity into actionable insights:

Temperature, humidity, and altitude sensors: Integrated into some bike computers and wearables, these help cyclists to analyse environmental stress on performance and pacing in extreme conditions.

Muscle oxygen sensors: Near-infrared spectroscopy (NIRS) devices measure local muscle oxygen saturation. Used primarily in labs or elite settings, they can fine-tune training intensity by showing when a muscle shifts to anaerobic energy systems.

Sweat sensors: Adhesive patches employ microfluidic technology to quantify sweat rate and sodium levels and analyse fluid loss and electrolyte content during exercise, offering immediate insights into hydration requirements. In hot weather or during prolonged training or racing this enables the development of tailored hydration plans to mitigate dehydration, and performance decline.

Continuous glucose monitors (CGMs): These devices employ a small subcutaneous sensor to measure interstitial glucose, allowing cyclists to understand their personal carbohydrate response during training and competition. While currently prohibited in UCI races, CGMs are often used in training to optimize nutrition plans and avoid energy dips, particularly in endurance events.

Sleep sensors: Advanced smartwatches, rings, and wrist-worn devices monitor sleep duration and quality using accelerometery and heart rate variability (HRV) to estimate sleep stages and recovery. This can help identify early indicators of fatigue or illness, optimize training schedules, and manage the stresses of travel and competition.

THE FUTURE OF DATA

As technology continues its rapid advancement, cycling is entering a new era characterized by deeper physiological understanding, real-time intelligence, and the integration of hardware with sophisticated analytics.

The combination of miniaturized sensors, wearable devices, cloud computing, and artificial intelligence is transforming how cyclists train, race, and recover. The future will likely bring even faster innovation, both in the available tools and in the way data is captured, interpreted, and used.

*AI is **already being used** to **optimize pace strategies** for time trials*

The evolution of cycling technology is shifting from simply collecting data to providing decision intelligence. Future systems will not only report what occurred but will also recommend the next steps and explain the reasoning. Cyclists of all levels will gain from smarter sensors, real-time analytics, and AI-driven decision support.

Ultimately, the future belongs to those who can integrate technology with intuition. The goal is to use data not as a substitute for understanding, but as a partner in the pursuit of performance, health, and long-term development.

INTEGRATION AND MINIATURIZATION

A clear trend is the development of multi-functional, miniaturized sensors designed for seamless wear without hindering performance. Instead of using multiple devices, sensor fusion is leading to integrated platforms that combine metrics like power, heart rate, temperature, motion, lactate, glucose, and sweat into a single wearable or embedded system (smart apparel).

Future wearables may integrate, core body temperature, skin hydration, muscle oxygen saturation, blood lactate quantification and real-time glucose monitoring within a single patch or incorporated into jersey fabric. This paves the way for continuous metabolic monitoring, rather than isolated measurements, allowing cyclists and coaches to make dynamic adjustments during training or competition.

REAL-TIME DATA PROCESSING

As sensor capabilities increase, the volume and speed of data collected are growing rapidly. To manage this, there's a move towards edge computing, processing data directly on the device or head unit instead of relying solely on post-ride uploads. This enables real-time fatigue alerts, in-ride fuelling recommendations, live aerodynamic optimization prompts, and auto-adjusted pacing based on wind, slope, and rider condition.

The development of AI-powered bike computers that interpret data, rather than just displaying it, is expected to be transformative, changing data's role from passive recording to active guidance.

AI FOR COACHING

Artificial Intelligence is already used in training platforms to automatically create workouts and adjust plans based on cyclist input and performance trends. The future is likely to see AI advance into predictive performance modelling, using past and present sensor data to forecast peak performance windows, risk of overtraining or injury, early detection of illness and race day performance under various condition.

With access to extensive datasets and machine learning algorithms, AI could guide fine-tuned adjustments in tapering, rest days, nutrition, and travel arrangements, all personalized and adaptive.

DATA COLLECTION

The use of AI will become more commonly used by cyclists to gather and interpret information that can be used to improve training and performance.

CYCLING TECHNOLOGY | The Future of Data

PERSONALIZATION OF EQUIPMENT

Sensor-driven bike fitting and equipment selection are set to become more advanced. While motion sensors, saddle pressure mapping, and pedal stroke analysis are already used in high-quality bike fits, future systems may offer dynamic fit adjustments based on riding conditions.

For example, saddles that change stiffness depending on vibration levels, aerobars that alter their tilt for climbing versus flat roads, smart insoles that provide real-time guidance on cleat positioning. These advancements will help cyclists maintain biomechanical efficiency and comfort across different race conditions and levels of fatigue.

HOLISTIC MONITORING PLATFORMS

The future of cycling technology lies in integrated platforms that unify training, health, nutrition, and recovery data. These platforms will enable coaches, sports scientists, and medical staff to operate from a shared dashboard with input from multiple sensor data sources.

Such systems will not only improve performance but also support well-being, helping to identify warning of factors contributing to underperformance. In fact, mainly professional cycling teams already use bespoke platforms to achieve this at the highest level of the sport.

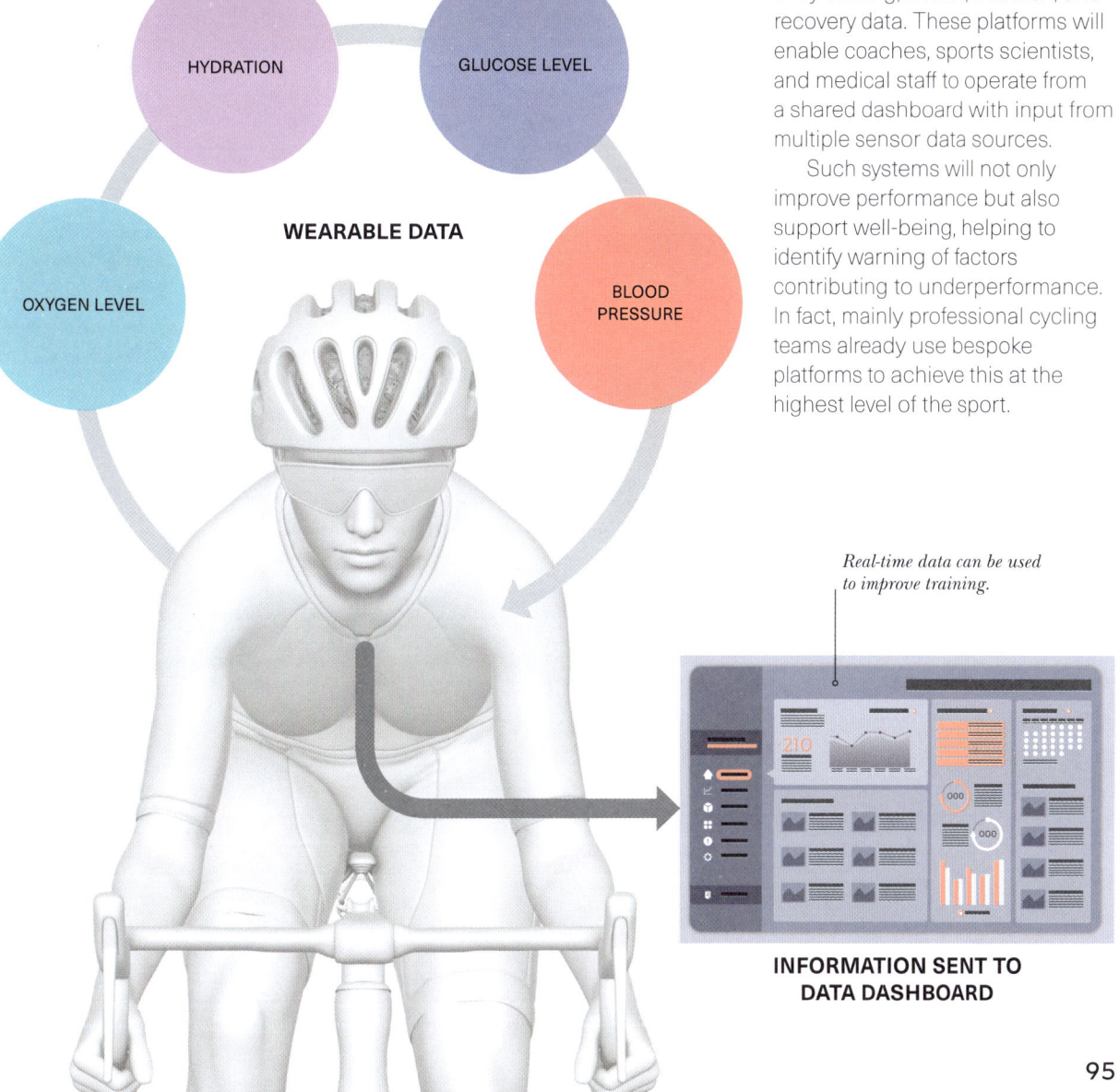

Real-time data can be used to improve training.

INFORMATION SENT TO DATA DASHBOARD

COMPETITIVE CYCLING

Competitive cycling uniquely fuses physical endurance, tactical intelligence, technical skill, and psychological resilience. At the professional level, it transcends a simple test of fitness, evolving into a complex team sport where strategic acumen, meticulous preparation, and the exploitation of marginal gains frequently prove as crucial as raw power in the pursuit of victory.

THE ROAD RACING SCENE

Road cycling is one of the most physically demanding and tactically complex sports in the world. At the elite level, it is a highly structured global sport driven by tradition, evolving technology, data analytics, and commercial interests. The World Tour calendar, governed by the Union Cycliste Internationale (UCI), forms the backbone of the competitive season and showcases the sport's top talent across various terrains and race formats.

STRUCTURE AND GOVERNANCE

The elite road racing scene is primarily centred around the UCI WorldTour, a collection of top-tier races including one-day classics, stage races, and Grand Tours.

These events are contested by UCI WorldTeams, the highest level of professional teams, alongside selected ProTeams (second-tier teams). The WorldTour calendar spans from late January, starting with events like the Tour Down Under in Australia, to October, culminating in prestigious one-day races in Italy. Supporting this top tier is the UCI ProSeries and a global calendar, which serve as proving grounds for developing riders and smaller teams aiming to move up the ranks. This structure allows for promotion and relegation between tiers based on performance and financial criteria.

3,400 KM is the *average* distance of the *Tour de France*

Race types and specializations

RACE TYPE	TOP EVENTS	FEATURES
Stage races Multi-day events ranging from three days to three weeks.	Giro d'Italia, Tour de France, and Vuelta a España.	These races test endurance, recovery, climbing, time trialling, and tactical acumen across diverse terrain and weather conditions.
One-day classics Single-day races over long distances.	Paris-Roubaix, Tour of Flanders, and Milan–San Remo	Often feature unpredictable weather, cobblestones, short steep climbs, and chaotic racing. These events are won by riders with toughness, race craft, and a deep understanding of positioning.
Time trials Individual or team races against the clock.	Some stages of the Tour de France, World Championships, and the Olympics.	Success depends on aerodynamics, pacing, and power output over specific durations. Time trials are standalone events but also feature within stage races, sometimes proving decisive for general classification outcomes.

TEAM DYNAMICS AND TACTICS

Each race type favours different rider profiles: sprinters, climbers, time triallists, "puncheurs" (cyclists specializing in short, steep climbs), and all-rounders. The most successful riders either specialize heavily or develop well-rounded skill sets to contend across multiple terrains.

Cycling is an individual sport executed in a team context. Each professional team fields six to eight riders per race, assigned specific roles including domestiques, sprint leaders, and road captains.

Winning at the elite level demands not only physical excellence but also precise strategy, communication, and timing. Teams deploy radios and follow detailed race plans devised by team directors but must constantly adapt to evolving race scenarios.

Modern professional cycling in recent years has undergone a data-driven transformation and integration of sports science and medicine. Teams now employ sports scientists, performance directors, nutritionists, biomechanists, engineers, and medical specialists to optimize every element of performance. Reconnaissance of race routes, wind tunnel testing, and individualized pacing strategies are now standard among top-tier teams.

TEAM FORMATION
In a typical team the leader will be protected by the team for most of the race. The domestiques may be at the front of the group setting the pace and leading out sprints.

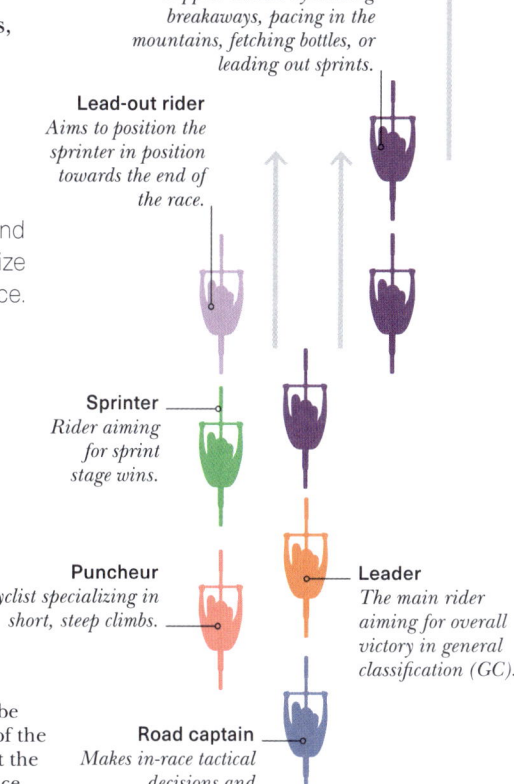

Domestiques
Support leaders by chasing breakaways, pacing in the mountains, fetching bottles, or leading out sprints.

Lead-out rider
Aims to position the sprinter in position towards the end of the race.

Sprinter
Rider aiming for sprint stage wins.

Puncheur
Cyclist specializing in short, steep climbs.

Leader
The main rider aiming for overall victory in general classification (GC).

Road captain
Makes in-race tactical decisions and manages positioning.

CULTURE, COMMERCIALIZATION, AND ANTI-DOPING

Cycling is deeply rooted in European culture, especially in countries like France, Italy, Belgium, and Spain.

However, the sport has globalized significantly in recent decades. The inclusion of races in the Middle East, Australia, and North America has expanded cycling's reach, while rider nationalities now span every continent.

At the elite level, commercial sponsorship is vital. Teams are primarily funded by corporate sponsors, and the return on investment is based on brand exposure during televised races, social media, and in-person events. Sponsorships can be volatile, leading to frequent team name changes and financial instability if performance declines.

The sport continues to battle historical associations with doping. However, biological passports, more stringent testing, and cultural shifts have gradually rebuilt credibility. The current generation of riders is largely seen as clean and extremely professional in their approach to training and recovery. Meanwhile, sustainability, rider safety, and mental health have become key areas of focus, with growing calls for reform in race organization, equipment standards, and rider representation.

DRAFTING AND THE PELOTON

In professional road cycling, the main group of riders, called the peloton, is more than a simple assembly of riders. It functions as a fluid entity, governed by the interplay of aerodynamic principles, rider physiology, and race tactics. To the casual viewer, the peloton may look chaotic with dozens of riders in tight formation, jostling for position. But to those who ride within it, the peloton is a marvel of energy flow, decision-making, and efficiency.

PELOTON DYNAMICS

At the core of peloton formation is the concept of drafting, sometimes referred to as slipstreaming. This fundamental aerodynamic mechanism allows riders to achieve substantial energy savings by positioning themselves within areas of reduced air resistance created by the riders ahead.

Drafting within the peloton is not merely a means of conserving energy; it influences the outcome of races and dictates the tactical approaches adopted by both riders and teams across every element of the competition.

ENERGY AND AERODYNAMICS
When a rider moves through the air, they must overcome two major resistive forces: rolling resistance and aerodynamic drag. While rolling resistance remains relatively constant, aerodynamic drag increases with speed. At a typical professional race speed of 40–50kph (25–32mph), a rider expends up to 90 per cent of their energy overcoming air resistance.

Drafting mitigates this by exploiting the low-pressure wake behind another rider. When one rider takes up a position behind another, they face significantly reduced air resistance. Scientific studies have quantified this benefit: a rider drafting directly behind another rider can save 27–50 per cent of their energy.

Riders in the centre of the peloton are shielded. This aerodynamic efficiency allows them to conserve energy and delay fatigue, a critical advantage in a sport where energy management over several hours is the difference between winning and falling out of contention.

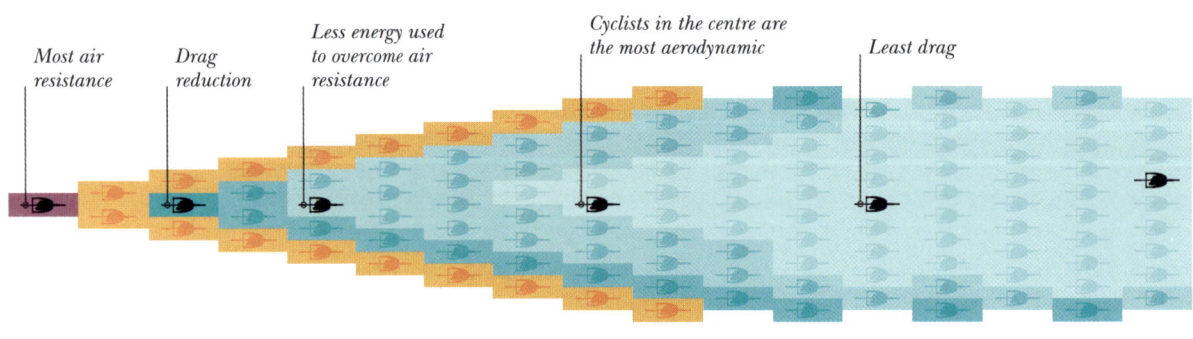

AERODYNAMIC DRAG REDUCTION
Where riders are shielded on all sides in the peloton, energy savings can exceed 90 per cent, particularly in the centre and back.

Percentages of energy being used to overcome air resistance

PELOTON TACTICS

Domestiques expend energy riding on the front and set the pace so their leaders can remain shielded in the draft. This sacrificial energy transfer is fundamental to professional cycling: a leader's victory is made possible by teammates absorbing wind load earlier in the race.

The peloton wields enormous collective power. A small group of riders may gain several minutes of advantage if they break away from the main group, but if the peloton commits to the chase, it is almost always faster due to superior drafting and resource sharing. A paceline in a breakaway rotates to minimize fatigue, but the peloton can spread the load across multiple riders. Teams with an interest in a sprint finish, for instance, will position their riders at the front of the peloton to maintain a controlled pace. Drafting ensures they can ride fast but efficiently, pulling back time without exhausting themselves. Maintaining a forward position in the peloton is a perpetual battle. It reduces crash risk, provides better access to the race's tactical cues, and is essential approaching narrow roads, climbs, or technical sectors. However, moving up can cost energy, especially in headwinds or open terrain. Drafting becomes a negotiation as riders tuck in behind others to save energy, but everyone wants a better spot. The friction of this constant positioning is where physical prowess meets tactical sharpness. Teams use lead-out trains, shields of domestiques, or even borderline-aggressive positioning to preserve their leaders' places.

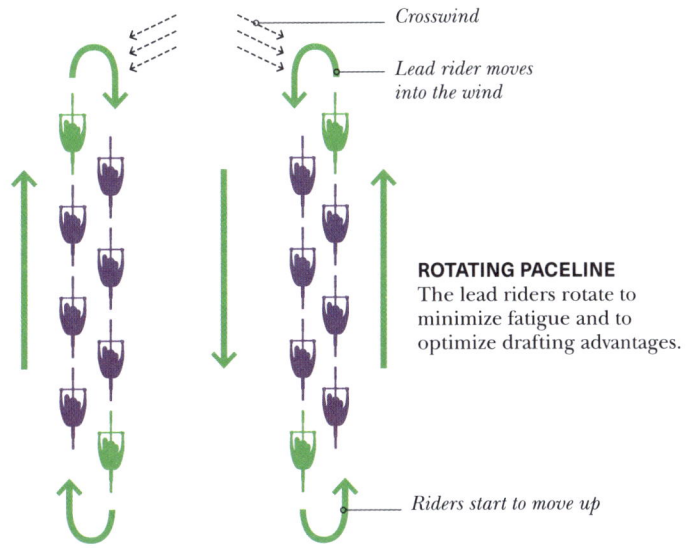

Crosswind

Lead rider moves into the wind

ROTATING PACELINE
The lead riders rotate to minimize fatigue and to optimize drafting advantages.

Riders start to move up

STRUCTURE AND MOVEMENT

The peloton is often described as a single unit, but it behaves more fluidly, with constantly shifting density, direction, and flow.

The shape of the peloton changes according to road conditions, race situations, wind direction, and the decisions of individual riders.

The centre and rear interior of the peloton provide the greatest aerodynamic shelter, but are also the most chaotic. Braking, surging, and cornering ripple through the group like waves. Constant surging is metabolically inefficient and mentally fatiguing. The front third of the peloton, especially the outer edge, is where control happens. This zone allows quick response to attacks and crashes and maintains tactical flexibility. However, it requires higher energy output due to partial wind exposure and the effort needed to maintain position.

Climbing and descending

Drafting is less effective at lower speeds. However, riders still draft when they can, particularly on moderate gradients and in large groups. Shielding from headwinds and psychological pacing benefits still apply. On longer climbs, tactical pacing and drafting within climbing groups can make marginal but crucial differences. On descents, the peloton becomes looser and more dynamic. Riders often regroup at the bottom of climbs or descents, where drafting helps to bring dropped riders back.

ECHELONS AND BREAKAWAYS

Echelons and breakaways reflect the complex interplay of physics, positioning, and team dynamics in professional cycling. Echelons leverage wind and organization to punish inattention and poor positioning, while breakaways test collaboration, endurance, and tactical nous against the collective strength of the peloton.

DEALING WITH CROSSWINDS

Echelons are diagonal formations created in response to crosswinds. When wind strikes from the side, the traditional straight-line drafting formation becomes inefficient. Instead, riders stagger diagonally across the road, each seeking shelter on the downwind side of the rider ahead.

This positioning creates a rotating line that offers optimal aerodynamic benefit while maintaining high speeds. However, road width limits the number of riders who can fit in the protected echelon. Those pushed into the "gutter" must work significantly harder, often leading to gaps and splits in the peloton.

In a crosswind, the effective wind angle increases the aerodynamic drag experienced by riders who are not well positioned. Because of this, riders who are unprepared or out of position when the peloton hits a crosswind section may lose minutes in just a few kilometres.

Tactically, echelons are often initiated by well-drilled teams looking to isolate rivals or reduce the size of the peloton. Classics-oriented teams or those from windy countries like the Netherlands and Belgium tend to excel at this. Directeurs Sportifs use course reconnaissance and real-time weather data to identify crosswind sectors and deploy riders to the front in anticipation. Echelons are effective on narrow roads or after high-speed sections where riders are fatigued. They are race-defining and can alter general classification standings in stage races like the Tour de France.

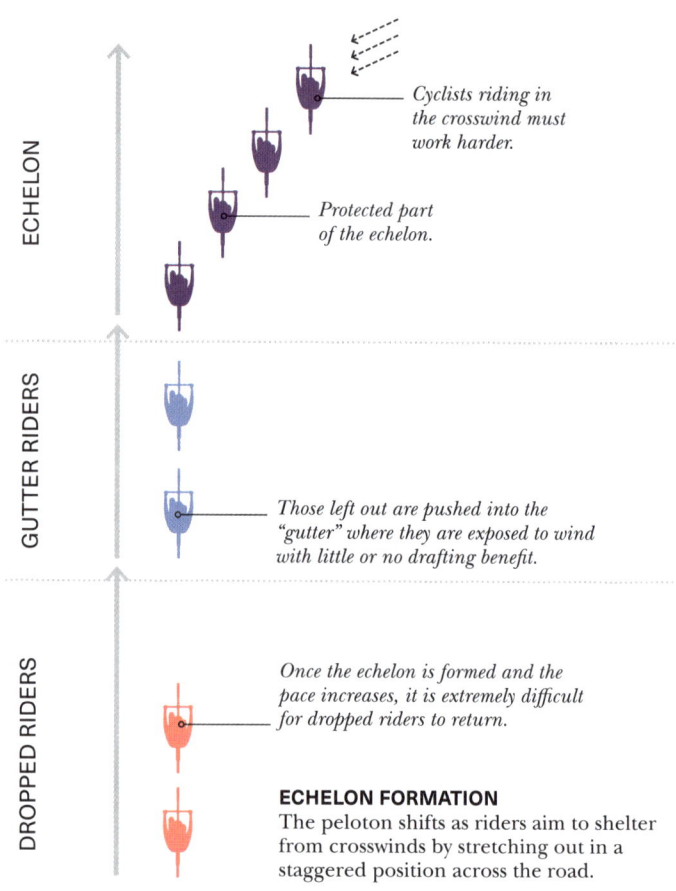

Cyclists riding in the crosswind must work harder.

Protected part of the echelon.

Those left out are pushed into the "gutter" where they are exposed to wind with little or no drafting benefit.

Once the echelon is formed and the pace increases, it is extremely difficult for dropped riders to return.

ECHELON FORMATION
The peloton shifts as riders aim to shelter from crosswinds by stretching out in a staggered position across the road.

ESCAPING THE PELOTON

Breakaways are attacks launched by individuals or small groups who escape the peloton, typically early or mid-race.

They serve various purposes: winning the stage, gaining exposure for sponsors, softening the peloton for later attacks, or acting as satellite riders for teammates.

From a scientific standpoint, breakaways are less efficient than riding in the peloton. The reduced number of riders means less drafting benefit, and each rider must spend more time at the front. The average power output of a breakaway rider is significantly higher than that of those in the peloton, especially during the first hour of racing, where average power output can exceed 5W/kg.

To counteract this, breakaway riders use rotating pacelines, where each rider takes a short pull and then returns to the draft line. The goal is to distribute the workload evenly and keep the pace steady. In ideal conditions, a cooperative breakaway can sustain a high tempo for hours, especially on terrain that disrupts the chasing rhythm of the peloton.

Success, however, is rare. The peloton, benefitting from deeper drafting and more manpower, can often close the gap in the final 20km (12 miles), especially when teams with sprinters or general classification contenders organize a chase.

Breakaways may be allowed to form when they pose no threat to the overall race outcome, and the composition is favourable (there are no dangerous riders or limited team representation).

Factors for success

Successful breakaways usually involve the following elements:

- **Strong, evenly matched** riders willing to collaborate for the good of the team as a whole rather than putting their own success first.
- **Tactical miscalculations** or lack of motivation in the peloton meaning that the breakaway isn't chased down and brought back in.
- **Favourable terrain** in the form of undulating, narrow, early climbs or technical roads can mean that the peloton are unable to pull the breakaway back.
- **They take place late** in the race, leaving little opportunity for the main peloton to catch up.

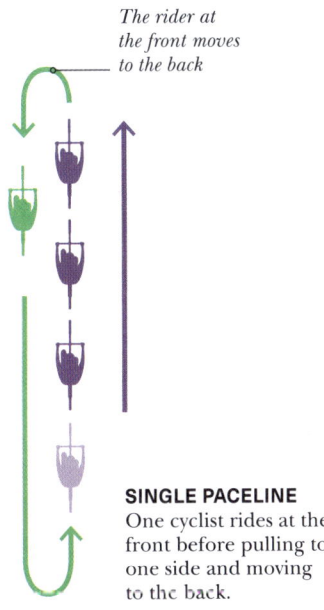

SINGLE PACELINE
One cyclist rides at the front before pulling to one side and moving to the back.

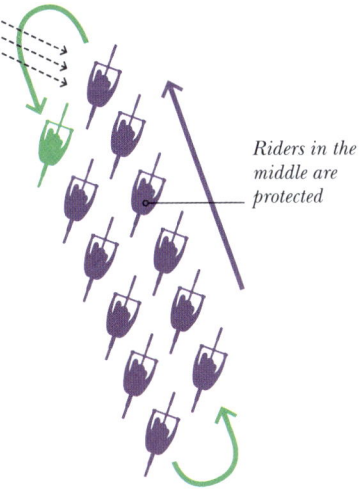

ECHELON PACELINE
In this formation riders draft at an angle against the wind. This can be difficult in narrow spaces.

Up to **30%** *more energy* is needed by cyclists riding in the *crosswind*

HOW TO RACE A CLIMB

In professional cycling, effective pacing and climbing are crucial to success, especially in stage races where cumulative fatigue and terrain intricacy determine outcomes. Mastering these elements involves a combination of physiological understanding, race craft, and strategic execution tailored to individual rider profiles and team objectives.

PACING PRINCIPLES

Effective pacing is the strategic distribution of a rider's energetic resources across a race segment and is especially critical during climbing. Mistakes in pacing, such as starting too aggressively, can lead to premature depletion of glycogen stores, elevated blood lactate levels, and a decline in neuromuscular efficiency.

Getting pacing right allows riders to maximize the time they spend at or near physiological limits while minimizing the accumulation of fatigue-related metabolites.

Accurate modelling of pacing strategies involves understanding the interplay between a climb's length, gradient profile, and altitude. Coaches often use power–duration curves, derived from historical training and race data, to estimate sustainable targets. For example, a 20-minute climb typically requires an effort around 100–105 per cent of Functional Threshold Power (FTP), depending on the rider's anaerobic work capacity and tolerance for over-threshold efforts. In contrast, a 40-minute climb necessitates a more conservative output, generally in the 95–100 per cent FTP range to avoid unsustainable lactate accumulation and maintain muscular endurance.

THE EFFECT OF ALTITUDE

Altitude further complicates pacing because reduced oxygen availability impairs aerobic metabolism. At elevations above 2,000m (6,562ft), maximal aerobic power and FTP can decrease by more than 10 per cent, even in acclimatized cyclists. Pacing targets must be adjusted downwards to account for diminished oxygen delivery and slower recovery from high-intensity efforts.

Performance in climbing

Climbing stresses the aerobic system more than almost any other cycling demand. Power-to-weight ratio (W/kg) is all-important on inclines. A rider with a high VO_2 max, excellent lactate clearance, and efficiency at threshold intensities will generally perform well uphill. Great climbers also possess a refined sense of pacing strategy, often guided by a deep awareness of their own physiological limitations, terrain nuances, and the behaviours of rivals.

RESISTANCE

On steep climbs of more than 8 per cent, gravitational resistance is more important than aerodynamics. On moderate gradients of 3–6 per cent, however, aerodynamic drag is still an important factor.

DRAFTING

Drafting can reduce energy expenditure by up to 40 per cent so needs to be factored into pacing strategies.

FTP

Cyclists on a 20-minute climb require an effort of 100–105 per cent of Functional Threshold Power (FTP) while a 40-minute climb requires a more conservative output of 95–100 per cent.

COMPETITIVE CYCLING | How to Race a Climb

TACTICAL CLIMBING

In race scenarios, pacing can be unpredictable, especially on climbs where terrain, team tactics, and race dynamics converge to create repeated surges in intensity.

At the base of a climb teams may surge to gain advantage. Cyclists must rapidly decide whether to respond to these surges, maintain a self-selected steady-state effort, or temporarily drop and rely on measured pacing to rejoin the group. These decisions are informed by a combination of real-time power output, internal load indicators such as rating of perceived exertion (RPE), and communication with team staff and teammates.

Attacking on a climb involves operating well above critical power (CP) or functional threshold power (FTP) for brief period, engaging anaerobic capacity and glycolytic reserves, before recovering at high aerobic intensities.

Team dynamics further shape pacing strategies. Domestiques often set a high but submaximal tempo just below threshold to suppress opportunistic attacks and control group composition, essentially creating a "pacing shield" for their team leader. This tactic serves both to regulate the peloton's velocity and to fatigue rivals through consistent high-intensity output. Conversely, general classification contenders must adopt a pacing strategy that allows them to remain within their sustainable physiological zones (for example, at or just below Critical Power), minimizing unnecessary surges while remaining alert to decisive moves by key rivals.

USING DATA

Modern technology plays a central role in pacing regulation during races. Power meters provide real-time data of external workload while heart rate monitors offer insights into internal load and cardiovascular strain. Coupled with atmospheric or GPS-based gradient data displayed on head units, these tools allow riders to modulate effort with a high degree of precision, particularly on variable terrain. However, data should inform rather than dictate pacing. Environmental factors such as altitude, temperature, wind, and road surface and race dynamics can affect the relationship between power output and physiological cost.

PSYCHOLOGICAL AND TACTICAL READINESS

Climbing effectively under race pressure demands significant mental resilience and cognitive control. Riders must regulate emotional responses during surges, manage pain perception, and keep focus on factors such as cadence, breathing rhythm, and pedalling mechanics. The ability to remain composed in volatile moments allows a rider to avoid unnecessary anaerobic spikes.

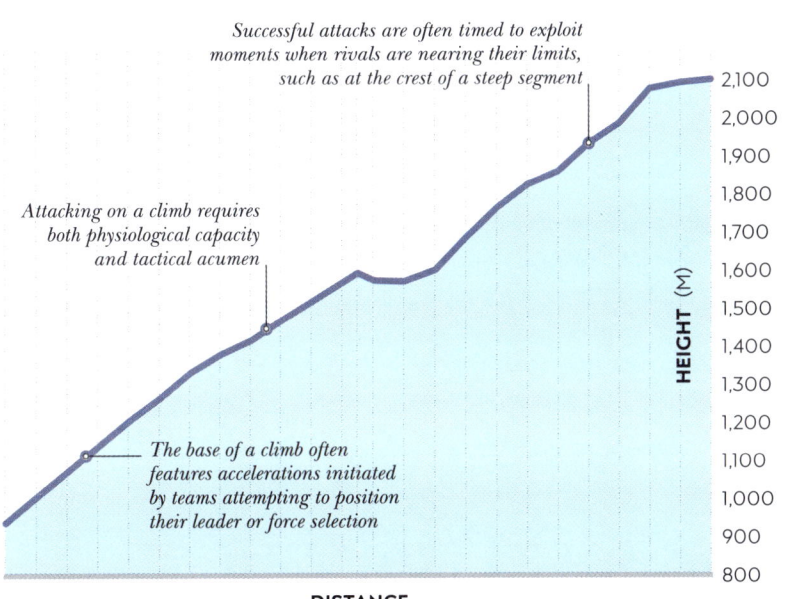

Successful attacks are often timed to exploit moments when rivals are nearing their limits, such as at the crest of a steep segment

Attacking on a climb requires both physiological capacity and tactical acumen

The base of a climb often features accelerations initiated by teams attempting to position their leader or force selection

HILL CLIMB TACTICS
Experienced cyclists know when to conserve energy and when to attack on a hill climb.

LEAD-OUTS AND SPRINT FINISHES

Sprint finishes in professional cycling represent some of the sport's most critical, technically intricate, and physically taxing phases. Success demands a unique combination of peak anaerobic power, refined neuromuscular control, astute tactical awareness, and coordinated team execution.

> Top road sprint cyclists can produce over **1,500** watts

THE END OF THE RACE

Although the final, explosive 200m (656ft) of a race often captures the attention, the result is the culmination of a meticulously orchestrated series of choices and efforts that typically commence more than 10km (6 miles) from the finish.

Central to this process is the lead-out train, a precisely paced formation of teammates collaborating to position their sprinter for the decisive launch with optimal speed, shelter, and track position.

Elite sprinters possess exceptional neuromuscular power, with peak outputs ranging from 1,600 to over 2,000 watts, and the ability to sustain 1,000+ watts for 10–20 seconds. These high-power efforts are underpinned by well-developed phosphagen and glycolytic energy systems. However, sprint performance does not occur in isolation. Sprints often follow 4–6 hours of racing at high average intensities, including multiple surges, positioning battles, and terrain variability.

Success thus depends not only on absolute power output, but also on fatigue resistance, or the ability to produce near-maximal efforts after significant prior workload. A robust aerobic base, reflected in high VO_2 max and lactate clearance capacity, is essential for conserving anaerobic reserves and enabling recovery between repeated high-intensity efforts, especially in the final 10km (6 miles) when speed and volatility increase dramatically.

Team dynamics

Trust and cohesion within the team are fundamental. Riders who commit themselves in the lead-out must trust the sprinter to deliver a result; conversely, the sprinter must respect and honour the team's work by executing their role with precision. Clear pre-race communication, real-time radio updates, and shared tactical understanding are critical for effective collaboration.

> 66 99
>
> *Every successful sprint is based on structured training, detailed planning, and flawless execution.*

THE ROLE OF THE LEAD-OUT TRAIN

The lead-out train functions as the tactical engine behind the final sprint. Typically composed of three to five riders, the train begins asserting control 10km (6 miles) from the finish.

EARLY LEAD-OUT — Skilled in positioning and holding speed.

LEAD-OUT SPRINTER — Able to make sustained sprints.

SPRINTER — Highest peak power to finish the race.

Roles within the train are clearly defined. Each transition between lead-out riders must be timed to keep kinetic momentum while minimizing exposure to the wind. These exchanges are usually planned with reference to course reconnaissance, wind direction, and road features such as curves or narrowing sections. Positioning the train along the sheltered side or the barriers in a crosswind scenario helps reduce aerodynamic drag and control lateral movement from rivals.

TACTICAL COMPLEXITY

The final 3km (1.9 miles) are a chaotic, high-risk environment. Riders must navigate narrow sections, rapidly changing speeds, road furniture, and the constant jostling for position. Teams contesting the sprint converge, and the lead-out trains must aggressively protect their line, often engaging in subtle contact and shoulder-to-shoulder positioning to defend themselves. The last kilometre sees speeds rise to over 65kmh (40mph), stringing out the peloton and reducing the likelihood of interference or crashes, a crucial goal given the high speeds involved.

MENTAL RESLIENCE

Mental acuity is just as important as physical strength. Sprinters must process environmental cues, make snap judgements under pressure, and commit to high-risk decisions, often threading through narrow spaces or choosing to launch early under uncertain conditions.

Sprint finishes in professional cycling epitomize the convergence of sophisticated physiology, strategic brilliance, and perfectly synchronized teamwork. They showcase a team's ability to orchestrate energy expenditure, spatial awareness, and precise timing within a chaotic, high-velocity context.

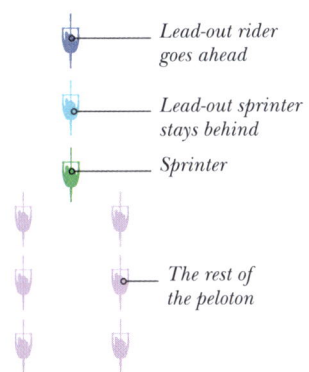

EARLY LEAD-OUT RIDERS
With a short distance to go, early lead-out riders maintain high tempo to reduce bunching and deter attacks.

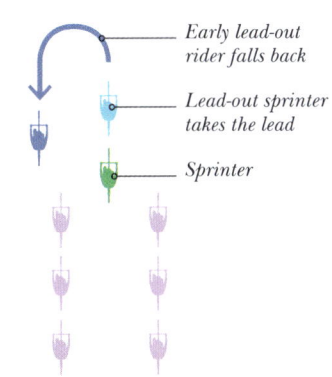

MIDDLE RIDERS
During the last 30 seconds of the race lead-out riders help position and shelter the sprinter.

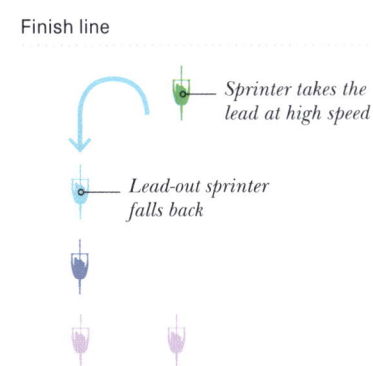

FINAL LEAD-OUT RIDER
A lead-out sprinter positions and launches the team's main sprinter for the final sprint to the finish.

TIME TRIALS

Often called "the race of truth", time trialling in professional cycling is a highly demanding and technically intricate discipline. In contrast to mass-start events where drafting and tactics are key, the individual time trial (ITT) focusses purely on a cyclist's power output, aerodynamic efficiency, pacing strategy, and mental resilience. Excelling in time trialling requires a synthesis of sports science principles, meticulous equipment setup, and the cyclists' capacity for excellent pacing to manage their effort.

PHYSIOLOGY AND POWER OUTPUT

Fundamental to time trial performance is the ability to sustain a high-power output over a given duration.

This requires a balance between aerobic and anaerobic energy systems (see p.48). Aerobic capacity (VO_2 max) is critical, as it determines the rider's ability to deliver oxygen to working muscles and maintain steady-state effort. The point at which lactate production exceeds clearance is another key metric, as cyclists must sustain efforts at or just below this threshold for the duration of the time trial. Critical power (or functional threshold power, FTP, see p.57) is a commonly used benchmark for time trial performance. Additionally, power-to-weight ratio is relevant, particularly in hilly time trials, where lighter cyclists have an advantage.

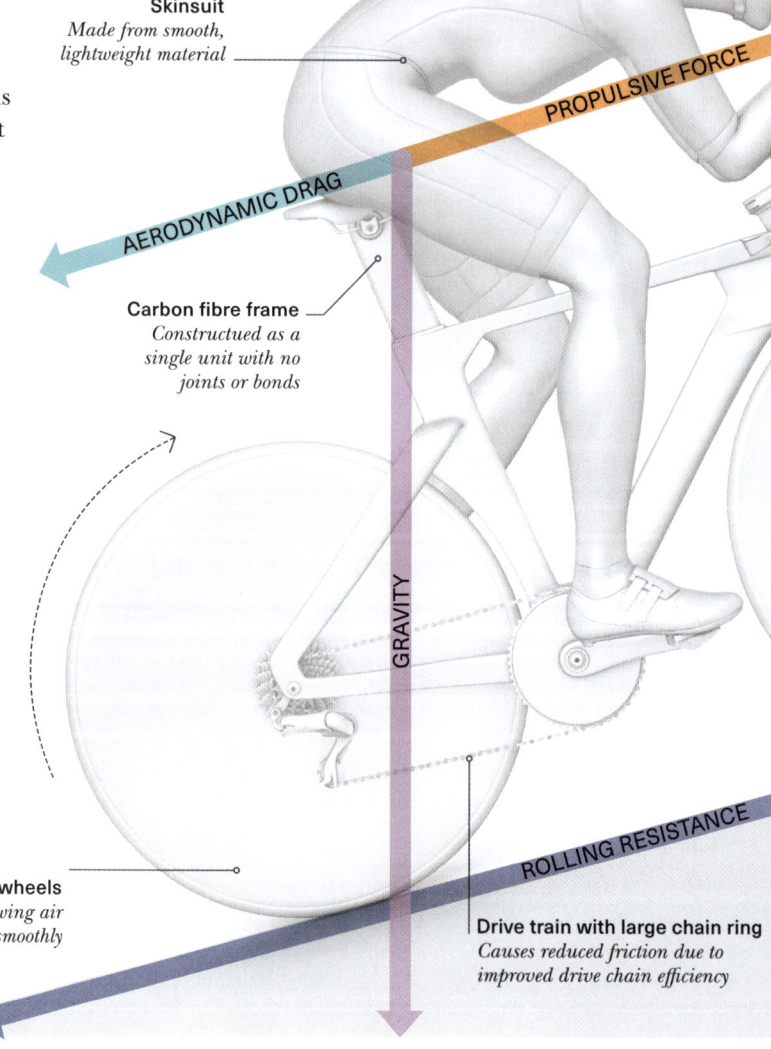

Aero helmet
Designed to allow unrestricted air flow over the top of the head and shoulders

Skinsuit
Made from smooth, lightweight material

Carbon fibre frame
Constructued as a single unit with no joints or bonds

Disc wheels
Reduce drag, allowing air to flow over them smoothly

Drive train with large chain ring
Causes reduced friction due to improved drive chain efficiency

COMPETITIVE CYCLING | Time Trials

> *The rider's position on the bike is the single most critical factor in reducing aerodynamic drag.*

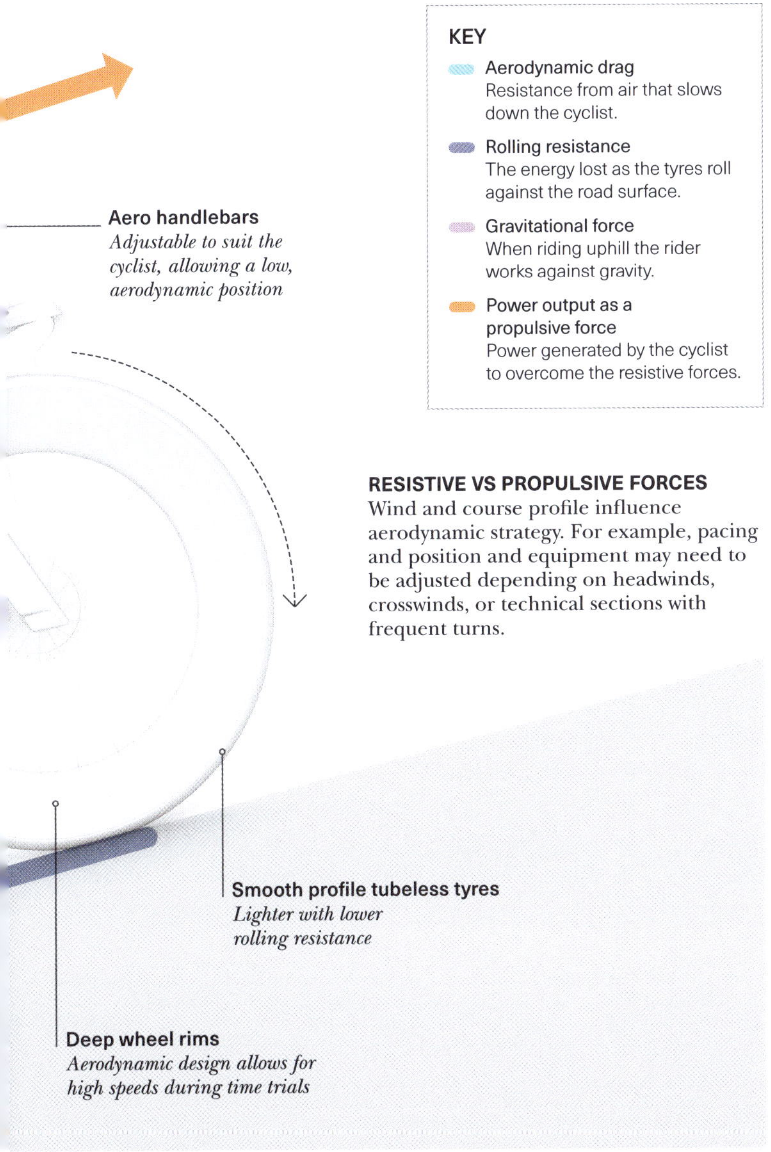

Aero handlebars
Adjustable to suit the cyclist, allowing a low, aerodynamic position

KEY
- Aerodynamic drag
  Resistance from air that slows down the cyclist.
- Rolling resistance
  The energy lost as the tyres roll against the road surface.
- Gravitational force
  When riding uphill the rider works against gravity.
- Power output as a propulsive force
  Power generated by the cyclist to overcome the resistive forces.

RESISTIVE VS PROPULSIVE FORCES
Wind and course profile influence aerodynamic strategy. For example, pacing and position and equipment may need to be adjusted depending on headwinds, crosswinds, or technical sections with frequent turns.

Smooth profile tubeless tyres
Lighter with lower rolling resistance

Deep wheel rims
Aerodynamic design allows for high speeds during time trials

THE KEY TO SPEED

The most important element in time trial performance is aerodynamics or power-to-drag ratio, sometimes referred to as watts per coefficient of drag area (W/CdA).

While power-to-weight ratio (W/kg) is important, time trialling rewards riders who can generate high absolute power (W) while maintaining a highly aerodynamic position. CdA, defined as the product of drag coefficient (Cd) and frontal area (A), is a central performance variable. It is a measure of how much aerodynamic resistance a rider or object presents to the air. The lower the CdA, the less resistance the rider experiences, and so the higher the velocity for a given power output. Reductions in CdA by as little as 0.01 can lead to time savings of several seconds over short courses.

The best rider position is a balance between aerodynamics and biomechanical efficiency. Time trial position is often based on a rider's body shape, flexibility, and sustained output capabilities.

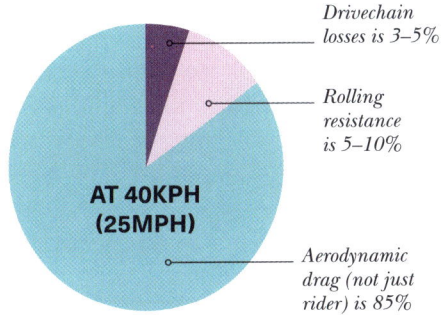

Drivechain losses is 3–5%

Rolling resistance is 5–10%

AT 40KPH (25MPH)

Aerodynamic drag (not just rider) is 85%

DIFFERENT RESISTIVE FORCES
The contributions from different forces depend on whether the cyclist is on a flat surface or climbing, moving fast or slow. This shows the effect of aerodynamic drag at 40kph (25mph) on a flat surface.

PACING STRATEGY

Effective pacing is critical in time trials. Cyclists must distribute their effort evenly or adjust for specific course demands to avoid fatiguing too early or finishing with unused reserves.

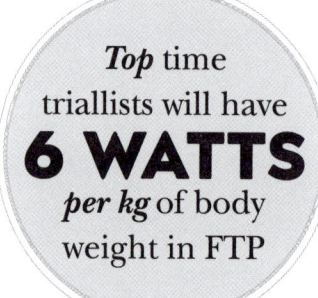

Top time triallists will have **6 WATTS** *per kg* of body weight in FTP

There are three main pacing strategies employed in time trials. These include even pacing with constant power close to FTP; negative splitting with a more conservative start; and variable pacing, which is often dictated by a course with variable terrain.

Advanced pacing strategies incorporate specific power output targets for individual parts of a given course. For example, accounting for higher power outputs on uphill or headwind sections and using downhill or tailwind sections to allow some recovery at lower power output. The most straightforward pacing strategies are those for flat, straight courses, whereas a course with rolling terrain or significant up and downhill sections requires more specific approaches.

WHAT ARE NEGATIVE SPLITS?

A negative split occurs when a cyclist completes the second half of a time trial with greater speed or intensity than the first half. This strategy involves starting the time trial at a controlled, conservative pace to avoid early overexertion and ensure maximum power output in the later stages. Importantly, a negative split doesn't always mean the second half is objectively faster in terms of time, especially when factors like terrain or weather make this impractical. Instead, it reflects a measured pacing approach, saving energy early to deliver a stronger performance later in the race.

PACING FOR A MIXED TERRAIN TIME TRIAL
Pace conservatively on climbs to avoid spikes, maintain effort on false flats, and maximize speed on descents and tailwind sections, balancing intensity with terrain to sustain consistent output.

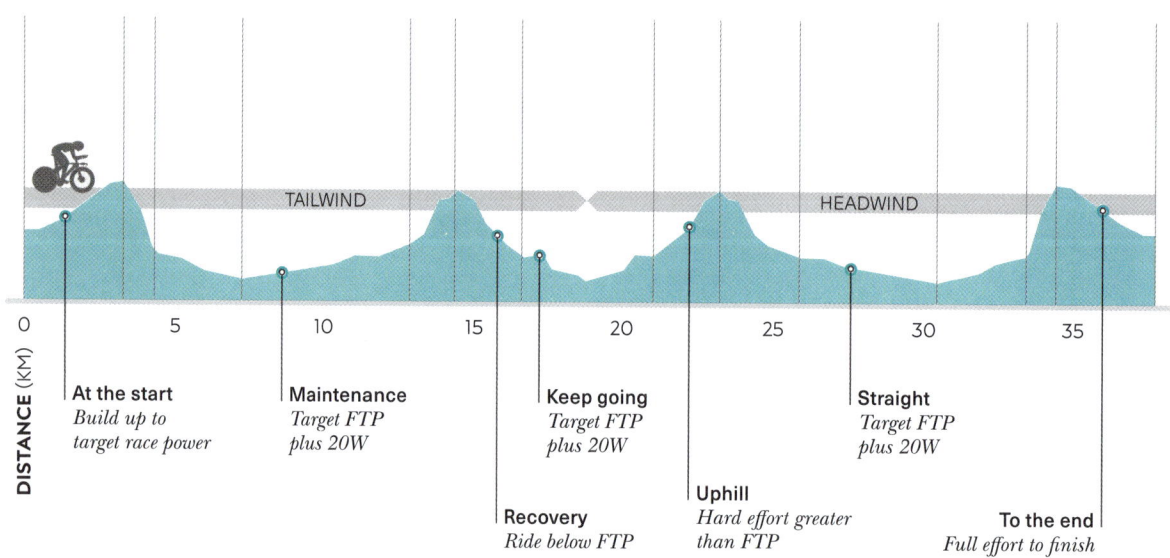

At the start — *Build up to target race power*
Maintenance — *Target FTP plus 20W*
Keep going — *Target FTP plus 20W*
Recovery — *Ride below FTP*
Uphill — *Hard effort greater than FTP*
Straight — *Target FTP plus 20W*
To the end — *Full effort to finish*

COMPETITIVE CYCLING | Time Trials

MENTAL PREPARATION

Time trials require exceptional mental resilience. Unlike mass-start races, cyclists face the challenge alone, without teammates or tactical dynamics to lean on. Success hinges on maintaining intense focus, managing physical discomfort, and sustaining motivation throughout the effort.

To prepare mentally, athletes often employ techniques such as visualization, goal setting, and mindfulness (see pp.114–115), which help them stay composed and engaged under pressure.

RACE PREPARATION

In stage racing, time trials can be decisive. A strong performance can create gaps that are nearly insurmountable in mountainous or sprint-heavy stages. For this reason, general classification (GC) contenders often develop their time trial capacity, even if they are not natural specialists. This includes tailoring training programs to include threshold intervals, motor-pacing sessions, aerodynamic drills, and "race simulation" efforts. Modern training approaches for time triallists emphasize polarized training structures, high volumes of low-intensity riding paired with targeted high-intensity efforts near critical power thresholds. Neuromuscular coordination and position-specific strength work (such as low-cadence intervals in aero position) are also integrated to improve output.

Team time trials

In team time trials, teams race together over a fixed course, and the finishing time is usually taken when the fourth or fifth rider crosses the line. This format emphasizes group cohesion, as all riders must take structured turns at the front, shielding teammates from the wind before peeling off in a smooth rotation. Precise pacing strategy, aerodynamic harmony among riders of different sizes and abilities, and seamless communication are all important. The strongest riders must moderate their efforts to avoid dropping teammates prematurely, while weaker riders must maximize drafting benefit and contribute effectively within their limits.

PACING FOR A MOUNTAIN TIME TRIAL
Keep pace near lactate threshold (FTP) to sustain aerobic power without early fatigue. Maintain consistent cadence, smooth force application, and controlled breathing.

TIME TRIAL BIKE — **ROAD BIKE**

Building up — *Start then move up to FTP*

On the flat — *Recover at below FTP*

Going up — *Build to FTP plus 30W*

Keep going — *Hold FTP plus 20W*

Keep it steady — *Hold FTP plus 20W*

Changeover — *Fast, efficient transition between bikes needed*

Maintain effort — *Hold FTP plus 30W*

The finish line — *Full finish effort*

GAME THEORY

Cycling races, especially road races, are often seen as strategic games where riders must decide when to expend energy on attacks, when to conserve energy in the peloton, and how to react to the moves of other riders.

A THEORETICAL FRAMEWORK

Game theory is a mathematical tool for modelling strategic interactions and is used by professional cycling teams to make strategic decisions in races.

STRATEGIC COOPERATION

Drafting is the core driver of strategic cooperation and competition in road racing. It creates different energy costs between leading and following riders, shaping the incentives and payoffs that affect every part of the sport.

Game theory classifies these interactions into cooperative and non-cooperative structures. For instance, breakaway groups form cooperative coalitions with the goal of staying ahead of the peloton, yet within the group, individual riders compete for the stage win. This structure can be seen as a game in which groups that include non-cooperative subgames work together. Each rider must assess the trade-off between contributing to the breakaway's success (increasing the probability of collective escape) and conserving energy for an individual outcome (increasing personal payoff if the breakaway survives).

	RIDER/TEAM B	
RIDER/ TEAM A	Collaborate	Don't collaborate
Collaborate	1, 1	0, 1
Don't collaborate	1, 0	0, 0

PLAYING THE GAME

The Nash Equilibrium (1,0 or 0,1) appears in a breakaway when each rider does just enough work to keep the break going, knowing that changing effort alone won't improve their outcome. It reflects a stable balance between cooperation and self-interest.

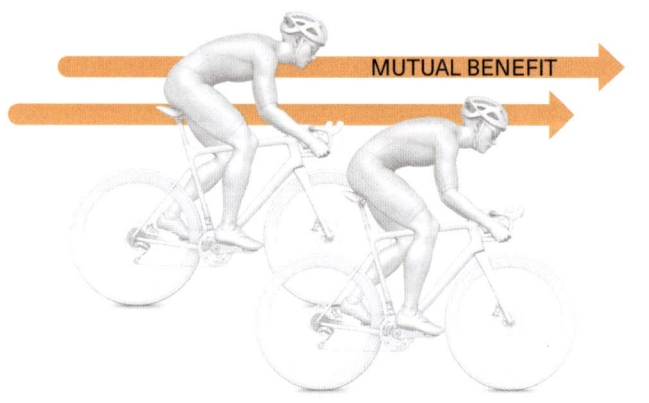

COLLABORATE
A non-zero-sum game (public good) is when riders can all benefit by working together (1,1) to stay ahead of the peloton. Cooperation creates mutual gain.

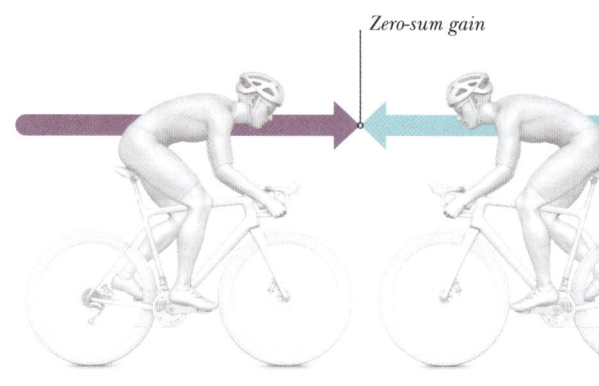

DON'T COLLABORATE
As the finish of the race nears, it shifts to a zero-sum game (personal good). Only one rider can win, so one's gain is another's loss (1,0).

COMPETITIVE CYCLING | Game Theory

In *cycling teams* **RISK VS REWARD** is constantly evaluated

VOLUNTEER'S DILEMMA

Game theory also provides a framework for looking at the problems encountered by the peloton when chasing breakaways.

Several teams may benefit from bringing a breakaway back into the peloton, but if each assumes another will perform the work, the break may succeed. This is a variant of the volunteer's dilemma, where the cost of action is borne by the few, and the benefit is distributed across many.

Decision-making in these situations can be made by teams making educated guesses about each other's intentions based on current rider deployment, past behaviour, and race context (for example, stage profile or GC standings). The likelihood of a successful chase increases when teams share common objectives and when the perceived threat from the breakaway is high.

THE PUBLIC GOOD

The cooperation dilemma in breakaway groups mirrors a behavioural experiment called the public goods game, which is a way of looking at the balance between contributing to the collective good, and the self-interest of individuals. In the cycling world, the "public good" is the collective maintenance of high speed, which all members benefit from, but which each rider contributes to at a cost.

Another model, known as the Nash equilibrium (see left), where no player has an incentive to unilaterally change their strategy, can be useful in situations where there are multiple potential strategies. This theory often predicts lower cooperation levels due to free-riding tendencies. However, data gathered from real-life situations and mathematical models suggest that cooperation develops through mutual support, damaged reputation if it fails, and punishment for non-cooperative riders. All of these variables are constantly changing over the course of a race and must be reevaluated.

SEQUENTIAL GAMES AND BACKWARD INDUCTION

In the final kilometres of a race, especially during sprint finishes, game-theoretic concepts such as sequential games and backward induction become relevant.

Cycling races involve a sequence of moves and countermoves. Riders must anticipate how their actions will affect other riders and how other riders will respond. Backward induction is when riders consider the end-game and work backwards to determine the best course of action. Decisions such as when to initiate their lead-out, how to position their sprinter, and whether to respond to rival strategies are shaped by preceding actions. Riders and teams predict and adapt tactics, often relying on snap decisions or learned behavioural patterns to get the best results possible.

TEAM TIME TRIALS

Teams must optimize collective output by distributing effort among riders of varying abilities. The goal is to minimize team time, often measured on the speed of the fourth or fifth rider.

The challenge in team time trials lies in constructing pacing strategies that avoid overburdening weaker riders while using the stronger riders to their maximum power and potential. This is done using precise modelling of each rider's individual power–duration curve, aerodynamic efficiency (CdA), and recovery kinetics.

Game-theoretic approaches in these situations emphasize role specialization and rotational strategy design. Computational simulations are often used to decide optimal pulling durations (the amount of time each rider should spend at the front) and the order of riders.

MENTAL PERFORMANCE

Performance in elite sport arises from intricate connections between the central nervous system, psychological states, and physiological functions. The brain serves as both the command centre and a regulator, integrating sensory information, generating motor outputs, and mediating cognitive and emotional responses to environmental challenges.

In cycling, an endurance discipline with substantial cognitive and emotional demands, the dynamic interplay between executive functions, the regulation of emotions, and a cyclist's awareness of their body's signals is particularly crucial for success.

The prefrontal cortex plays a central role in goal setting, inhibitory control, and decision-making under stress. It interacts with the amygdala, which processes threat and emotional salience, and the anterior cingulate cortex (ACC), which monitors performance and detects conflict or error. These neural circuits influence attention, arousal, and effort regulation during racing. Simultaneously, neuroendocrine systems, particularly the hypothalamic-pituitary-adrenal (HPA) axis, modulate stress reactivity through cortisol release, impacting both mental state and physical output.

PSYCHOLOGY AND CYCLING

Psychological performance in cycling is best understood as a biopsychosocial phenomenon, where mental states influence physiological output and where social–environmental cues shape emotional responses and behavioural regulation. This framework underpins the idea of developing psychological capacities like mental toughness and resilience in elite cyclists.

TOUGHNESS AND RESILIENCE

In competitive cycling mental toughness and resilience are foundational components of elite performance. Mental toughness refers to a cyclist's ability to sustain optimal performance under pressure, maintain focus, and execute skill despite fatigue or adversity. Resilience complements this by defining the capacity to recover quickly from setbacks, such as crashes, illness, poor results, or tactical errors. In professional cycling, psychological attributes often determine who succeeds when physical capabilities are otherwise evenly matched.

Cycling presents a uniquely taxing mental landscape in which cyclists contend with isolation, unpredictable conditions, and an ever-present threat of mechanical failure or injury. They must navigate complex team dynamics, constant changes in strategy, and physical discomfort. Mental toughness enables a cyclist to push through the final kilometres of a mountain stage in a highly fatigued state or to maintain composure when rivals attack unexpectedly.

TRAINING

Mental toughness and resilience are malleable psychological elements that can be developed. Psychological skills training (PST), encompassing techniques such as cognitive restructuring via self-talk modification, mental imagery for performance enhancement and stress reduction, and hierarchical goal setting to foster motivation and self-efficacy, are commonly integrated into high-performance cycling training regimens. PST over time enables neuroplastic adaptations, helping cyclists to enhance their perceptual tolerance of physiological discomfort, develop more adaptive cognitive appraisals of failure experiences, and improve their capacity for self-regulation when under pressure.

Team enviroments

The culture within a cycling team also affects cyclists' mental toughness and resilience. Teams that normalize psychological challenges, promote transparent communication, and prioritize process-oriented goals tend to cultivate more robust psychological frameworks in their cyclists. Conversely, environments characterized by rigid perfectionism, a culture of blame attribution, or a pervasive fear of failure can impede psychological development and amplify stress responses.

COMPETITIVE CYCLING | Mental Performance

CONTROL CENTRE

The brain is integral in helping us manage our emotions and physical performance. Mental resilience and toughness can be developed and improved to help cyclists get the edge.

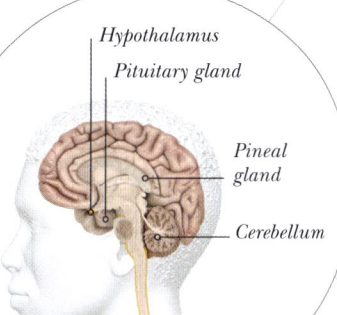

Hypothalamus
Pituitary gland
Pineal gland
Cerebellum

ATTENTIONAL CONTROL
Enables cyclists to focus on relevant cues, such as cadence, breathing, positioning, while filtering out distractions such as crowd noise, negative self-talk, or tactical missteps by competitors.

COGNITIVE RESTRUCTURING
Cognitive restructuring helps cyclists reinterpret stress as challenge rather than threat, reinforcing adaptive appraisals that support confidence and perseverance.

BUILDING MOTIVATION
Neurotransmitters such as dopamine and noradrenaline, which influence motivation and arousal regulation, interact with stress hormones like cortisol to affect a cyclist's emotional and cognitive states.

GOAL PERSISTENCE
Goal persistence involves long-term motivation and the maintenance of intentional effort even in the absence of immediate feedback or success.

PST
Psychological skills training (PST) is used in high-performance cycling training. It can help cyclists build tolerance and self-regulate.

EMOTIONAL FLEXIBILITY
Emotional flexibility allows cyclists to process emotional experiences (fear, anger, disappointment) without becoming dysregulated or disengaged.

SOCIAL SUPPORT
Strong social support from teammates, coaches, and psychologists provides practical guidance and emotional support during recovery.

RESILIENCE
Resilient cyclists mentally prepare for challenges, employing pre-developed coping mechanisms like imagery, pre-performance routines, or mindfulness to lessen the impact of setbacks.

STRESS REDUCTION
Training can be a social activity, which can ease stress. Exercise, including cycling, has been shown to significantly improve stress- and anxiety-related symptoms.

MENTAL TOUGHNESS
Mental toughness in this context is about sustained attention to task, pain tolerance, and emotional self-regulation.

MANAGING ANXIETY

Performance anxiety is a psychobiological response to a perceived threat related to performance, manifesting as heightened physiological arousal, disrupted cognitive processes, and inflexible behaviour. In the demanding arena of professional cycling, where riders frequently compete in high-stakes environments, this anxiety can significantly impair performance.

HORMONAL CHANGES

From a neurobiological perspective, performance anxiety triggers activation of the sympathetic–adrenomedullary system and the hypothalamic–pituitary–adrenal axis.

This results in an increased release of catecholamines (adrenaline and noradrenaline) and cortisol. While these hormonal changes can enhance cardiovascular and metabolic function in the short-term, they can also disrupt the fine motor control and cognitive processing crucial for cycling performance if this activation becomes excessive or poorly regulated. Performance anxiety can be triggered when cyclists focus on what they perceive is threatening, such as crowds, competitors, or technicals issues and internalizing thoughts. This type of anxiety can be seen in increased monitoring of internal states such as heart rate, breathing, or perception of effort, which can lead to hyperawareness and impaired ability to carry out physical actions.

HORMONES AND THE BRAIN
The cognitive–behavioural model of performance anxiety theorizes that beliefs such as a fear of failure or perfectionism interact with situational stressors to generate cognitive distortions and affective dysregulation.

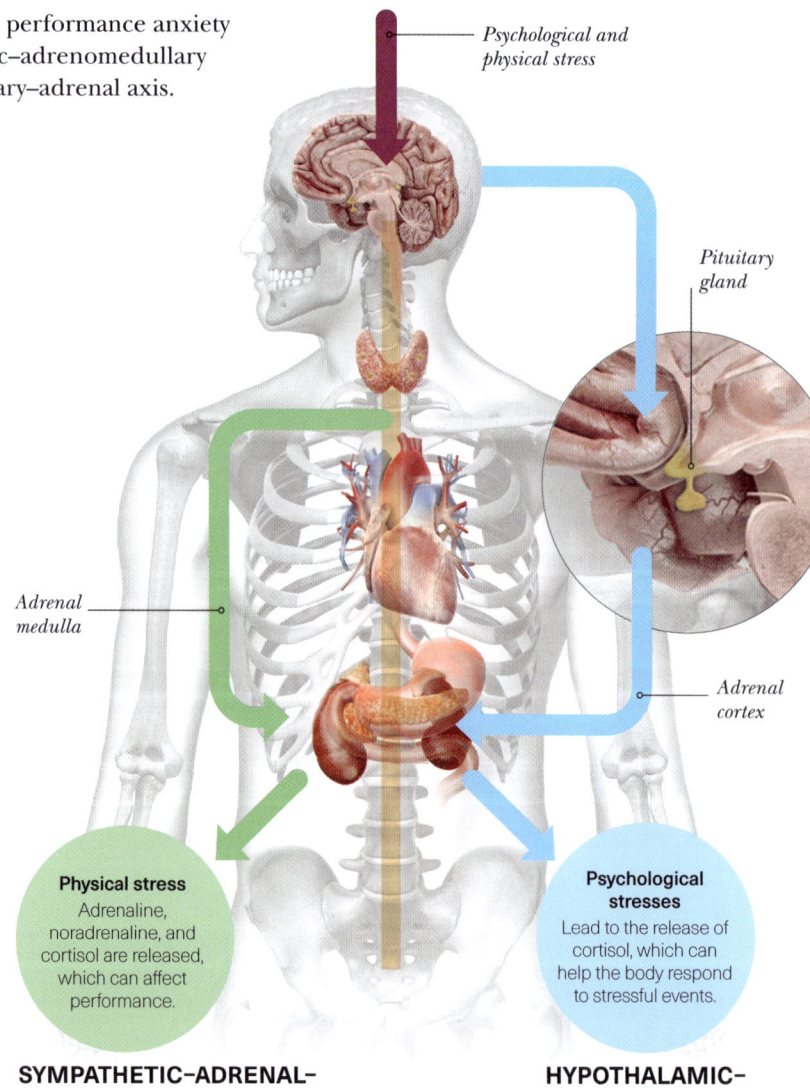

Psychological and physical stress

Pituitary gland

Adrenal medulla

Adrenal cortex

Physical stress
Adrenaline, noradrenaline, and cortisol are released, which can affect performance.

Psychological stresses
Lead to the release of cortisol, which can help the body respond to stressful events.

SYMPATHETIC–ADRENAL–MEDULLARY AXIS

HYPOTHALAMIC–PITUITARY–ADRENAL AXIS

STRESS INOCULATION

Exposure-based training, also known as performance simulation under controlled stress conditions, is another way of managing performance anxiety.

By systematically exposing cyclists to anxiety-eliciting scenarios within a safe and monitored environment, such as simulated time trials replicating race pressures or internal selection events, psychologists can help cyclists desensitize fear responses and build their tolerance to anxiety. This form of stress inoculation contributes to increased performance confidence and improved cognitive resilience, and is particularly beneficial for younger cyclists or those with less competitive experience.

Below is an example of how a cyclist, in this case a time triallist, may use these methods to mitigate performance anxiety before a race.

" "

I ride to test my limits, not prove my worth.

Treating anxiety

Case: A time triallist experiences anticipatory anxiety before major events, with elevated heart rate, shallow breathing, and intrusive thoughts of failure.

Intervention sequence:

1. Baseline assessment: complete psychometric profiling with tools such as the Competitive State Anxiety Inventory (CSAI-2) or Sports Anxiety Scale (SAS-2) or similar assessments of performance anxiety that can be interpreted by a sports psychologist.

2. Cognitive restructuring: reframe perfectionism underpinning perceived anxiety, for example, "If I don't win, I'm a failure" is transitioned to "Although winning is great, I've prepared well, and my goal is to execute my pacing strategy. This is one race in a season of opportunities."

3. Somatic control training: use of diaphragmatic breathing techniques, progressive muscle relaxation, and biofeedback-assisted methods using assessment of heart rate or breathing to assist relaxation prior to the race and before TT training sessions or race simulations.

4. Mindfulness training: before the race (or hard interval training) spend 10 minutes seated with eyes closed, focussing on breathing and gently returning attention from distraction. Repeat this during competition to stay focussed on effort and execution rather than outcomes or discomfort.

5. Pre-performance routine: inclusions of standard task such as a warm-up, listening to a specific music playlist and diaphragmatic breathing all anchored to core values: "I ride to test my limits, not prove my worth."

6. Post-event debriefing: taking time after the race to review performance (physical and mental) individually and with others to reinforce self-efficacy and adaptive interpretation.

Mindfulness

Mindfulness-based interventions, particularly Mindfulness-Acceptance-Commitment (MAC) training, are increasingly used in elite sport to reduce experiential avoidance and enhance present-moment focus. In cyclists, these programmes improve attentional control during long-duration efforts and decrease the reactivity to negative performance-related thoughts. Neuroimaging studies support these effects, demonstrating reduced amygdala activation and increased prefrontal modulation following mindfulness training.

Medication

In some instances, the use of medication may be considered, particularly when anxiety symptoms are chronic and significantly impede daily functioning beyond performance. Beta-adrenergic antagonists can be used in the short term to help improve the physical effects of anxiety. Selective serotonin reuptake inhibitors (SSRIs) may be used in cases where performance anxiety and generalized anxiety disorder are experienced. It is imperative that any pharmacological strategy adheres strictly to anti-doping regulations and is implemented under the guidance of a qualified psychiatrist, ensuring a holistic and ethical approach to athlete care.

HOW TO TRAIN

Cycling is a highly demanding endurance sport requiring the integration of physical, psychological, and tactical capacities.
To prepare effectively, training must be grounded in sound physiological principles tailored to the specific demands of competition and continually adapted based on emerging scientific insights. This chapter outlines the key principles, current practices, and scientific evidence guiding effective training in modern cycling disciplines.

QUANTIFICATION OF TRAINING

Effective training in cycling is guided by the fundamental biological principle that the body adapts in response to imposed stress. The General Adaptation Syndrome (GAS), first described in the 1930s, provides a conceptual framework for understanding how training stimuli elicit physiological responses that, with appropriate recovery, result in performance improvements.

INTERNAL AND EXTERNAL MEASURES

In cycling, measuring and evaluating training includes both external load (what the cyclist does) and internal load (how the cyclist responds). External load refers to measures such as power output, duration, distance, and cadence, while internal load includes heart rate (HR), rate of perceived exertion (RPE), blood lactate, and hormonal or neuromuscular markers of stress. The interaction between these two domains provides insight into training effectiveness and readiness.

TRAINING ZONES

Performance testing forms the foundation for individualizing training and interpreting load metrics. Standardized field or laboratory tests, such as incremental ramp tests, lactate threshold tests, and critical power testing protocol are used to determine key physiological benchmarks (for example, VO_2 max, functional threshold power, and anaerobic capacity). These metrics help define training zones, enabling coaches and cyclists to prescribe workloads that target specific adaptations such as aerobic endurance, glycolytic capacity, or neuromuscular power.

MONITORING TRAINING

With individualized zones established, monitoring of day-to-day training can be done using power meters and heart rate monitors, supported by assessments like session RPE (rate of perceived exertion). Common external load metrics include a training stress or load, kilojoules of work, average power output, and measuring how much time cyclists spend in each zone, while internal metrics may include heart rate variability (HRV), perceived fatigue, mood state, and sleep quality. Modern cycling software platforms

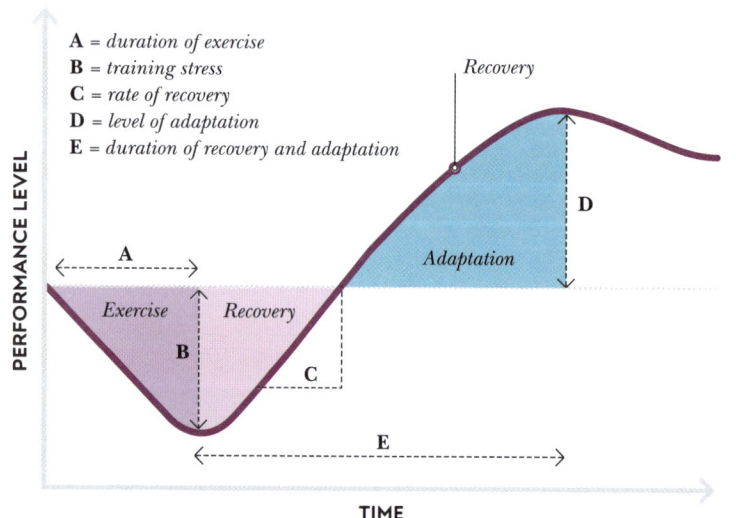

A = duration of exercise
B = training stress
C = rate of recovery
D = level of adaptation
E = duration of recovery and adaptation

ADAPTATION TO TRAINING

According to the General Adaptation Syndrome (GAS) model, a cyclist exposed to a training stress experiences a temporary reduction in performance (fatigue), followed by recovery and supercompensation (adaptation). If the training stress is insufficient, no adaptation occurs; if it is excessive or recovery is inadequate, maladaptation or non-functional overreaching may result.

HOW TO TRAIN | Quantification of Training

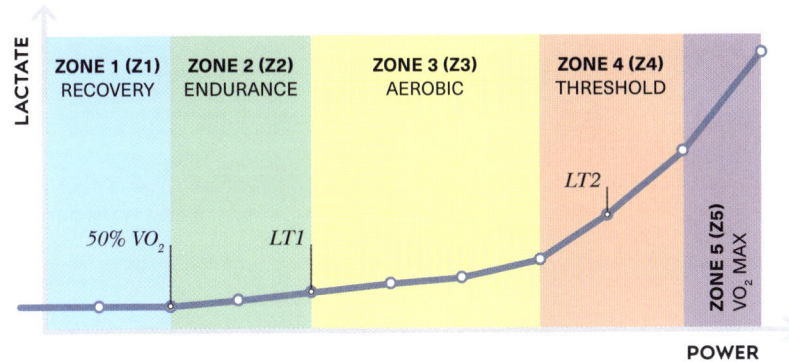

TRAINING ZONES AND LACTATE
This example chart shows how training zones relate to the blood lactate profile test.

ZONE	% OF FTP	PHYSIOLOGICAL EFFECT
1	<55%	Promotes recovery via increased blood flow for metabolite clearance and nutrient delivery; minimal physiological stress.
2	55–75%	Enhances endurance capacity through mitochondrial biogenesis, improved fat oxidation, and increased capillary density; builds cardiovascular efficiency.
3	76–90%	Develops muscular endurance and lactate shuttle capacity; improves sustained moderate-to-high power output.
4	91–105%	Elevates FTP; increases maximal metabolic steady state by improving lactate production/clearance balance.
5	106–120%	Increases maximal oxygen uptake (VO_2 max); targets cardiovascular output and muscle oxygen utilization for high-intensity efforts (suprathreshold with anaerobic contribution).

often combine this information to provide rolling measures of acute and chronic training load, highlighting a cyclist's fatigue and fitness over time.

Advanced monitoring systems integrate long-term data, which helps cyclists analyse training responsiveness. For example, trends in power–heart rate decoupling (cardiovascular drift), improvements in submaximal power at fixed heart rates, or shifts in lactate thresholds are used to evaluate training effectiveness.

Training load inputs can also be used to predict changes in fitness and fatigue, although this requires careful individual calibration and has limitations under certain conditions. Examples of this include the Banister Training Impulse model (TRIMP), which combines exercise duration, heart rate, and intensity into a single score to estimate the internal training load of a session. Another measure, Training Stress Balance (Training Peaks) is calculated as the difference between Chronic Training Load (CTL) – your long-term fitness – and Acute Training Load (ATL) – your short-term fatigue. A positive TSB suggests you're well-rested and ready to perform, while a negative TSB indicates accumulated fatigue and potential need for recovery. The Critical Power/W' balance model and power–duration curves (see pp.50–51) are used to determine how much power a cyclist can sustain and for how long.

A MULTIDIMENSIONAL APPROACH

Effective training quantification must be contextualized. The same power output or training stress may elicit different responses depending on the cyclist's recovery status, environmental conditions, or psychological state. Therefore, integrating subjective and objective data, such as pairing RPE (rate of perceived exertion) with power output or tracking mood alongside HRV (heart rate variability), provides a more complete picture of a cyclist's status. This approach supports early detection of maladaptation, informs training adjustments, and enhances communication between cyclists and their coaches.

PERIODIZATION

Training periodization is the systematic organization of training over time, designed to optimize performance, manage fatigue, and support long-term development. In cycling, periodization provides a framework for sequencing training stimuli to enhance physiological adaptation and prepares cyclists to perform at their best during key competitions.

In basic terms, periodization is based on two core principles: progressive overload and variation. Progressive overload ensures that training stress increases over time to stimulate physiological improvement, while variation prevents stagnation and overtraining by altering the type, volume, and intensity of training. These principles are built into a structure of macrocycles (long-term planning over a season or year), mesocycles (medium-term blocks lasting several weeks), and microcycles (short-term periods, typically one week). Together, these timeframes help organize training phases that shift focus from general to specific preparation, ultimately leading to a peak in performance.

MODELS OF PERIODIZATION

The traditional model of periodization begins with a preparatory phase focussed on general conditioning. This phase emphasizes high training volume at low to moderate intensity, aimed at developing the aerobic base, muscular endurance, and technical proficiency on the bike. Strength training and cadence drills are often included to enhance muscular resilience and pedalling efficiency.

Training during the build phase becomes more specific to the physiological demands of competition. Intensity is gradually increased through the inclusion of structured interval training targeting the lactate threshold, VO₂ max, and race-specific power profiles. While training volume may decrease slightly, the focus turns towards higher-intensity sessions, such as threshold efforts, repeated intervals at VO₂ max intensity, and race simulation rides.

During the next phase, the competition phase, the goal is to translate the accumulated fitness into race form. The focus is on maintaining intensity while reducing overall volume to preserve freshness and avoid excessive fatigue. The use of high-intensity, low-volume sessions, including short sprints, race-specific intervals, and tactical group sessions, helps sharpen performance.

STRUCTURE
Breaking training down into long-, medium-, and short-term cycles helps maintain focus on long-term goals.

Types of periodization

MODEL	KEY FEATURE
Traditional	Progresses from high-volume, low-intensity to low-volume, high-intensity
Reverse	Starts with high-intensity, low-volume work; base volume is added later
Block	Concentrated blocks of training focussed on a single physiological target
Polarized	More low-intensity training with small amounts of high-intensity work
Pyramidal	Mostly Z1, with a moderate amount in Z2, and a small portion in Z3

HOW TO TRAIN | Periodization

The taper is a planned reduction in training volume over one to two weeks, while maintaining training intensity and frequency. The goal is to reduce accumulated fatigue without compromising fitness, allowing for supercompensation and peak performance. Effective tapering is highly individualized,

Ultimately, individualization remains the most important aspect of periodization. The same structure will not suit everyone, as training responses are influenced by age, training history, discipline, competition schedule, and individual recovery capacity. Effective periodization requires continuous assessment and plans must remain dynamic, adjusting to illness, injury, travel, and shifts in motivation or form.

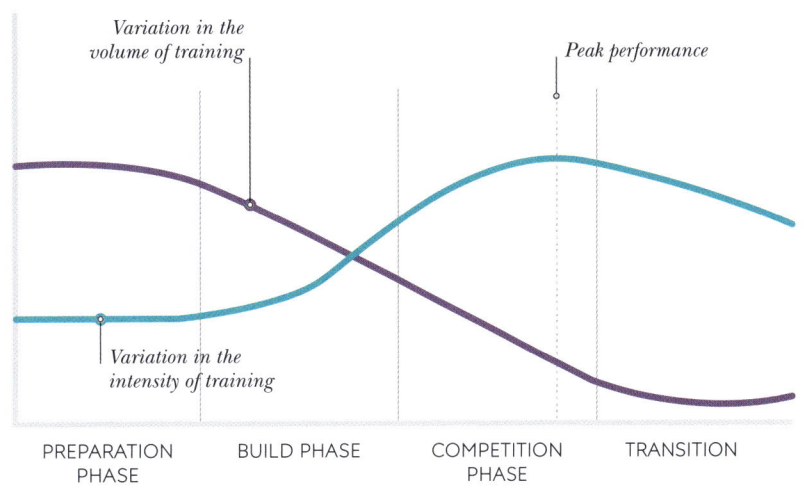

TRAINING PHASES
This traditional periodization model is used to build up to a competition. It should be followed by two weeks of taper during which training is reduced.

KEY
— Volume (training)
— Intensity (quality)

INTENSITY DISTRIBUTION	FOCUS	BEST FOR	LIMITATION
Early: Around 80% Z1–3, <10% Z4/Z5 Late: increased Z4/Z5	Builds general base fitness before developing specificity	Riders with a long season build-up and clear target events	Can lack early high-intensity stimulus
Early: >30% Z4/5, low Z1–3 Later: more Z1–3 added	Neuromuscular readiness, anaerobic power, high-intensity conditioning	Track cyclists, criterium racers, or cyclists with limited time	Risk of insufficient endurance base; harder to sustain long-term
Varies by block: typically ≥50% of one intensity zone	Maximizes adaptation to specific training targets	Advanced or elite cyclists with focussed objectives or multiple peaks	High fatigue risk; requires monitoring and planning
Around 70% Z1–2, around 5% Z3, 10–15 % Z4/5	Aerobic efficiency and peak power development with minimal fatigue	Endurance cyclists aiming for long-term development	Hard to apply to cyclists needing threshold work or frequent racing
Around 70% Z1/2, around 20–25% Z3, around 5–10% Z4/5	Threshold development, sustained power, and endurance base	Stage races, road cyclists, gravel racers and all-rounders preparing for variable events	Less emphasis on high-end intensity; may undertrain sprint/anaerobic power

INTERVAL TRAINING

Interval training is a cornerstone of cycling performance, enabling cyclists to improve various physiological systems and enhance their ability to sustain high-intensity efforts. This structured approach to training alternates between periods of intense effort and recovery, targeting specific energy systems, including aerobic, anaerobic, and neuromuscular capacities.

> *High-intensity intervals benefit all cyclists through greater efficiency and enhanced fatigue resistance.*

Interval training is a cornerstone of cycling performance, enabling cyclists to improve physiological systems and their ability to sustain high-intensity efforts. This approach to training alternates between periods of intense effort and recovery, targeting aerobic, anaerobic energy systems, and neuromuscular capacities.

During high-intensity intervals, the cardiovascular and muscular systems are selectively overloaded, increasing maximal oxygen uptake (VO_2 max), lactate threshold, and muscular metabolic efficiency. The recovery periods allow partial replenishment of energy stores and clearance of metabolic byproducts, enabling high-intensity efforts later.

The intensity, duration, and number of intervals determine the training outcome. These variables influence which energy systems are targeted and how the body adapts. For example, short, intense intervals of 30-second all-out sprints develop anaerobic capacity, while longer intervals at moderate intensity, such as 5–10 minutes at threshold power, enhance aerobic endurance and lactate clearance.

TEMPO, SWEAT SPOT OR ZONE 3
These intervals are performed at 85–95 per cent of FTP, a zone where both intensity and sustainability are balanced. These intervals efficiently build aerobic fitness and muscular endurance without excessive fatigue.

25 minutes

95%

THRESHOLD
Threshold intervals target the lactate threshold, the point at which lactate production exceeds clearance. Performed at 90–105 per cent of FTP, these intervals usually last 10–20 minutes and improve a cyclist's ability to sustain near-maximal effort without fatiguing prematurely.

10–20 minutes

105%

VO_2 MAX
VO_2 max intervals, typically lasting 4–6 minutes at 105–120 per cent of functional threshold power (FTP), aim to maximize aerobic capacity. These intervals are at a high intensity but crucial for improving the ability to sustain high power outputs for extended periods.

4–6 minutes

120%

ANAEROBIC CAPACITY
Short, intense efforts of 30 seconds to 2 minutes at 120–150 per cent of FTP train the anaerobic system. These intervals are essential for cyclists involved in sprint finishes, breakaways, or punchy climbs.

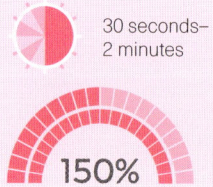

30 seconds– 2 minutes

150%

HOW TO TRAIN | Interval Training

TRAINING STRUCTURE

A typical interval training session begins with a warm-up to increase heart rate, blood flow, and muscle temperature. This is then followed by the main set that includes the intervals, interspersed with recovery periods that vary in length depending on the goal of the session. They may be longer for maintaining the quality of intervals and shorter to focus on building tolerance and fatigue resistance. A cooldown phase at the end of the session helps the body recover before ending the session.

Interval training improves performance across all types of cycling, whether road racing, time trials, or mountain biking. It increases efficiency, enhances fatigue resistance, and develops the ability to perform repeated high-intensity efforts. However, correct periodization of these interval sessions to suit an individual's fitness levels and requirements is essential to avoid overtraining and underperformance. Cyclists should monitor training load using metrics like power, heart rate, and perceived exertion, ensuring sessions align with their goals and individual fitness.

Intervals are performed at up to **200%** of normal functional *threshold* power

SPRINTS
All-out efforts lasting 10–20 seconds at maximum power output improve neuromuscular power and explosive strength. These are critical for track cyclists and sprinters but also beneficial for road riders who need to contest sprints or respond to attacks.

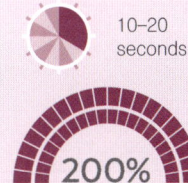
10–20 seconds
200%

Examples of training sessions
The examples below show typical interval training sessions from a range of training programmes. Each should begin with a warm-up and end with a cool-down with recovery sessions in between.

VO₂ MAX SESSION

Z5 Zone 5

5 x 5 min @ 110% of FTP with 5 min recovery

THRESHOLD SESSION

Z4 Zone 4

3 x 15 min @ 95% of FTP with 10 min rest

ANAEROBIC CAPACITY

Z5 Zone 5

8 x 1 min efforts @ 150% of FTP with 5 min recovery between each effort

SPRINTS

Z6 Zone 6

3 x 15 min @ 95% of FTP with 10 min rest

TEMPO, SWEAT SPOT, OR ZONE 2

Z3 Zone 3

5 x 5 min @ 110% of FTP with 5 min recovery

Z2 Zone 2

5 x 5 min @ 110% of FTP with 5 min recovery

STRENGTH TRAINING

Strength training is an essential component of a cycling training programme. Once viewed as non-essential or even detrimental, it is now widely accepted as a performance-enhancing and injury-preventing tool for cyclists across all disciplines.

When implemented correctly, strength training supports improvements in power, efficiency, fatigue resistance, and overall robustness. These are all qualities that benefit every cyclist, from sprinters to stage racers and technical off-road specialists.

The basic principles of effective strength training in cycling mirror those used in other sports: specificity, progressive overload, individualization, and recovery. While the goal is not to maximize muscle size (hypertrophy), cyclists aim to improve maximal strength, rate of force development, and neuromuscular coordination. Training programmes typically emphasize compound movements, such as squats, deadlifts, lunges, and step-ups, that target the glutes, quadriceps, hamstrings, and core. These exercises are selected for their transfer to pedalling dynamics and the ability to be loaded progressively.

A well-structured programme generally includes three phases: an adaptation phase to introduce strength work and develop proper movement patterns, a maximum strength phase with heavier loads, and an in-season maintenance phase to retain strength with reduced volume. Strength training is ideally performed 1–3 times per week, depending on the time of year, training load, and racing schedule.

DIFFERENT DISCIPLINES

Application of strength training varies by cycling discipline, as shown below. Across all disciplines, strength training supports better pedalling mechanics, resilience to crashes or overuse injuries, and long-term development. When integrated appropriately with on-bike training, it is a powerful tool that contributes to sustained performance gains and rider health throughout a cycling career.

Example session

This sample strength training session (maximum strength phase) uses a combination of exercises to build up muscle without maximizing muscle size.

EXERCISE	SETS/REPS	LOAD	FOCUS
Barbell front squat (pp.154–55)	4 × 5	80–85% 1RM (1 repetition maximum)	Maximal lower-body strength
Trap bar deadlift (pp.160–61)	3 × 8	Moderate	Hamstring and glute development
Split squat (pp.156–57)	3 × 6 per leg	Bodyweight or light	Single-leg strength and stability
Pull up (pp.176–77)	3 × 8	Bodyweight or moderate	Upper body pulling strength
Glute bridge (pp.192–93)	3 × 10	Moderate	Hip extension and posterior chain activation
Plank with shoulder taps (pp.184–85)	3 × 30 secs	Bodyweight	Core stability and anti-rotation
Russian twist (pp.188–89)	3 × 12 per side	Light to moderate (medicine ball)	Rotational core control

ROAD CYCLISTS

Road cyclists benefit from improved fatigue resistance, sprint performance, and musculoskeletal durability, particularly during long stages or when attacking under load.

WHOLE BODY TRAINING

While cycling strength training is focussed on the lower body, exercises that work on the upper body are also important in supporting maximal performance and injury prevention.

A strong upper body helps maintain good posture on the bike, especially during long rides or races where fatigue can lead to slouching and discomfort. In addition to posture, upper body strength contributes to better bike handling and resilience. A stronger upper body helps cyclists absorb vibrations from rough terrain, stabilize the handlebars during sprints or descents, and manoeuvre more effectively through corners or in a peloton. Strengthening muscles around the shoulders, arms, and trunk can also help prevent overuse injuries, particularly in the neck, shoulders, and wrists, which often result from prolonged static positioning. Overall, balanced upper body training can enhance comfort, control, and durability – key factors in both performance and long-term cycling health.

Stretching

Long periods in a fixed, forward-leaning position can lead to muscle tightness and imbalances, particularly in the hip flexors, hamstrings, lower back, and chest. Regular stretching can help counteract these effects, promoting better flexibility and joint mobility. Stretching also plays a role in injury prevention and recovery and can reduce the risk of strains and overuse injuries common in cyclists, such as lower back pain or knee issues. Post-ride stretching helps relax muscles and in the removal of metabolic waste products. It also supports faster recovery.

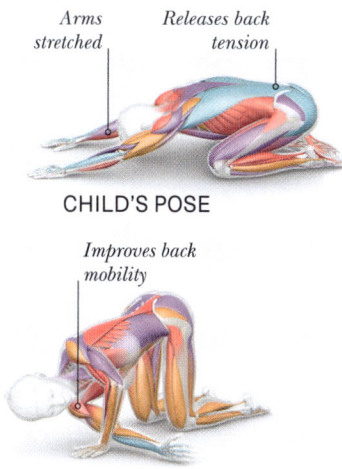

Arms stretched — *Releases back tension*

CHILD'S POSE

Improves back mobility

THREAD THE NEEDLE

Pectoralis major
Biceps
Abdominal muscles

BUILDING THE CORE
Core and upper back strength is essential for keeping the torso stable, which helps maintain efficient pedalling.

RUSSIAN TWIST (PP.188–189)

TRACK CYCLISTS AND BMX RIDERS

Track cyclists and BMX riders, who rely heavily on explosive power, require greater emphasis on maximal strength, Olympic lifts, and plyometrics to maximize acceleration and torque.

MOUNTAIN BIKERS AND CYCLOCROSS

Mountain bikers and cyclocross riders must also develop upper-body and core strength to handle technical terrain and off-bike efforts, with added focus on grip strength, shoulder stability, and dynamic core control.

GRAVEL AND ULTRA-ENDURANCE

For gravel and ultra-endurance or multi-stage cyclists, strength training improves efficiency, injury resistance, and posture during prolonged efforts.

THE TRAINING PROGRAMMES

Cycling training programmes must be structured according to a cyclist's experience, current fitness, and the demands of their event or discipline. Variables such as training history, technical ability, and time availability all influence programme design, while the physiological and tactical requirements of different events demand tailored preparation.

The programmes in this book are designed for amateur cyclists preparing for a specific event in their chosen discipline. Each plan is grounded in key training principles including progressive overload, specificity, and periodization, with targeted work across training zones to develop aerobic endurance, lactate threshold, and neuromuscular power. Recovery, nutrition, and practical considerations are included to support sustainable progress.

THE WORKOUTS

Each training plan includes workouts designed to target specific physiological capacities and skill components relevant to the demands of the target event. Workout intensities are prescribed using predefined training zones, which are derived from the performance testing protocols outlined in this book. These zones are typically anchored to an individual's Functional Threshold Power (FTP) and expressed as percentages of FTP to standardize training stimuli across different cyclist's profiles. This individualized approach enables more precise control of training load and physiological stress.

Key training intensities

ZONES	NAME	INTENSITY DESCRIPTION	POWER (%FTP)
Z1 Zone 1	Recovery	Very easy, active recovery	< 55%
Z2 Zone 2	Endurance	Easy aerobic	56–75%
Z3 Zone 3	Aerobic	Moderate aerobic (minimized in polarized)	76–90%
Z4 Zone 4	Threshold	Hard, just below or at FTP	91–105%
Z5 Zone 5	VO$_2$ max/ anaerobic	Very hard, race-specific burst	>105%
Z6 Zone 6	Neuromuscular	Maximal sprint	>150%

OTHER SYMBOLS

- Group rides
- Surges
- Race simulation
- Race starts
- Track exercises
- @ Cycle at a given pace
- R Rest

BELOW FTP

Training zones below and around FTP are relatively well-defined and correspond to distinct metabolic responses. These adaptations are critical for sustainable power output and fatigue resistance during prolonged efforts.

ABOVE FTP

Training zones above FTP are less defined, as they engage overlapping and variable physiological systems depending on the intensity and duration of the effort. High-intensity intervals target different combinations of central and peripheral adaptations.

ADJUSTMENTS

As adaptations occur training zones must be recalibrated to ensure the prescribed intensities continue to elicit the intended stimulus. Failure to update these zones may result in underloading or overreaching, both of which can compromise the effectiveness of training and the accuracy of load management. Zone adjustments can be achieved through periodic re-assessment using standardized performance testing protocols.

One effective approach is the continuous monitoring of power–duration curves, which reflect a cyclist's peak power outputs across a range of durations. Shifts in the 20–60-minute range of the curve are particularly relevant for estimating changes in FTP or threshold power. By integrating these updates, the training programme remains appropriately challenging and continues to target the cyclist's evolving physiological profile and optimize the rate and quality of adaptation.

FTP is the *highest average power a rider can sustain for* **60 MINUTES**

TRAINING VOLUME

Across all programmes, progressive overload is systematically applied by manipulating volume, intensity, and frequency across the training weeks. Training load typically increases over two to three weeks before a recovery week is introduced, aligning with block periodization models. This wave-like pattern of loading and unloading supports long-term progression and minimizes the risk of overtraining. Supportive elements such as strength training off-the-bike are also included in each plan, with a focus on core stability, neuromuscular control, and lower-limb strength, all of which contribute to improved force transmission, injury prevention, and sustained posture on the bike.

TRAINING PHASES
- Base phase
- Recovery
- Race-specific work
- Taper
- Race week

LOAD MANAGEMENT
Balancing hard and easy days, incorporating rest weeks, and tapering appropriately before key events is important for effective training and recovery.

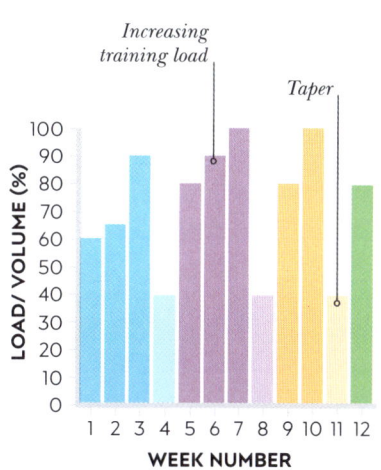

TRAINING PHASES AND PERIODIZATION

The training plans are grounded in a shared set of core principles derived from contemporary endurance training science. Despite differences in race duration, terrain, and tactical demands, all plans adhere to a coherent framework that balances intensity, specificity, and recovery to optimize physiological adaptation and prepare for peak performance.

GENERAL ENDURANCE FITNESS

Focus on developing overall cycling fitness by incorporating a mix of aerobic endurance rides, anaerobic intervals, and technique-focussed sessions. This balanced approach helps build a more complete and well-rounded cyclist.

INCREASED INTENSITY

Workouts are progressed by either increasing the intensity; extending the duration; increasing the ratio of hard efforts to recovery time; or riding the recovery intervals at a slightly higher intensity.

INCREASE VOLUME

Increase cycling volume towards a planned peak over the course of the training programme, with periodic lower-volume weeks to allow the body to recover and adapt to the accumulated training load.

PROGRESSIVE ADAPTATION

Introduce varied training stimuli to drive physical adaptations by adjusting the volume, intensity, or frequency of cycling sessions throughout the programme.

OPTIMIZE TRAINING LOAD

Training load should increase at a rate the body can adapt to. Watch for signs of overtraining and adjust the plan if necessary to ensure continued progress and recovery.

POLARIZED INTENSITY

Each programme is designed around polarized training intensity distribution. This emphasizes the accumulation of training time at low intensities (Zones 1–2), typically accounting for 80–90 per cent of total volume, with the remaining 10–20 per cent allocated to high-intensity efforts (Zones 4–5). This approach is supported by a growing body of evidence demonstrating that frequent, high-quality work at both ends of the intensity spectrum leads to superior endurance adaptations compared to training that relies heavily on moderate-intensity efforts. The result is a training stimulus that promotes both aerobic efficiency and high-end fitness while reducing the risk of maladaptation, excessive fatigue, or stagnation.

All plans are periodized into sequential training phases that progressively develop and refine the cyclist's capacity. They begin with a base phase in which aerobic conditioning, muscular endurance, and foundational strength are developed through extended Zone 2 rides and introductory threshold or VO_2 max efforts. This is followed by a recovery week in which volume is reduced to promote physiological consolidation and repair. A subsequent build or race-specific phase introduces greater intensity and specificity, incorporating race-simulation workouts, higher training loads, and terrain-relevant demands. The final one to two weeks of each plan serve as a taper phase, reducing overall training volume while maintaining intensity to allow for full recovery, glycogen replenishment, and the expression of peak fitness on race day.

EFFORTS

Each plan incorporates a variety of structured workouts designed to elicit specific physiological responses aligned with the demands of the target event. For instance, VO$_2$ max intervals (for example, 4–6 x 3–5 minutes at Zone 5 with equal recovery) are used to stimulate maximal oxygen uptake and central cardiovascular adaptations. Threshold intervals (for example, 2–3 x 10–20 minutes at Zone 4) are used to increase sustainable aerobic power and metabolic efficiency. Sprint and anaerobic intervals (e.g., 8–12 x 30 seconds all-out) are included in plans where explosive power and repeated high-intensity efforts are critical, such as criterium and cyclocross races. Long rides at Zone 2 form the backbone of all plans and help develop fatigue resistance, capillarization, and the ability to mobilize and oxidize fatty acids as a fuel source.

RECOVERY

Recovery periods are not merely passive gaps between efforts, they actively determine the metabolic demands, neuromuscular loading, and cardiovascular stress of the session. In all the training plans, rest days are typically scheduled once a week depending on the structure of the week. These are either complete rest days or very low-intensity rides. Their purpose is to promote recovery, support adaptation, and maintain freshness for key training sessions. During recovery weeks and taper phases, additional rest or recovery days are included to reduce cumulative fatigue and optimize performance ahead of competition.

SLEEP AND NUTRITION

Sleep supports growth hormone secretion, muscle repair, immune function, and cognitive performance. Poor sleep impairs recovery regardless of other interventions and should be a primary focus for all cyclists. After sleep, correct nutrition is the most important element of recovery. Timely intake of carbohydrates and protein following training and racing accelerates glycogen replenishment and muscle repair (see pp.62–63). Hydration is equally important, especially in hot or high-altitude environments.

> *Recovery between intervals is a fundamental component of the training stimulus.*

Further strategies

Additional recovery methods may be useful in different circumstances. Active recovery or low-intensity riding enhances blood flow and reduces perception of muscle soreness. Massage, compression garments, cold water immersion, and mobility work may support recovery, although their effects can vary between individuals. These strategies are most effective when used during high-demand phases of training and racing. Chronic use of anti-inflammatory based recovery modalities (cold water immersion, contrast bathing, cryotherapy) may, however, blunt training adaptations as the inflammatory response is crucial to training adaptation.

Warm up and cool down

Establish a regular routine for warming up and cooling down around each training session. Ensure you reserve enough time for each stage. Start your workouts with a dynamic warm-up involving stretches and drills, which helps to prepare your body for the demands of cycling, prevents injury, and improves performance. It is especially important on race day that you warm up adequately, so you can be confident that you are physically and mentally prepared to do your best. End each run with an adequate cool-down session to begin the recovery process.

8-WEEK TRAINING PLAN FOR A CRITERIUM RACE

This is an 8-week polarized training plan designed for amateur cyclists preparing for short criterium races (approximately 1 hour). The objective is to enhance aerobic capacity, improve repeatability of high-intensity efforts, and increase anaerobic power, while maintaining a polarized distribution.

The plan is structured into progressive phases. Weeks 1–3 build aerobic fitness with VO₂ max intervals and sprint work. Week 4 is a recovery week. Weeks 5–6 introduce race-specific efforts, including anaerobic intervals, sprint starts, and simulations. Week 7 tapers volume, leading into race preparation and the target event in Week 8. Core sessions include VO₂ max intervals, short sprints, and threshold work, supported by longer Zone 2 rides. Intervals are placed on key training days, with lighter sessions and rest days ensuring recovery. By progressively increasing training intensity and specificity, this plan prepares cyclists to meet the demands of criterium racing, frequent accelerations, high-speed corners, surges, and sprint finishes, with fitness and tactical resilience.

Phase		WEEK	MONDAY	TUESDAY	WEDNESDAY
Base phase	Aerobic base + VO₂ max	1	R	1.5 hours @ Z2	1 hour @ Z1
	Sprint power + endurance	2	R	1.5 hours @ Z2 + 5 x 20s sprints (2 min recovery)	1 hour @ Z1
	VO₂ max intervals	3	R	1.5 hours @ Z2 + 5 x 3 min @ Z5 (5 min recovery)	1 hour @ Z1
	Recovery	4	R	1 hour @ Z2	45 mins @ Z1
Specificity phase	Threshold + criterium work	5	R	2 hours @ Z2 + 3 x 10 min @ Z4 (10 min recovery)	1 hour @ Z1
	Anaerobic power	6	R	2 hours @ Z2 + 4 x 30 s all-out (3 min recovery)	45 mins @ Z1
	Taper start	7	R	1.5 hours @ Z2 +2 x 5 min @ Z5 (5 min recovery)	1 hour @ Z1
	Race week	8	R	1 hour @ Z1 or Z2	R

TRAINING PLAN | 8-Week Training Plan for a Criterium Race

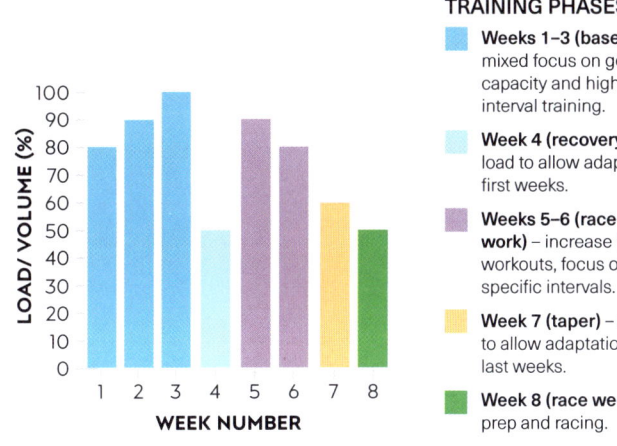

TRAINING PHASES

- **Weeks 1–3 (base phase)** – mixed focus on general aerobic capacity and higher-intensity interval training.
- **Week 4 (recovery)** – reduce load to allow adaptation to the first weeks.
- **Weeks 5–6 (race-specific work)** – increase high-intensity workouts, focus on race specific intervals.
- **Week 7 (taper)** – reduce load to allow adaptation to the last weeks.
- **Week 8 (race week)** – race prep and racing.

Training notes

Strength training (1–2 x per week)
Focus on core, stability, and leg strength/power.

Nutrition and hydration
Practise fuelling strategy during long rides.

Taper effectively
Reduce volume in the last two weeks to peak on race day.

THURSDAY	FRIDAY	SATURDAY	SUNDAY
1.5 hours @ Z2 + 4 x 3 min @ Z5 (5 min recovery)	R	2 hours @ Z2 + 3 x 30s sprints	1.5 hours @ Z1
1.5 hours @ Z2 + 4 x 4 min @ Z5 (5 min recovery)	R	2.5 hours @ Z2	1.5 hours @ Z1
1.5 hours @ Z2 + 4 x 2 min @ Z5 (4 min recovery)	R	2 hours @ Z2 + 4 x15s race starts	1.5 hours @ Z1
1 hour @ Z2	R	1.5 hours @ Z2	1 hour @ Z1
1.5 hours @ Z2 + 6 x 2 min @ Z5 (5 min recovery)	R	2 hour group ride + surges	1.5 hours @ Z1
1.5 hours @ Z2 + 5 x 3 min @ Z5 (5 min recovery)	R	1 hour race simulation	1.5 hours @ Z1
1 hour @ Z2	R	1.5 hours @ Z2 + 3 x 1min @ Z5 (full recovery)	1.5 hours @ Z1
1.5 hours @ Z2 + 2 x 1 min @ Z5 (full recovery)	R	Criterium race (1 hour)	1 hour @ Z1

8-WEEK TRAINING PLAN FOR A CROSS-COUNTRY MOUNTAIN BIKE RACE

This 8-week polarized training plan is for an amateur cyclist preparing for a cross-country (XC) mountain bike race lasting approximately one hour. The plan focusses on enhancing aerobic efficiency, technical consistency under fatigue, and high-intensity tolerance.

The plan begins with three base weeks focussed on aerobic conditioning, VO₂ max efforts, and MTB handling skills. Week 4 is a recovery week to support adaptation. Weeks 5 and 6 increase specificity with off-road intervals, race simulations, and technical climbs. Week 7 tapers training volume to reduce fatigue, leading into race preparation and the target event in Week 8. Core sessions include Zone 5 intervals, threshold efforts on undulating terrain, sprint starts, and technical drills. These are supported by longer Zone 2 MTB endurance rides. The structure balances intensity, volume, and recovery to prepare riders for the repeated surges, variable terrain, and technical demands of the sport.

Phase		WEEK		MONDAY	TUESDAY	WEDNESDAY
Base phase	Base endurance + skills	1		R	1.5 hours @ Z2 + MTB skills	1 hour @ Z1
	VO₂ max development	2		R	1.5 hours @ Z2 + 4 x 4 min @ Z5 off-road (5 min recovery)	1 hour @ Z1
	Threshold strength	3		R	1.5 hours @ Z2 + 2 x 10 min @ Z4	1 hour @ Z1
	Recovery week	4		R	1 hour @ Z2	45 mins @ Z1
Specificity phase	XC-specific intensity	5		R	2 hours @ Z2 + 5 x 3 min @ Z5 (3 min recovery)	1 hour @ Z1
	Anaerobic repeatability	6		R	2 hours @ Z2 + 6 x 2 min @ Z5 (2 min recovery)	45 mins @ Z1
	Taper start	7		R	1.5 hours @ Z2 + 2 x 5 min @ Z5 (5 min recovery)	1 hour @ Z1
	Race week	8		R	1 hour @ Z1 or Z2	R

HOW TO TRAIN | 8-Week Training Plan for a Cross-Country Mountain Bike Race

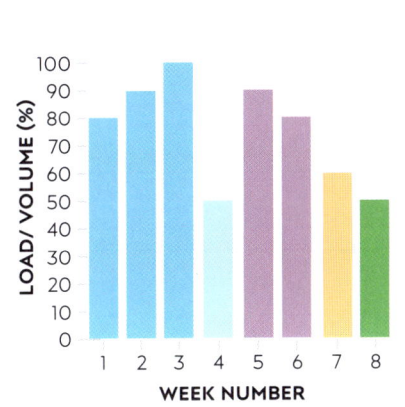

TRAINING PHASES

- **Weeks 1–3 (base phase)** – mixed focus on general aerobic capacity and higher-intensity interval training combined with MTB technical skills.
- **Week 4 (recovery)** – reduce load to allow adaptation to the first weeks.
- **Weeks 5–6 (race-specific work)** – increase high-intensity workouts, focus on race specific intervals on MTB race terrain.
- **Week 7 (taper)** – reduce load to allow adaptation to the last weeks.
- **Week 8 (race week)** – race prep and racing.

Training notes

Strength training (1–2 x per week)
Focus on core, stability, and leg strength/power.

Nutrition and hydration
Practise fuelling strategy during long rides.

Taper effectively
Reduce volume in the last two weeks to peak on race day.

THURSDAY	FRIDAY	SATURDAY	SUNDAY
1.5 hours @ Z2 + 3 x 3 min @ Z5 off-road (5 min recovery)	R	2 hours @ Z2 + 3 x 30s sprints on MTB	1.5 hours @ Z2
1.5 hours @ Z2 + MTB skills	R	2.5 hours @ Z2 + 4 x 1 min @ Z5 on MTB (full recovery)	2 hours @ Z1
1.5 hours @ Z2 + 3 x 6 min @ Z4 tech climbs (5 min recovery)	R	2.5 hours @ Z2 on MTB	1.5 hours @ Z1
1 hour @ Z2	R	1.5 hours @ Z2 on MTB	1 hour @ Z1
1.5 hours @ Z2 + 3 x 10 min @ Z4 MTB circuit (5 min recovery)	R	1.5 hours MTB race simulation	2 hours @ Z1
1.5 hours @ Z2 + 4 x 30s sprints + MTB skills	R	2 hours race pace simulation + 3 x 15 min on/off efforts	1 hour @ Z1
1 hour @ Z2	R	1.5 hours @ Z2 + 3 x 1 min @ Z5 (full recovery)	1.5 hours @ Z2
1.5 hours @ Z2 + 2 x 1min @ Z5 (full recovery)	R	MTB-XC race	1 hour @ Z1

8-WEEK TRAINING PLAN FOR A CYCLOCROSS RACE

This is an 8-week polarized training plan for an amateur cyclist preparing for a 1-hour cyclocross (CX) race. Cyclocross requires a combination of anaerobic power, repeatability, technical handling, and durability on various terrains.

The first three weeks of the plan focusses on aerobic endurance, VO_2 max development, and basic CX skills such as dismounts and remounts. Week 4 is a recovery phase. Weeks 5 and 6 increase specificity with intervals on muddy or grassy terrain, repeated short surges, and CX circuit simulations. Week 7 tapers the training load to reduce fatigue, leading into race preparation and competition in Week 8.

The plan includes Zone 5 intervals, threshold efforts, off-road sprints, and skill-focussed sessions. These are supported by longer Zone 2 rides and structured recovery. The progression develops the physical and technical demands needed for the frequent accelerations, variable surfaces, and complex race scenarios typical of cyclocross.

Phase	Focus	WEEK	MONDAY	TUESDAY	WEDNESDAY
Base phase	Base + CX skills	1	R	1.5 hours @ Z2 + dismount/remounts	1 hour @ Z1
Base phase	VO_2 max + CX handling	2	R	1.5 hours @ Z2 + 4 x 4 min @ Z5 muddy terrain (5 min recovery)	1 hour @ Z1
Base phase	Threshold build	3	R	2 hours @ Z2 + 2 x 10 min @ Z4 undulating (10 min recovery)	1 hour @ Z1
Base phase	Recovery	4	R	1 hour @ Z2	45 mins @ Z1
Specificity phase	CX-Specific Intensity	5	R	2 hours @ Z2 + 5 x 3 min @ Z5 (3 min recovery)	1 hour @ Z1
Specificity phase	Anaerobic repeatability	6	R	2 hours @ Z2 + 6 x 2 min @ Z5 (5 min recovery)	45 mins @ Z1
	Taper start	7	R	1.5 hours @ Z2 + 2 x 5min @ Z5 (5 min recovery)	1 hour @ Z1
	Race week	8	R	1 hour @ Z1 or Z2	R

HOW TO TRAIN | 8-Week Training Plan for a Cyclocross Race

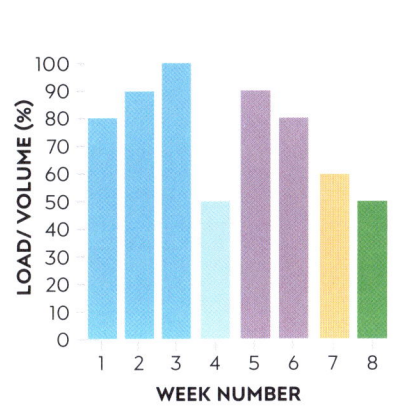

TRAINING PHASES

- **Weeks 1–3 (base phase)** – mixed focus on general aerobic capacity and higher-intensity interval training combined with CX skills.
- **Week 4 (recovery)** – reduce load to allow adaptation to the first weeks.
- **Weeks 5–6 (race-specific work)** – increase high-intensity workouts, focus on race specific intervals on CX race terrain.
- **Week 7 (taper)** – reduce load to allow adaptation to the last weeks.
- **Week 8 (race week)** – race prep and racing.

Training notes

Strength training (1–2 x per week)
Focus on core, stability, and leg strength and power.

Nutrition and hydration
Practise fuelling strategy during long rides.

Taper effectively
Reduce volume in the last two weeks to peak on race day.

THURSDAY	FRIDAY	SATURDAY	SUNDAY
1.5 hours @ Z2 + 3 x 5min @ Z4 on grass or dirt *(5 min recovery)*	R	2 hours @ Z2 + 3 x 30s sprints off road	1.5 hours @ Z2
1.5 hours @ Z2 + corners, barriers + run ups	R	2.5 hours @ Z2 + 5 x 1 min @ Z5 *(full recovery)*	2 hours @ Z1
1.5 hours @ Z2 + 3 x 6 min @ Z4 + CX skills *(5 min recovery)*	R	2 hours @ Z2 + 3 x 10 min @ Z4 CX circuit *(10 min recovery)*	1 hour @ Z1
1 hour @ Z2	R	1.5 hours @ Z2	1 hour @ Z1
1.5 hours @ Z2 + 3 x 10 min @ Z4 CX circuit *(5 min recovery)*	R	1.5 hours @ Z2 + 2 x 20 min @ Z4 race simulation	2 hours @ Z1
1.5 hours @ Z2 + 4 x 30s sprints + CX skills	R	1.5 hours @ Z2 + 2 x 12 min race efforts @ Z4 *(full recovery)*	2.5 hours @ Z1
1 hour @ Z2	R	1.5 hours @ Z2 + 3 x 1 min @ Z5 *(full recovery)*	1 hour @ Z2
1.5 hours @ Z2 + 2 x 1 min @ Z5 *(full recovery)*	R	Cyclocross race	1 hour @ Z1

8-WEEK TRAINING PLAN FOR A TIME TRIAL

This is an 8-week time trial (TT) training plan for an amateur cyclist preparing for a 20km (12 mile) TT race, which typically lasts around 30–40 minutes. The plan incorporates a 5-zone model aimed at enhancing aerobic capacity, sustainable threshold power, and tolerance to aero positioning. It includes a combination of polarized and race-specific threshold training.

The first three weeks focus on aerobic development, VO$_2$ max intervals, and threshold work to raise power output. Week 4 is a recovery phase. Weeks 5 and 6 increase race-specific efforts on the time trial bike, including intervals at and above race pace. Week 7 is a taper to reduce fatigue while maintaining intensity, leading into final race preparation and competition in Week 8.

Key workouts include Zone 4 threshold intervals, Zone 5 VO$_2$ max efforts, and sustained efforts in the aero position. These are supported by longer Zone 2 rides and active recovery. The plan builds the power, pacing, and aerodynamic resilience needed for time trial success.

Phase		Week	Monday	Tuesday	Wednesday
Base phase	Aerobic base	1	R	1.5 hours @ Z2	1 hour @ Z1
Base phase	Threshold build	2	R	2 hours @ Z2 + 2 x 10 min Z4 @ Z4 (10 min recovery)	1 hour @ Z1
Base phase	VO$_2$ max focus	3	R	1.5 hours @ Z2 + 4 x 4 min @ Z4 (5 min recovery)	1 hour @ Z1
Base phase	Recovery week	4	R	1 hour @ Z2	45 mins @ Z1
Specificity phase	Threshold Development	5	R	2 hours @ Z2 + 3 x 12 min @ Z4 (10 min recovery)	1 hour @ Z1
Specificity phase	VO$_2$ max + Race specific	6	R	1.5 hours @ Z2 + 5 x 3 min @ Z5 (5 min recovery)	45 mins @ Z1
Specificity phase	Taper start	7	R	1.5 hours @ Z2 + 2 x 6 min @ Z4 (5 min recovery)	1 hour @ Z1
Specificity phase	Race week	8	R	1 hour @ Z1 or Z2	R

TRAINING PLANS | 8-Week Training Plan for a Time Trial

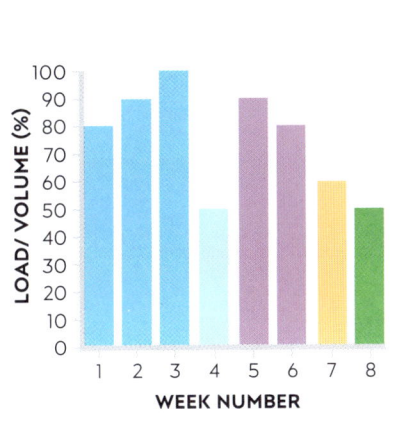

TRAINING PHASES

- **Weeks 1–3 (base phase)** – focus on building both power at threshold and increasing VO₂ max specific to time trial performance.
- **Week 4 (recovery)** – reduce load to allow adaptation to the first weeks.
- **Weeks 5–6 (race-specific work)** – increase high-intensity workouts, focus on race specific intervals on the time trial bike.
- **Week 7 (taper)** – reduce load to allow adaptation to the last weeks.
- **Week 8 (race week)** – race prep and racing.

Training notes

Strength training (1–2 x per week)
Focus on core, stability, and leg strength/power.

Nutrition and hydration
Practise fuelling strategy during long rides.

Taper effectively
Reduce volume in the last two weeks to peak on race day.

THURSDAY	FRIDAY	SATURDAY	SUNDAY
1.5 hours @ Z2 + 3 x 3 min @ Z5 (3 min recovery)	R	2.5 hours @ Z2 long ride	1.5 hours @ Z2
1.5 hours @ Z2 + 3 x 6 min @ Z4 (5 min recovery)	R	1 hour @ Z2 Long ride	1.5 hours @ Z2
1.5 hours @ Z2 + 6 x 2 min @ Z5 (5 min recovery)	R	1 hour @ Z2 + 3 x 30s sprints (full recovery)	2 hours @ Z2
1 hour @ Z2	R	1.5 hours @ Z2	1.5 hours @ Z2
1.5 hours @ Z2 + 2 x 15 min @ Z4 (in aero position)	R	40 mins @ Z4 race pace effort (TT bike)	1.5 hours @ Z2
1 hour @ Z2 + 20 min sustained @ Z4 (TT bike)	R	1.5 hours @ Z2 + 2 x 10 min @ Z4 TT race pace (full recovery)	2 hours @ Z2
1 hour @ Z2	R	1 hour @ Z2 + 3 x 1 min @ Z5 (full recovery)	20 min TT effort @ Z4
1 hour @ Z2 + 2 x 1min @ Z5 (full recovery)	R	20km Time Trial Race	1 hour @ Z1

8-WEEK TRAINING PLAN FOR A TRACK CYCLIST

This is an 8-week time trial training plan tailored for a track cyclist targeting the individual pursuit and bunch races (for example, scratch race, points race, and elimination). This version balances the demands of both: the sustained power and pacing of the pursuit, and the repeated high-intensity efforts and tactical surges of bunch racing.

The first three weeks focus on building aerobic capacity, threshold power, and sprint tolerance. Week 4 reduces load to support recovery and adaptation. Weeks 5 and 6 increase intensity and specificity, with anaerobic intervals, pursuit efforts, and race simulations. Week 7 tapers volume while maintaining intensity and sharpness. Week 8 focusses on final race preparation and execution. Key sessions include Zone 4 pursuit intervals, Zone 5–6 sprints and starts, bunch race simulations, and track-specific drills. Strength training and technical work are emphasized early, tapering before race week. The programme develops the power and control needed to perform in both timed and tactical track events.

Phase		WEEK		MONDAY	TUESDAY	WEDNESDAY
Base phase	Aerobic + sprint base	1		R	1.5 hours @ Z2 + 5 x 15s Z6 sprints (3 min recovery)	1 hour @ Z1
	Threshold + anaerobic capacity	2		R	2 hours @ Z2 + 3 x 8 min @ Z4 (5 min rec)	1 hour @ Z1
	VO₂ max focus	3		R	1.5 hours @ Z2 + 4 x 4 min @ Z5	1 hour @ Z1
	Recovery week	4		R	1 hour @ Z2	1 hour @ Z1
Specificity phase	Race-specific build	5		R	2 hours @ Z2 + 3 x 6 min @ Z4 (pursuit gear, aero)	1 hour @ Z1
	Anaerobic speed	6		R	1.5 hours @ Z2 + 6 x 1 min @ Z5 / Z6	1 hour @ Z1
	Taper	7		R	1.5 hours @ Z2 + 2 x 3 min @ Z4 + 2 x 30s @ Z6	1 hour @ Z1
	Race week	8		R	1 hour @ Z1 or Z2	R

TRAINING PLANS | 8-Week Training Plan for a Track Cyclist

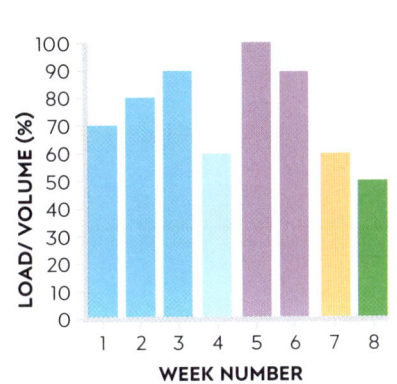

TRAINING PHASES

- **Weeks 1–3 (base phase)** – build threshold, aerobic capacity, and start repeated sprint tolerance.
- **Week 4 (recovery)** – reduce load to absorb gains.
- **Weeks 5–6 (race-specific work)** – intensify high-speed, anaerobic, and pursuit-specific work.
- **Week 7 (taper)** – reduce volume, maintain sharpness.
- **Week 8 (race week)** – event simulation and taper.

Training notes

Strength training (1–2 x per week)
Focus on core, stability, and leg strength/power.

Nutrition and hydration
Practise fuelling strategy during long rides.

Taper effectively
Reduce volume in the last two weeks to peak on race day.

THURSDAY	FRIDAY	SATURDAY	SUNDAY
1 hour @ Z2 + 4 x 4 min @ Z5 (5 min rec)	R	Track skills + 3 x 2km @ Z4 (race gear)	2 hours @ Z2
1 hour @ Z2 + 6 x 30s seated sprints (4 min rec)	R	Track: bunch race simulation + 5 x 1km pursuit efforts	1.5 hours @ Z2
1 hour @ Z2 + 4 x 3 min @ Z5 seated starts	R	Track: 2 x 3km pursuit + flying lap + race drills	1.5 hours @ Z2
1 hour @ Z2	R	1 hour skills + 3 x 30s @ Z6 (full rec)	1.5 hours @ Z2
1 hour @ Z2 + 5 x 30s sprint + 3 x 1 km pursuit pace	R	Track: Points race simulation 3 x 2km pursuit efforts	1.5 hours @ Z2
1 hour @ Z2 + flying 500m starts (5x)	R	Track: 4 x 1km pursuit scratch race efforts	1.5 hours @ Z2
1 hour @ Z2	R	1 hour track: 1 x 2km pursuit, 1x flying lap	1 hour @ Z2
1 hour @ Z2 + 2 x 1 min @ Z5	R	Track event: individual pursuit and bunch races	Recovery spin

12-WEEK TRAINING PLAN FOR A GRAN FONDO/CYCLOSPORTIVE

This plan is a more polarized training approach emphasizing spending most of the time in low-intensity (Zone 1–2) and high-intensity (Zone 4–5) while minimizing time in the moderate-intensity Zone 3 to optimize endurance and performance.

The plan is divided into four phases. Weeks 1–4 (base phase) focus on building aerobic capacity and muscular endurance. Weeks 5–8 (build phase) introduce higher-intensity efforts while maintaining volume. Weeks 9–11 (race-specific phase) simulate the demands of long-distance events with extended endurance rides and sharpening intervals. Week 12 is a taper to reduce fatigue and prepare for peak performance.

Key workouts include VO_2 max intervals, sustained threshold intervals, and long Zone 2 rides of up to 6 hours to develop endurance. Sprint efforts and hill surges are also used to prepare for challenging terrain. Weekly structure balances intensity with recovery, including regular Zone 1 days and full rest to support adaptation. This plan builds the aerobic durability, pacing control, and fuelling strategy required to complete a Gran Fondo.

Training notes

Strength training (1–2 x per week)
Focus on core, stability, and leg strength/power.

Nutrition and hydration
Practise fuelling strategy during long rides.

Taper effectively
Reduce volume in the last two weeks to peak on race day.

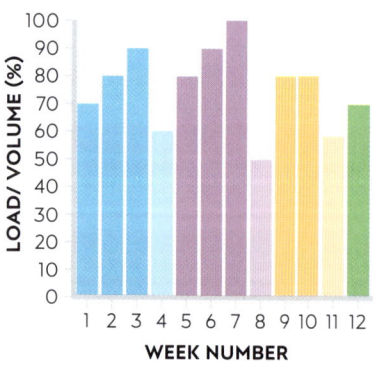

TRAINING PHASES

- **Weeks 1–4 (base phase)** – focus on building endurance with a small amount of intensity.
- **Weeks 5–8 (build phase)** – increase high-intensity workouts while maintaining endurance volume.
- **Weeks 9–11 (race-specific work)** – race simulations and sharpening efforts.
- **Week 12 (taper and race week)** – reduce volume, maintain intensity, and recover for peak performance.

> " "
>
> *Gran fondos are mass-participation long-distance cycling events.*

TRAINING PLANS | 12-Week Training Plan for a Gran Fondo/Cyclosportive

Phase		Week	MONDAY	TUESDAY	WEDNESDAY
Base phase	Recovery	1	R	2 hours @ Z2 + 4 x 4 min @ Z5 (5 min recovery)	2.5 hours @ Z2
Base phase		2	R	2 hours @ Z2 + 5 x 3 min @ Z5 (5 min recovery)	3 hours @ Z2
Base phase		3	R	2 hours @ Z2 + 4 x 5 min @ Z5 (5 min recovery)	3.5 hours @ Z2
Base phase	Recovery	4	R	2 hours @ Z2 + 2 x 8 min @ Z4 (5 min recovery)	2.5 hours @ Z2
Build phase		5	R	2 hours @ Z2 + 6 x 3 min @ Z5 (5 min recovery)	3.5 hours @ Z2
Build phase		6	R	2 hours @ Z2 + 7 x 3 min @ Z5 (5 min recovery)	4 hours @ Z2
Build phase		7	R	2 hours @ Z2 + 4 x 6 min @ Z5 (5 min recovery)	4.5 hours @ Z2
Build phase	Recovery	8	R	2 hours @ Z2 + 4 x 3 min @ Z5 (5 min recovery)	3 hours @ Z2
Specificity phase		9	R	2 hours @ Z2 + 4 x 8 min @ Z4 (5 min recovery)	4 hours @ Z2
Specificity phase		10	R	2 hours @ Z2 + 3 x 10min @ Z4 (10 min recovery)	4 hours @ Z2
Specificity phase	Taper	11	R	2 hours @ Z2 + 4 x 4 min @ Z5 (5 min recovery)	3.5 hours @ Z2
Specificity phase	Race week	12	R	2 hours @ Z2 + 3 x 5 min @ Z5 (5 min recovery)	2 hours @ Z2

WEEK	THURSDAY	FRIDAY	SATURDAY	SUNDAY
1	1.5 hours @ Z2 + 10 x 30s @ Z5 (30 sec recovery)	1.5 hours @ Z1	3 hours @ Z2	1.5 hours @ Z1
2	1.5 hours @ Z2 + 12 x 40s @ Z5 (20 sec recovery)	1.5 hours @ Z1	3.5 hours @ Z2	1.5 hours @ Z1
3	1.5 hours @ Z2 + 5 x 2min @ Z5 (2 min recovery)	1.5 hours @ Z1	4 hours @ Z2	1.5 hours @ Z1
4	1.5 hours @ Z2 + 12 x 20s @ Z5 (10 sec recovery)	1.5 hours @ Z1	3 hours @ Z2	1 hour @ Z1
5	1.5 hours @ Z2 + 5 x 2min @ Z5 (2 min recovery)	1.5 hours @ Z1	4.5 hours @ Z2	1.5 hours @ Z1
6	1.5 hours @ Z2 + 12 x 30s @ Z5 (30 sec recovery)	1.5 hours @ Z1	5 hours @ Z2	1.5 hours @ Z1
7	1.5 hours @ Z2 + 5x2min @ Z5 (2 min recovery)	1.5 hours @ Z1	5.5 hours @ Z2	1.5 hours @ Z1
8	1.5 hours @ Z2 + 8 x 40s @ Z5 (20 sec recovery)	1.5 hours @ Z1	4 hours @ Z2	1 hour @ Z1
9	1.5 hours @ Z2 + 5 x 2 min @ Z5 (2 min recovery)	1.5 hours @ Z1	6 hours @ Z2 (race simulation)	1.5 hours @ Z1
10	1.5 hours @ Z2 + 12 x 20s @ Z5 (10 sec recovery)	1.5 hours @ Z1	5 hours @ Z2	1.5 hours @ Z1
11	1.5 hours @ Z2 + 10 x 40s @ Z5 (20 sec recovery)	1.5 hours @ Z1	4 hours @ Z2	1.5 hours @ Z1
12	1 hour @ Z2	R	1.5 hours @ Z2	Race day

12-WEEK TRAINING PLAN FOR A ROAD CYCLIST

This 12-week plan for amateur cyclists focusses on polarized training to prepare for 3–4 hour road races. It uses 5 training zones with 80–90 per cent of training in low intensity and 10–20 per cent in high intensity, minimizing time in moderate intensity.

The plan begins with three base weeks to establish aerobic fitness, followed by a recovery week. The mid-phase introduces threshold and VO_2 max intervals while maintaining long endurance rides. Two race-specific phases focus on simulation rides, group surges, and high-intensity sharpening. Tapering occurs in the final two weeks to reduce fatigue and ensure peak performance on race day. Key sessions in the plan include VO_2 max intervals, sustained threshold efforts, and weekend group rides with surges. Structured recovery rides and complete rest days are included weekly.

Training notes

Strength training (1–2 x per week)
Focus on core, stability, and leg strength/power.

Nutrition and hydration
Practise fuelling strategy during long rides.

Taper effectively
Reduce volume in the last two weeks to peak on race day.

TRAINING PHASES

- **Weeks 1–3 (base phase)** – focus on building endurance with a small amount of intensity.
- **Week 4 (recovery)** – reduced load to allow adaptation to the first weeks
- **Weeks 5–7 (race-specific work)** – increase high-intensity workouts while maintaining endurance volume.
- **Week 8 (recovery)** – reduce load to allow adaptation to the last weeks
- **Weeks 9–10 (race-specific work)** – race simulations and sharpening efforts.
- **Weeks 11–12 (taper and race week)** – reduce volume, maintain intensity, and recover for peak performance

> *This plan builds the aerobic engine, intensity tolerance, and tactical preparation needed for successful road racing.*

		WEEK	MONDAY	TUESDAY	WEDNESDAY
Base phase	Base phase	1	1.5 hours @ Z2	1 hour @ Z1	2 hours @ Z2 + 3 x 3 min @ Z5 (3 min recovery)
	Aerobic endurance	2	2 hours @ Z2	1 hour @ Z1	2 hours @ Z2 + 4 x 4 min @ Z5 (4 min recovery)
	Base and threshold	3	2 hours @ Z2 + 2 x 10 min @ Z4 (10 min recovery)	1.5 hours @ Z1	2 hours @ Z2 + 3 x 6 min @ Z4 (5 min recovery)
	Recovery	4	1.5 hours @ Z2	1 hour @ Z1	1.5 hours @ Z2
Build phase	Threshold development	5	2 hours @ Z2 + 4 x 8 min @ Z4 (10 min recovery)	1.5 hours @ Z1	2 hours @ Z2 + 3 x 10 min @ Z4 (10 min recovery)
	VO₂ max build	6	2 hours @ Z2 + 4 x 4 min @ Z5 (5 min recovery)	1 hour @ Z1	2 hours @ Z2 + 5 x 3 min @ Z5 (3 min recovery)
	Mixed high intensity	7	2 hours @ Z2 + 3 x 10 min @ Z4 (10 min recovery)	1.5 hours @ Z1	2 hours @ Z2 + 4 x 3 min @ Z5 (3 min recovery)
	Recovery	8	1.5 hours @ Z2	1 hour @ Z1	1.5 hours @ Z2
Specificity phase	Specificity phase	9	2 hours @ Z2 + 5 x 5 min @ Z5 (5 min recovery)	1 hour @ Z1	2 hours @ Z2 + 2 x 15 min @ Z4 (10 min recovery)
	Specificity phase	10	2 hours @ Z2 + 6 x 3 min @ Z5 (5 min recovery)	1 hour @ Z1	2 hours @ Z2 + 3 x 8 min @ Z4 (8 min recovery)
	Taper	11	2 hours @ Z2 + 2x 10 min @ Z4 (10 min recovery)	1 hour @ Z1	1.5 hours @ Z2
	Race week	12	1.5 hours @ Z1 or Z2	(R)	1 hour @ Z2 + 2 x 1 min @ Z5 (5min recovery)

TRAINING PLANS | 12-Week Training Plan for a Road Cyclist

THURSDAY	FRIDAY	SATURDAY	SUNDAY
R	3 hours @ Z2	2 hours @ Z2	1.5 hours @ Z2
R	3.5 hours @ Z2	1.5 hours @ Z1 or Z2	2 hours @ Z2
R	3.5 hours @ Z2	2 hours @ Z2	2 hours @ Z2 + 2 x 10 min @ Z4 (10 min recovery)
R	2 hours @ Z2	1.5 hours @ Z1 or Z2	1.5 hours @ Z2
R	4 hours @ Z2	2 hours @ Z2	2 hours @ Z2 + 4 x 8 min @ Z4 (10 min recovery)
R	3-hour group ride or Z2 + surges	4 hours @ Z2	2 hours @ Z2 + 4 x 4 min @ Z5 (5 min recovery)
R	4-hour group ride or Z2 + surges	4 hours @ Z2	2 hours @ Z2 + 3 x 10 min @ Z4 (10 min recovery)
R	2 hours @ Z2	1.5 hours @ Z1 or Z2	1.5 hours @ Z2
R	4-hour group ride or Z2 + surges	4 hours @ Z2	2 hours @ Z2 + 5 x 5 min @ Z5 (5 min recovery)
R	3-hour race simulation	4 hours @ Z2	2 hours @ Z2 + 6 x 3 min @ Z5 (5 min recovery)
R	2 hours @ Z2 + 3 x 1 min @ Z5 (5 min recovery)	2.5 hour race simulation + surges	2 hours @ Z2 +2 x 10 min @ Z5 (10 min recovery)
R	Race day	1 hour @ Z1	1.5 hours @ Z1 or Z2

147

STRENGTH AND STRETCH EXERCISES

Strength training has become an essential part of modern cycling, not just for injury prevention but for complementing on-bike performance. By targeting the muscles most involved in pedalling, strength training can help cyclists generate more power, improve endurance, and handle the physical demands of training and racing. This section introduces a series of exercises specifically chosen to support cycling performance. The stretches work to reduce tightness, build resilience to injury, and aid recovery.

WHY WORK OUT OFF-BIKE?

For both amateur and professional cyclists, strength exercises can help build power and injury resilience. Each movement in this section is explained step by step, with options to modify or progress. The goal is to help you build a stronger, more resilient body that can withstand the repetitive stress of riding, all while minimizing the risk of injury.

MAINS AND VARIATIONS

The exercises are organized by muscle group and further categorized into "main" and "variation" movements. Each main exercise is selected for its effectiveness in targeting specific muscle groups relevant to cycling, often using multi-joint compound movements to build strength and coordination. Variation exercises complement the main ones, offering alternative ways to challenge the same muscles and improve functional strength. For each main exercise, the key muscle groups involved are illustrated anatomically, with clear step-by-step guidance on how to perform the movement safely and effectively. Safety reminders are included throughout to ensure that you can train with confidence and consistency.

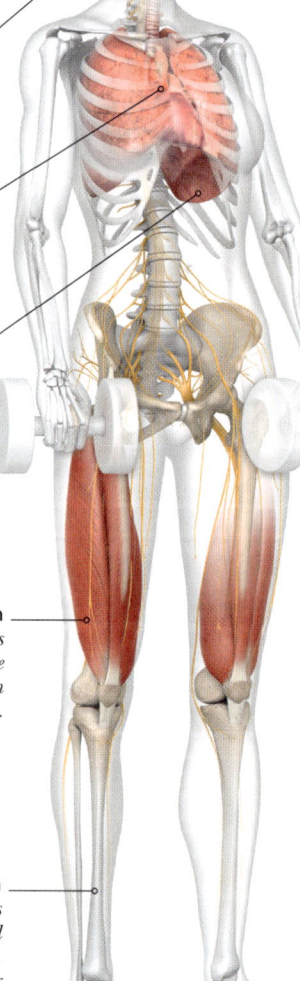

Brain and nervous system
The mind-to-body connection helps target the intended muscles and improve coordination.

Cardiovascular system
This carries blood enriched with oxygen to fuel the body and energize the muscles.

Respiratory system
Proper breathing helps to increase available oxygen; use the rhythm of your breath to aid form.

Muscular system
Performing these exercises correctly helps put the tension and stress on designated muscles.

Skeletal system
Muscles attach to bones and contract and relax to pull on them, causing movement. Performing exercises with proper form places stress in the right areas, helping to prevent injuries.

> **! Caution boxes**
> Caution boxes in this book highlight potential risks. They draw attention to critical warnings or guidelines, ensuring users are aware of hazards and necessary precautions to prevent accidents or errors.

CORRECT EXERCISE EXECUTION
Correct exercise execution involves proper form, controlled movements, and appropriate weight distribution. Focus on technique, alignments, and breathing to maximize benefits.

STRENGTH AND STRETCH EXERCISES | Why Work Out Off-Bike?

THE IMPORTANCE OF BREATHING

Breathing is a key component of strength training for cyclists, supplying oxygen to the muscles and supporting efficient energy production during lifts. Using proper breathing techniques improves focus, endurance, and the ability to generate force under load. It also helps regulate heart rate, manage fatigue, and prevent light-headedness, especially during heavy or complex movements.

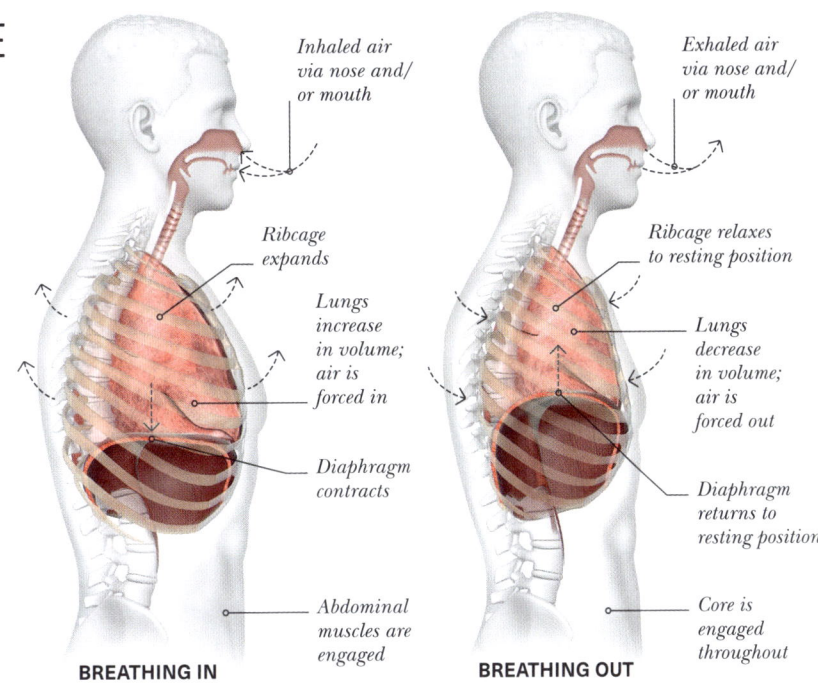

BREATHING IN
- Inhaled air via nose and/or mouth
- Ribcage expands
- Lungs increase in volume; air is forced in
- Diaphragm contracts
- Abdominal muscles are engaged

BREATHING OUT
- Exhaled air via nose and/or mouth
- Ribcage relaxes to resting position
- Lungs decrease in volume; air is forced out
- Diaphragm returns to resting position
- Core is engaged throughout

Training equipment

Many of the exercises require minimal equipment, making them practical and easy to complete at home using the items listed below. A few of the exercises, such as trap bar and leg press machine, will require gym access.

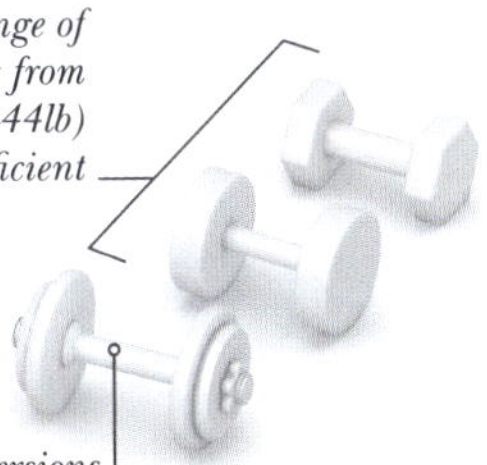

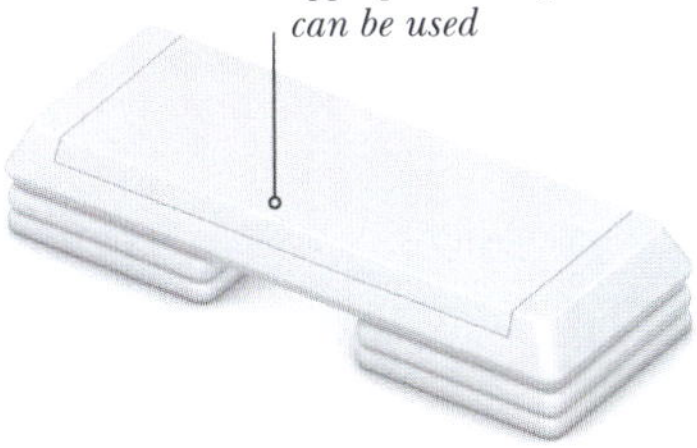

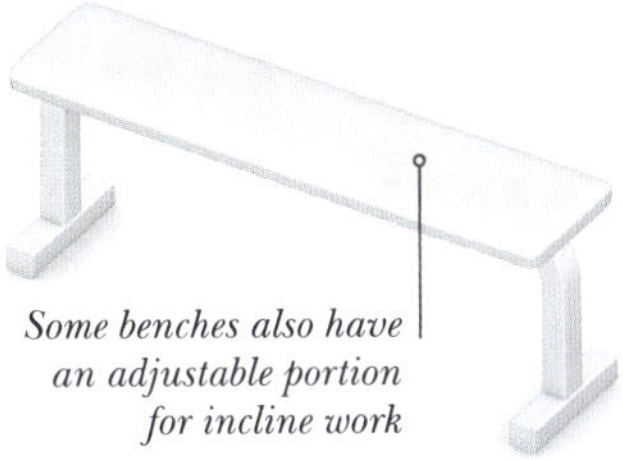

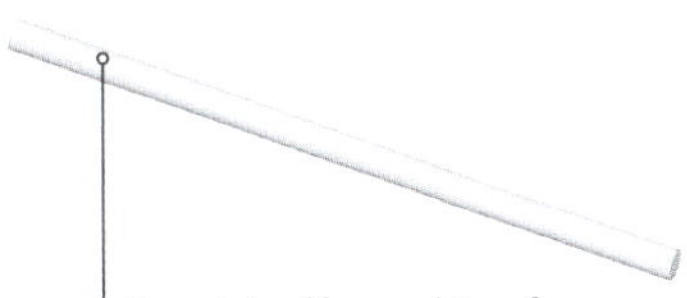

DUMBBELLS
- A range of weights from 5–20kg (10–44lb) is sufficient
- Adjustable versions can carry a wide range of weights

BOX OR STEP
- Anything at an appropriate height can be used

MEDICINE BALL
- 10–20kg (5–44lb) depending on fitness levels

ROLL MAT
- Choose a non-slip mat that is easy to roll up and store

BASIC BENCH
- Some benches also have an adjustable portion for incline work

PULL-UP/CHIN-UP BAR
- A variety of bars exists – from telescopic to those needing fixing to a wall or ceiling

BEST PRACTICES

When starting out, it's essential to begin with light weights and focus on mastering proper form. Begin with a warm-up and gradually build up the weights and load range as your become more experienced.

CHOOSING THE RIGHT WEIGHT

Technique should always take priority over load. One effective approach to guide your effort level is the Reps in Reserve (RIR) method, where you aim to finish each set with 2–3 reps left in the tank. This helps avoid excessive fatigue that could interfere with your cycling training.

As your technique improves and you build confidence, gradually increase the weight to apply progressive overload. Your training goal will determine your rep and load range. For building strength, aim for 3 to 6 reps at around 80–90 per cent of your one-rep max. For power development, use moderate loads moved with speed, typically 3–5 reps at 60–75 per cent of your max. For either stability work or rehabilitation, use lighter weights with higher reps, generally in the 10–15 range.

LIFTING TECHNIQUE

Every session should begin with a proper warm-up that activates key muscle groups, especially the glutes and core. When lifting, control the speed, lower the weight slowly and lift with strength and control. This not only improves muscle engagement but reduces injury risk. Using a full range of motion is also important, unless you're limited by mobility or dealing with an injury. Good posture underpins everything: maintain a neutral spine, keep your shoulders back, and stay engaged through the core during all movements.

USE OF EQUIPMENT

While most of the exercises can be done at home, having access to a full range of dumbbells, a trap bar or barbell can expand your training options. Incorporating unilateral exercises like split squats or single-leg step-ups is beneficial for cyclists, as these exercises help address strength imbalances between legs. When doing power-based exercises, such as box jumps or explosive movements, it's best to perform them at the beginning of your session while you're fresh, keeping reps low to preserve quality and speed.

WORKING ON MACHINES

Each machine in the gym should be adjusted to fit your body structure. If you're new to using these machines, consider having a session with a trainer to learn how they work and what settings are optimal for you. Common adjustments include the seat pad, back pad, and thigh pad. Pay attention to the machine's axis of rotation to align your legs correctly. If anything feels uncomfortable during your first repetitions, adjust the settings until the exercise feels right.

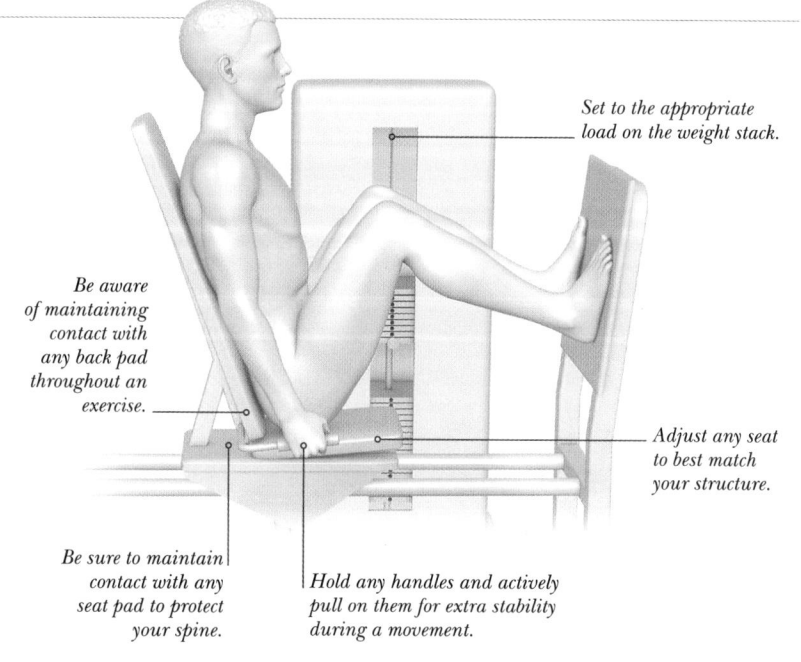

Set to the appropriate load on the weight stack.

Be aware of maintaining contact with any back pad throughout an exercise.

Adjust any seat to best match your structure.

Be sure to maintain contact with any seat pad to protect your spine.

Hold any handles and actively pull on them for extra stability during a movement.

SAMPLE WORKOUTS

The following sample programmes are designed to help cyclists integrate strength training effectively alongside their on-bike training.

Each programme aligns with a different phase of the training year, whether you're just getting started, building strength in the off-season, or maintaining gains during the competition phase. The sessions are structured to be time-efficient and focussed on movements that support cycling-specific performance, including lower body power, postural strength, and core stability. Depending on your equipment access and experience level, exercises can be scaled or substituted to fit your individual needs.

3–6 reps at **80–90%** of your one-rep max will *build strength*

Introductory phase (2 x week)

Goal: Build movement competency and core stability

Session 1: lower body + core

- Goblet squat – 3 x 10
- Step-ups (bodyweight or light dumbbells) – 3 x 8 per leg
- Glute bridge – 3 x 15
- Plank with shoulder taps – 3 x 30s
- Deadbugs – 3 x 10 each side

Session 2: Upper body + mobility

- Push up – 3 x 10
- Bent-over dumbbell row – 3 x 10
- Shoulder press (5–10kg dumbbells) – 3 x 8
- Cat cow / Figure 4 stretch – 2 x 30s
- Half-kneel hip flexor stretch – 2 x 30s per side

Development phase/off season (2–3 x week)

Goal: Build strength and core control

Session 1: strength focus

- Goblet squat – 3 x 10
- Front squat or trap bar deadlift – 4 x 5
- Split squat – 3 x 6 per leg
- Hip thrust – 3 x 10
- Hanging knee raise – 3 x 10
- Plank hold – 3 x 45s

Session 2: Power + upper body

- Push-ups – 3 x 10
- Box jump – 3 x 5
- Single-leg box jump – 3 x 3 per leg
- Pull-ups – 3 x 8
- Shoulder press – 3 x 10
- Russian twists – 3 x 20
- Thread the needle stretch – 2 x 30s per side

Optional Session 3: Core + mobility

- Bear plank – 3 x 20s
- Glute bridge – 3 x 12
- Alternating foot switch – 3 x 10
- Static hamstring stretch – 2 x 30s

In-season maintenance (1–2 x week)

Goal: Maintain strength and support recovery without impacting cycling performance

Total-body routine (45 mins)

- Goblet squat – 3 x 10
- Dumbbell goblet squat – 3 x 8
- Split squat – 3 x 6 per leg
- Pull up or row – 3 x 8
- Push up – 2 x 15
- Low plank hold – 2 x 45s
- Child's pose and hip flexor stretch – 2 x 30

" "

Strength training for cyclists can be done effectively at home or in a gym.

BARBELL FRONT SQUAT

The barbell front squat is a compound lower-body strength exercise emphasizing the quadriceps, core stability, and upper back engagement. For cyclists, it builds leg strength, improves hip mobility, and enhances pedalling efficiency under load.

This exercise develops lower-body strength in the quadriceps. The front-loaded position encourages a more upright posture and greater core activation, which translates to better spinal alignment and stability on the bike. Improved hip and ankle mobility from front squatting can enhance pedal stroke efficiency and comfort during long rides. Beginners can do 3 sets of 8–10 reps.

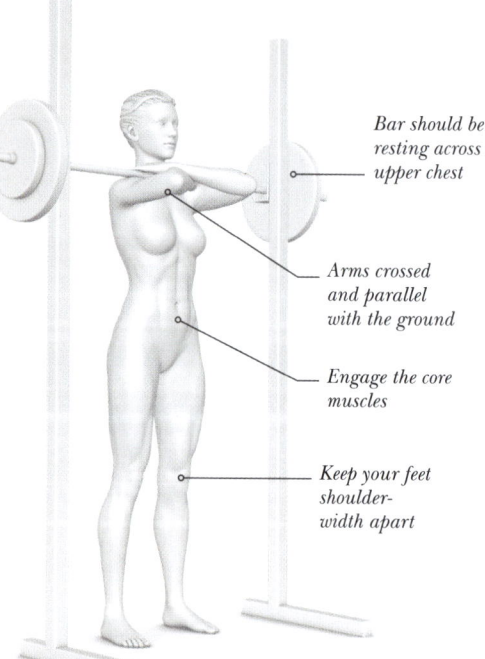

- Bar should be resting across the upper chest
- Arms crossed and parallel with the ground
- Engage the core muscles
- Keep your feet shoulder-width apart

PREPARATORY STAGE
Stand tall with your feet roughly shoulder-width apart. Lift the barbell and allow it to settle at the top of your shoulders in line with your collar bones.

Mid deltoid
Latissimus dorsi
Serratus anterior
Quadratus lumborum
Internal oblique
Rectus abdominis

Upper body and core
This exercise engages the **trapezius**, **deltoids, upper back**, and **core** muscles to stabilize the torso and support the barbell, promoting posture and control throughout the lift.

STAGE ONE
Breathe in, engage your core, and "sit" into a deep squat, keeping your torso as upright as possible. Hold a forward gaze and keep the barbell still.

STRENGTH AND STRETCH EXERCISES | Barbell Front Squat

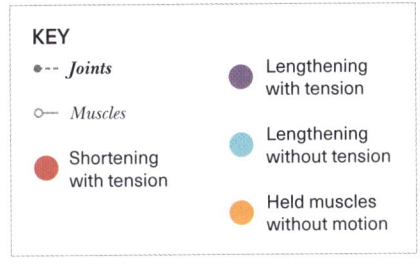

KEY
- Joints
- Muscles
- Shortening with tension
- Lengthening with tension
- Lengthening without tension
- Held muscles without motion

Maintain arms parallel with the ground

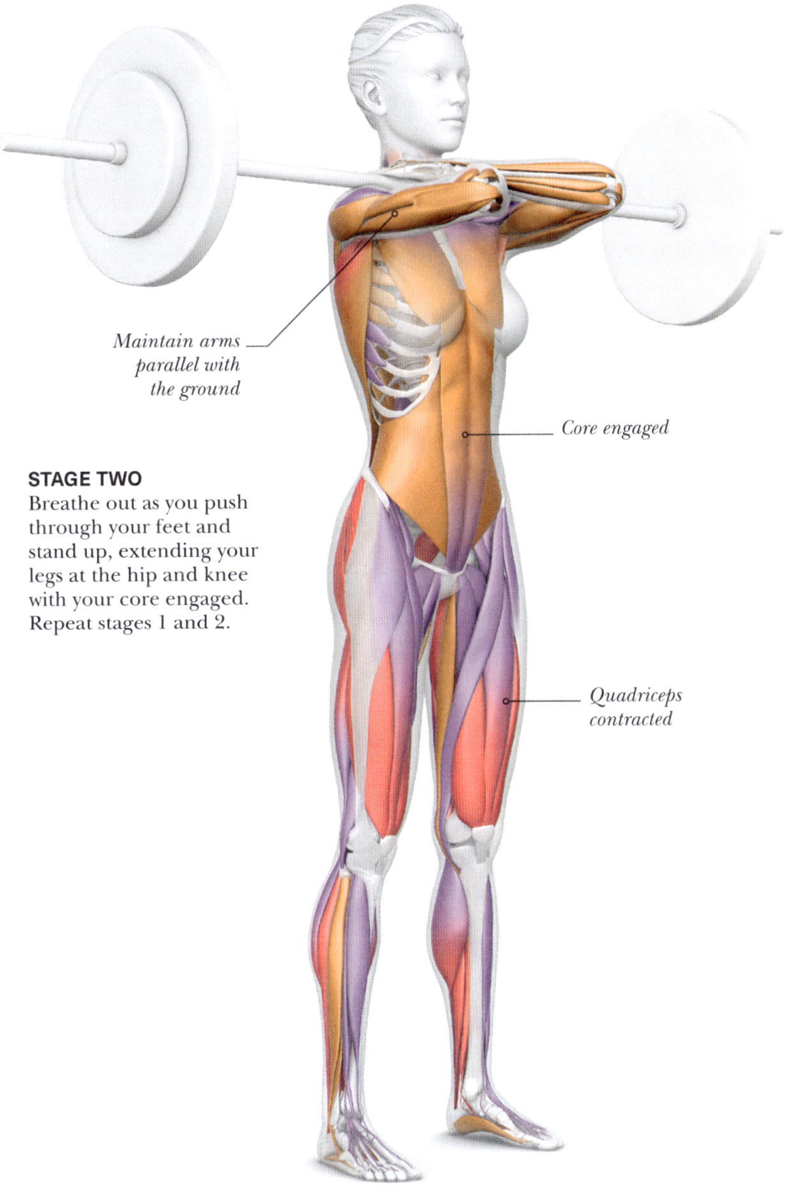

STAGE TWO
Breathe out as you push through your feet and stand up, extending your legs at the hip and knee with your core engaged. Repeat stages 1 and 2.

Core engaged

Quadriceps contracted

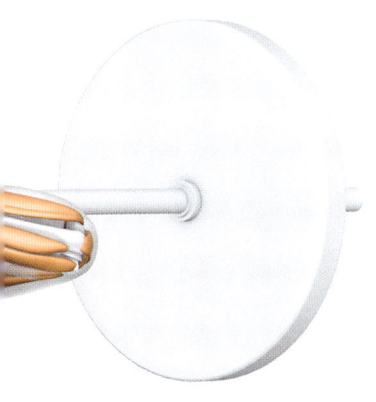

Rectus femoris
Vastus medialis
Gracilis
Knee
Gastrocnemius
Tibialis anterior
Soleus
Abductor hallucis

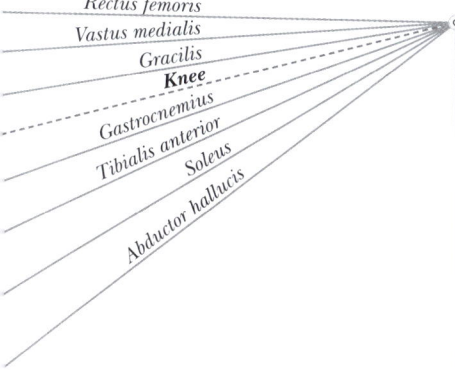

Lower body
This exercise primarily targets the **quadriceps**, with support from the **glutes**, **hamstrings**, and **calves**, while also engaging the hip flexors and adductors to control movement and generate power.

❝ ❞

The barbell front squat builds the strength and mobility to ride harder, longer, and more efficiently.

155

SPLIT SQUAT

The split squat is a single-leg exercise that builds strength and stability, and enhances hip mobility, all of which benefit cycling performance. It engages the hamstrings, quadriceps, glutes, and core. The hamstrings are especially active during the lowering phase, supporting balance and stability while contributing to muscle strength and growth.

Focussing on one leg at a time makes it more specific to cycling movement and will reduce the risk of overuse injuries in these muscle groups. The correct stance is crucial – avoid placing your legs too close together or too far apart. Keep your shoulders back, avoid hunching, engage your core, and ensure the front knee stays aligned with the toes, not beyond them. When starting, beginners can do 1 set of 8–12 reps per leg and gradually progress to 3 or 4 sets as they become more comfortable with the movement. To increase the load for further strength gains, hold a dumbbell (less than 15kg/33lb) in each hand.

Upper body
The abdominals, particularly the **obliques** and **rectus abdominis**, play a key role in stabilizing the core and bracing the spine during the squat. This support allows proper hip function and helps resist rotational forces from balance challenges. Holding dumbbells also activates the arms by adding tension and promoting full-body stability.

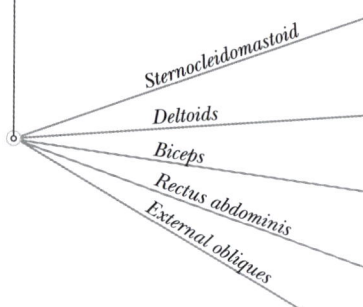

Lower body
The glutes drive hip extension and stabilize the pelvis in the split stance. The **quadriceps** extend the knee and help initiate the movement, while the **hamstrings** assist with balance, control, and the lowering phase. The calf muscles also engage to support ankle stability and overall posture.

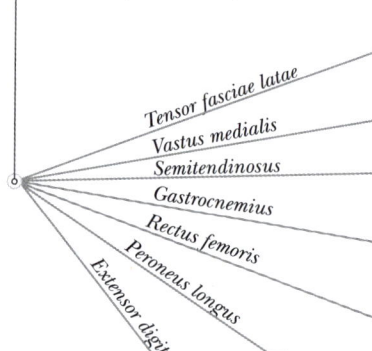

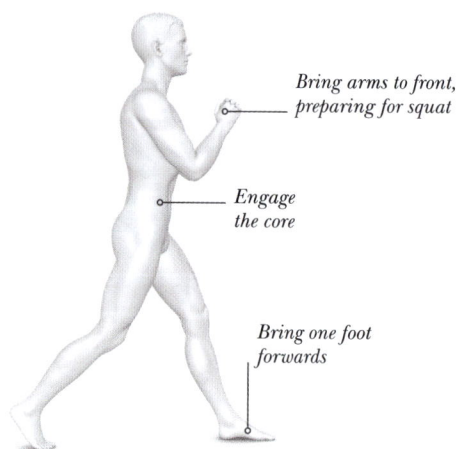

Bring arms to front, preparing for squat

Engage the core

Bring one foot forwards

PREPARATORY STAGE
Stand upright with your feet hip- to shoulder-width apart and toes pointing forwards. Lightly clasp your hands in front of your chest. Step forward with one leg, and once you feel stable and balanced, engage your core to prepare for movement.

STAGE ONE
Inhale as you slowly lower into a squat, keeping your chest lifted and torso upright. Lower until your back knee nearly touches the floor, allowing your back heel to rise so you're balanced on your toes. Maintain core engagement, and ensure your front knee stays at a 90-degree angle. Pause briefly at the bottom.

STRENGTH AND STRETCH EXERCISES | Split Squat

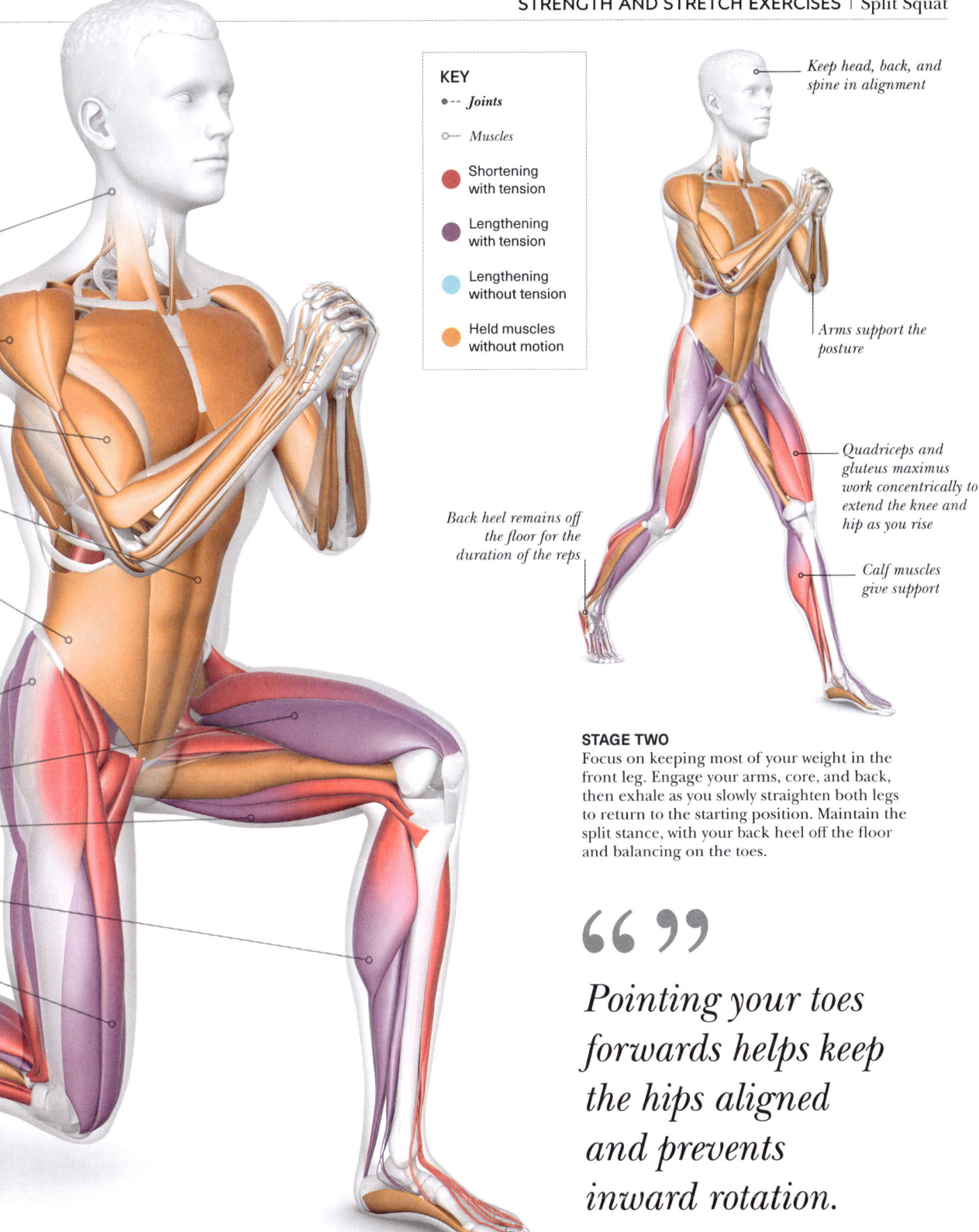

KEY
- Joints
- Muscles
- Shortening with tension
- Lengthening with tension
- Lengthening without tension
- Held muscles without motion

Keep head, back, and spine in alignment

Arms support the posture

Quadriceps and gluteus maximus work concentrically to extend the knee and hip as you rise

Back heel remains off the floor for the duration of the reps

Calf muscles give support

STAGE TWO
Focus on keeping most of your weight in the front leg. Engage your arms, core, and back, then exhale as you slowly straighten both legs to return to the starting position. Maintain the split stance, with your back heel off the floor and balancing on the toes.

> *Pointing your toes forwards helps keep the hips aligned and prevents inward rotation.*

» SQUAT VARIATIONS

Squat variations work the gluteal muscles, quadriceps, and hamstrings, but changing how the body position is held can shift emphasis to different muscles and movement patterns. Incorporating variations not only builds balanced strength and mobility but also helps reinforce proper technique and prevent overuse injuries.

> *Squat variations build balanced leg strength, mobility, and control.*

BULGARIAN SPLIT SQUATS

A single-leg strength exercise that targets the quads, glutes, and hamstrings while also challenging balance and core stability. For cyclists, it closely mimics the unilateral nature of the pedal stroke, helping to correct imbalances between legs, improve stability through the hips and knees, and develop force production in each leg independently.

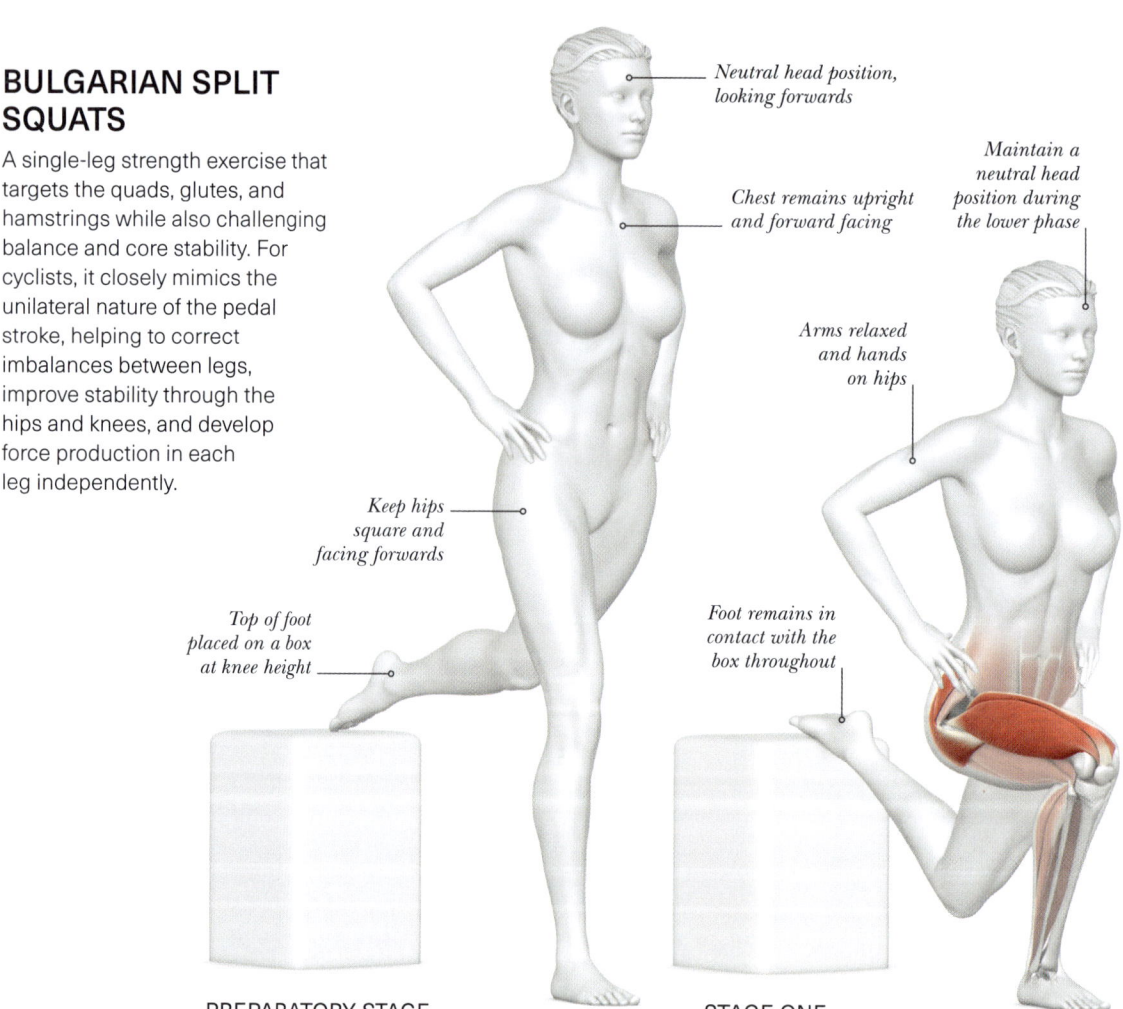

PREPARATORY STAGE

STAGE ONE

PREPARATORY STAGE
Stand a few feet in front of a bench or box, and place the top of one foot behind you on the bench. Keep your front foot flat and positioned so that your knee stays over your ankle when you lower down.

STAGE ONE
Bend your front knee to lower your body, keeping your torso upright and hips square. Lower until your front thigh is nearly parallel to the floor, or as far as your mobility allows.

STAGE TWO
Push through the heel of your front foot to return to the starting position, keeping balance and control throughout the movement.

STRENGTH AND STRETCH EXERCISES | Squat Variations

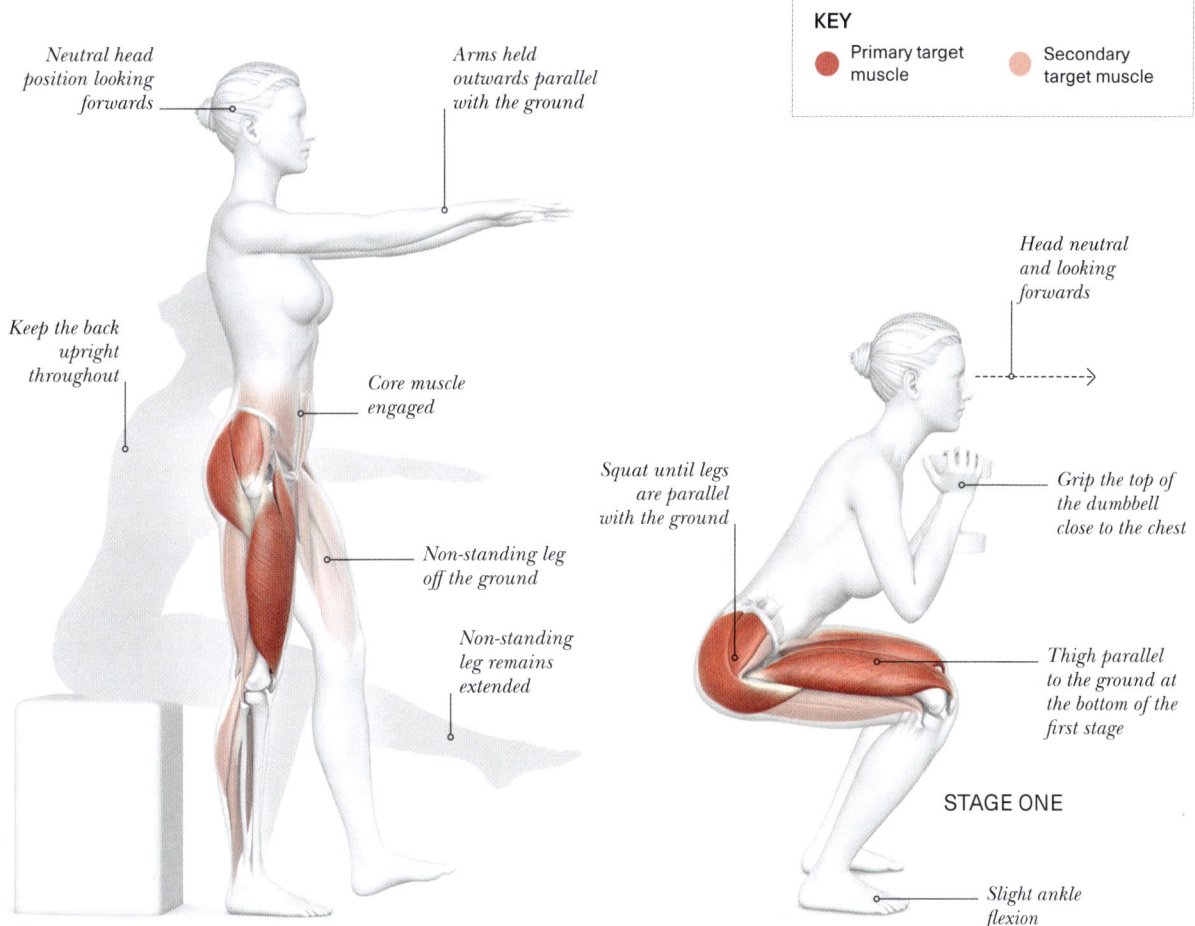

PISTOL SQUAT (WITH BOX)

The pistol squat is a challenging single-leg bodyweight exercise that builds strength, stability, and control in the quads, glutes, hamstrings, and core. For cyclists, it enhances joint stability around the knees and ankles.

PREPARATORY STAGE
Stand on one leg with the other leg extended straight in front of you. Keep your arms forwards for balance, chest upright, and core engaged.

STAGE ONE
Slowly bend the standing leg, lowering your hips down, keeping the extended leg off the ground. Maintain balance and control as you descend as low as your mobility allows.

STAGE TWO
Push through the heel of your standing leg to rise back up, keeping the extended leg and arms steady. Maintain alignment and stability throughout the movement.

DUMBBELL GOBLET SQUAT

The dumbbell goblet squat is a cyclist-friendly strength exercise that builds leg power, core stability, and hip mobility. Holding the weight at the chest encourages good posture and control, making it ideal for reinforcing form and developing balanced strength for efficient pedalling.

PREPARATORY STAGE
Hold a dumbbell vertically against your chest with both hands, feet shoulder-width apart, and chest upright.

STAGE ONE
Bend your knees and hips to squat down, keeping your elbows inside your knees and your torso tall.

STAGE TWO
Push through your heels to stand back up again, maintaining core engagement and balance throughout the movement.

TRAP BAR DEADLIFT

This deadlift is a compound lift using a hex bar that promotes a more upright torso and neutral grip, reducing lower back strain. It targets the glutes, hamstrings, quads, and core, with support from the traps, erector spinae, and forearms.

The trap bar deadlift builds powerful hip and leg drive, essential for sprinting and climbing in cycling. It strengthens the posterior chain while reinforcing core stability and joint alignment, helping prevent injuries and improve overall pedalling efficiency. Its balanced load makes it a safe, effective tool for developing cycling-specific strength. Beginners can start by doing 4 sets of 6–8 reps.

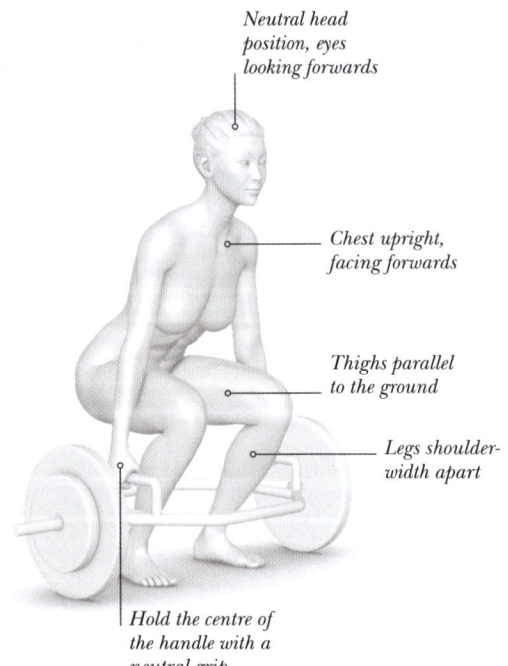

Neutral head position, eyes looking forwards

Chest upright, facing forwards

Thighs parallel to the ground

Legs shoulder-width apart

Hold the centre of the handle with a neutral grip

PREPARATORY STAGE
Set the weights and step into the trap bar (hex bar). Stand with feet angled outward slightly. Push your hips back to bend your knees and grip the handle.

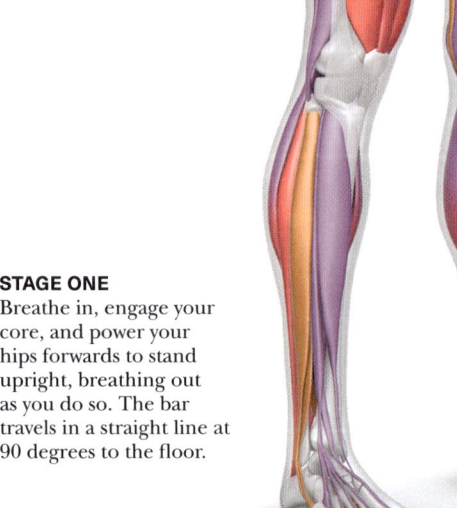

STAGE ONE
Breathe in, engage your core, and power your hips forwards to stand upright, breathing out as you do so. The bar travels in a straight line at 90 degrees to the floor.

STRENGTH AND STRETCH EXERCISES | Trap Bar Deadlift

Upper body and core

The **trapezius** and **rhomboids** help keep the shoulders retracted and support the upper back, while the **erector spinae** maintain spinal alignment. The forearms and grip muscles are heavily engaged to hold the handles, and the core – including the **abdominals** and **obliques** – stabilizing the torso throughout the lift.

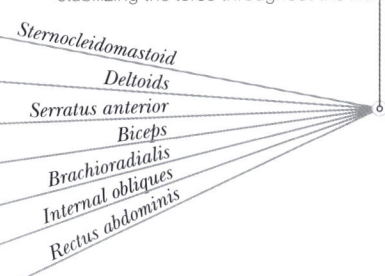

Sternocleidomastoid
Deltoids
Serratus anterior
Biceps
Brachioradialis
Internal obliques
Rectus abdominis

KEY
- •-- *Joints*
- ○— *Muscles*
- ● Shortening with tension
- ● Lengthening with tension
- ● Lengthening without tension
- ● Held muscles without motion

Lower body

This exercise engages the **glutes**, **hamstrings**, and **quadriceps**, driving hip and knee extension during the lift. The **calves** assist in stabilizing the ankles, while the **hip adductors** and **abductors** contribute to balance and control.

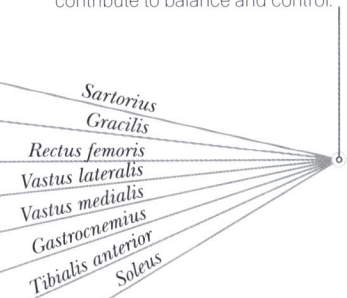

Sartorius
Gracilis
Rectus femoris
Vastus lateralis
Vastus medialis
Gastrocnemius
Tibialis anterior
Soleus

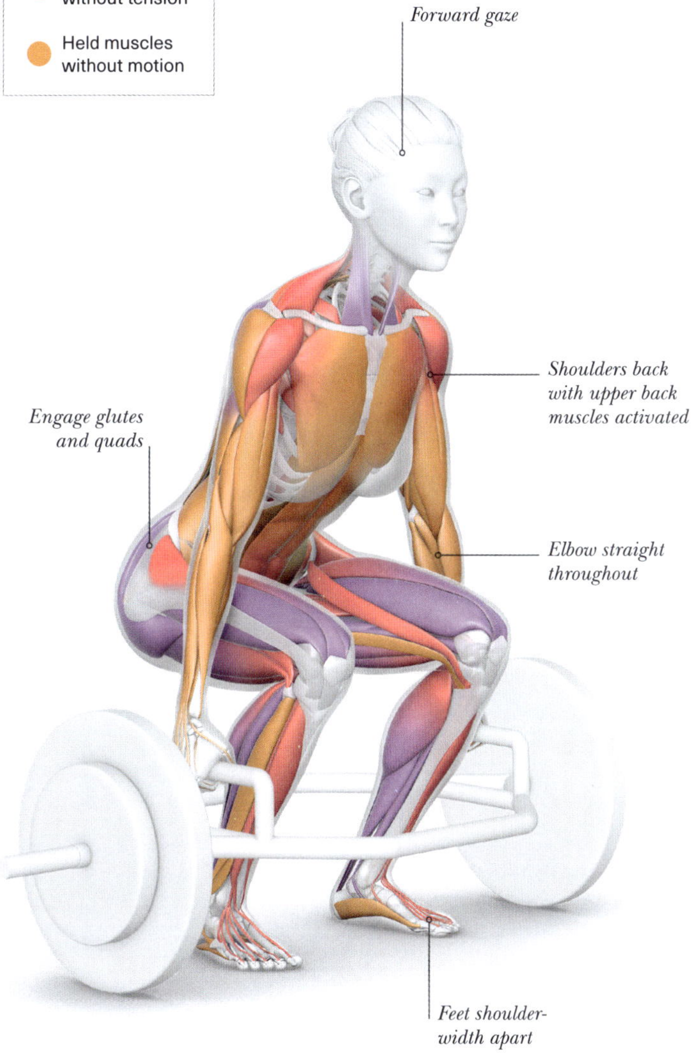

! Caution

Common trap bar deadlift mistakes include rounding the back, letting hips rise too early, and failing to engage the core. To avoid these, keep a neutral spine, lift smoothly with hips and shoulders together, and brace your core throughout the movement. Prioritize control over heavy loading.

Forward gaze

Shoulders back with upper back muscles activated

Engage glutes and quads

Elbow straight throughout

Feet shoulder-width apart

STAGE TWO
Push your hips back to return to the starting position, keeping your shoulders back and holding a forward gaze throughout. Repeat stages 1 and 2.

161

STEP UP WITH DUMBBELLS

This exercise engages the core and targets the glutes, quadriceps, hamstrings, and calves. It mimics the unilateral force production seen in cycling, improving leg strength, balance, and coordination.

Use an exercise step at least 30cm (12in) high. Focus on the front leg, which stays planted on the step, and keep your core engaged throughout. Make sure your entire foot is on the step and your feet are shoulder-width apart. Drive through the front leg to step-up, avoid pushing off with the back foot. Alternate legs each rep or complete full sets on one leg before switching. Beginners can start with 4 sets of 8–10 reps.

Hip and leg

As you step up, focus on engaging the glutes and quads. This concentric phase strengthens the **glutes**, **quadriceps**, and **proximal hamstrings** as you drive through the front leg, fully extending the hip and knee, with the calves aiding balance. During the lowering phase, control the movement by maintaining tension in the front leg to avoid dropping.

Upper body and core

The **core**, **upper back**, **arms**, and **shoulders** engage throughout the movement to help maintain a neutral spine and proper posture during both the lifting and lowering phases.

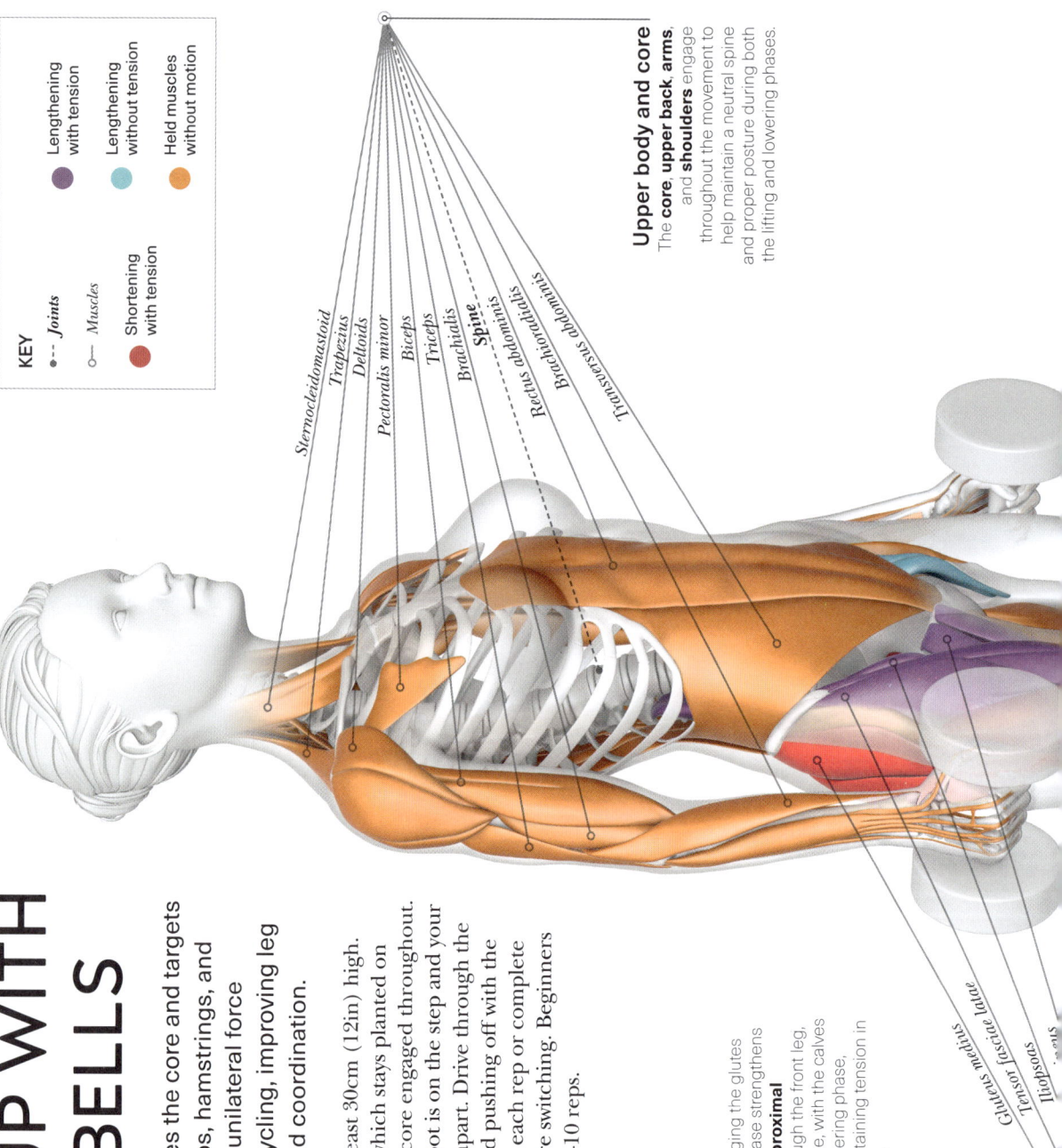

KEY
- --- Joints
- ○ Muscles
- ● Shortening with tension
- ● Lengthening with tension
- ● Lengthening without tension
- ● Held muscles without motion

Sternocleidomastoid
Trapezius
Deltoids
Pectoralis minor
Biceps
Triceps
Brachialis
Spine
Rectus abdominis
Brachioradialis
Transversus abdominis

Gluteus medius
Tensor fasciae latae
Iliopsoas

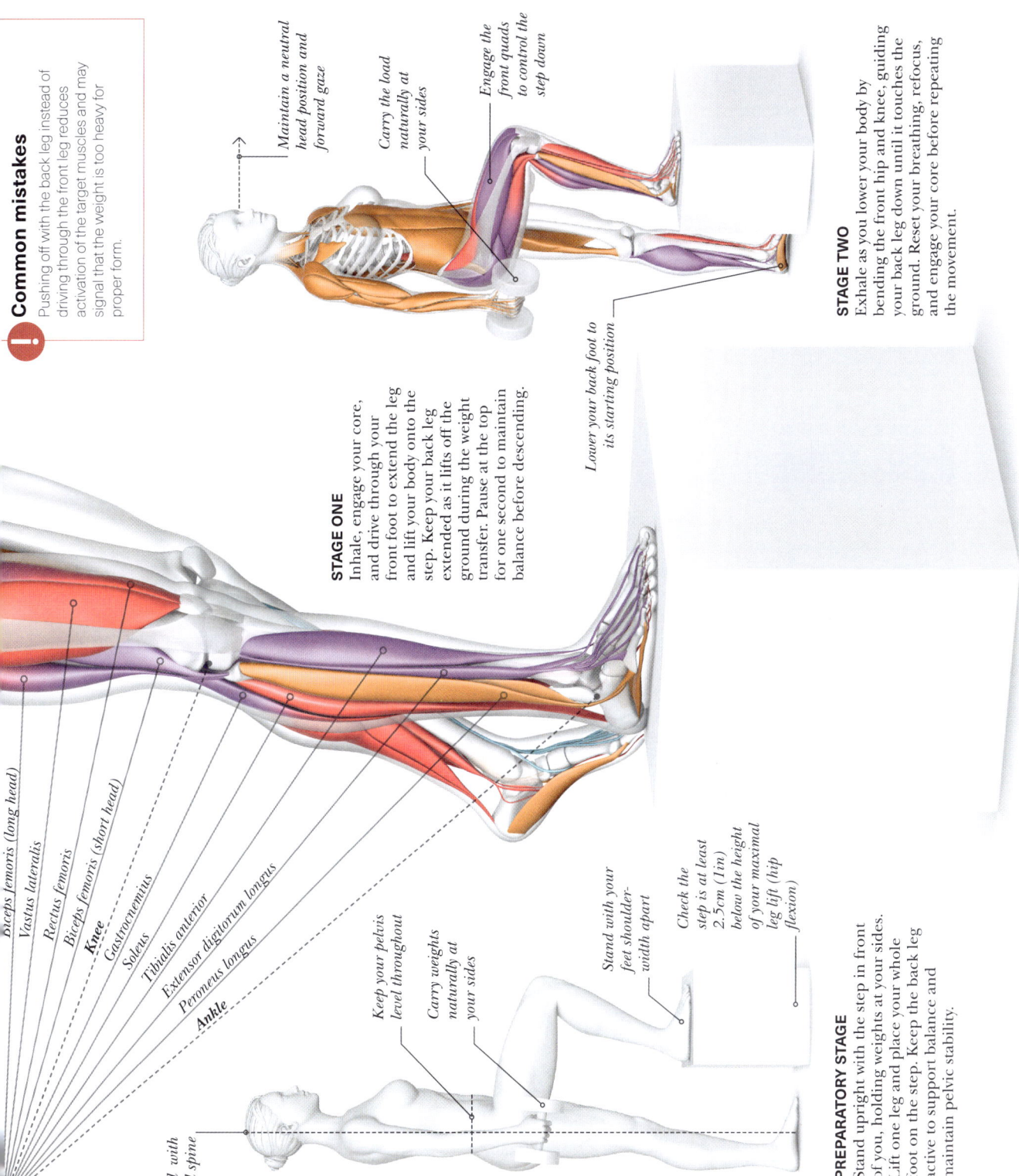

Common mistakes

Pushing off with the back leg instead of driving through the front leg reduces activation of the target muscles and may signal that the weight is too heavy for proper form.

Maintain a neutral head position and forward gaze

Carry the load naturally at your sides

Engage the front quads to control the step down

STAGE TWO
Exhale as you lower your body by bending the front hip and knee, guiding your back leg down until it touches the ground. Reset your breathing, refocus, and engage your core before repeating the movement.

STAGE ONE
Inhale, engage your core, and drive through your front foot to extend the leg and lift your body onto the step. Keep your back leg extended as it lifts off the ground during the weight transfer. Pause at the top for one second to maintain balance before descending.

Lower your back foot to its starting position

- Biceps femoris (long head)
- Vastus lateralis
- Rectus femoris
- Biceps femoris (short head)
- **Knee**
- Gastrocnemius
- Soleus
- Tibialis anterior
- Extensor digitorum longus
- Peroneus longus
- **Ankle**

Stand tall with a neutral spine

Keep your pelvis level throughout

Carry weights naturally at your sides

Stand with your feet shoulder-width apart

Check the step is at least 2.5cm (1in) below the height of your maximal leg lift (hip flexion)

PREPARATORY STAGE
Stand upright with the step in front of you, holding weights at your sides. Lift one leg and place your whole foot on the step. Keep the back leg active to support balance and maintain pelvic stability.

HIP THRUST

This is a lower-body exercise that primarily targets the gluteus maximus, with support from the hamstrings, quadriceps, and core muscles. The exercise enhances hip extension strength, improves pelvic stability, and supports powerful, efficient movement patterns, key for cycling performance, especially during sprints and climbs.

Use a sturdy bench or step to support your upper back. Place the barbell securely in your hip crease and move through hip extension and flexion to lift and lower your body. If the bar feels uncomfortable, use a pad for cushioning. Proper alignment of your feet, knees, and ankles is key for smooth movement and injury prevention. Beginners can start with 4 sets of 8–10 reps.

Upper body and arms
The **abdominal muscles** play a key role in stabilizing the spine and pelvis, ensuring smooth coordination between the upper and lower body. Meanwhile, the arms and shoulders help maintain control of the load throughout the movement.

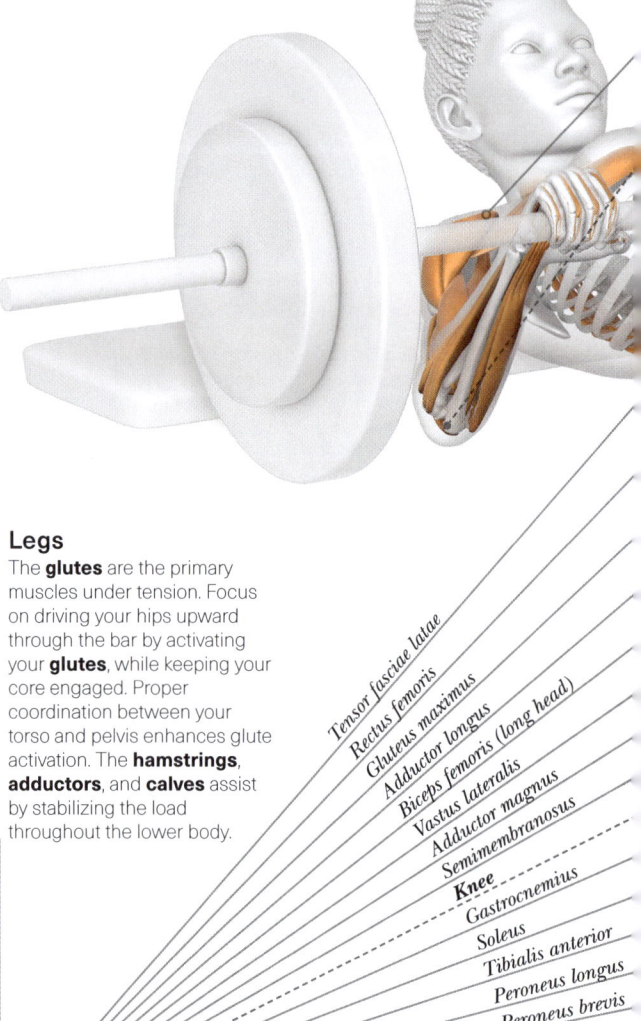

Legs
The **glutes** are the primary muscles under tension. Focus on driving your hips upward through the bar by activating your **glutes**, while keeping your core engaged. Proper coordination between your torso and pelvis enhances glute activation. The **hamstrings**, **adductors**, and **calves** assist by stabilizing the load throughout the lower body.

Make contact with the bench just below your shoulder blades

Tuck your chin in

Position your hands as wide as you like, as if holding the bar

Flex your knees (in stage 1, knees will align with ankles)

Tensor fasciae latae
Rectus femoris
Gluteus maximus
Adductor longus
Biceps femoris (long head)
Vastus lateralis
Adductor magnus
Semimembranosus
Knee
Gastrocnemius
Soleus
Tibialis anterior
Peroneus longus
Peroneus brevis
Ankle
Extensor d. lon
Extensor h. lor

PREPARATORY STAGE
Sit with your upper back resting on a bench, knees bent, and feet positioned just outside shoulder-width. Place the barbell securely in your hip crease. Engage your glutes to lift your hips off the floor into the starting position. Inhale to brace your core and prepare for the movement.

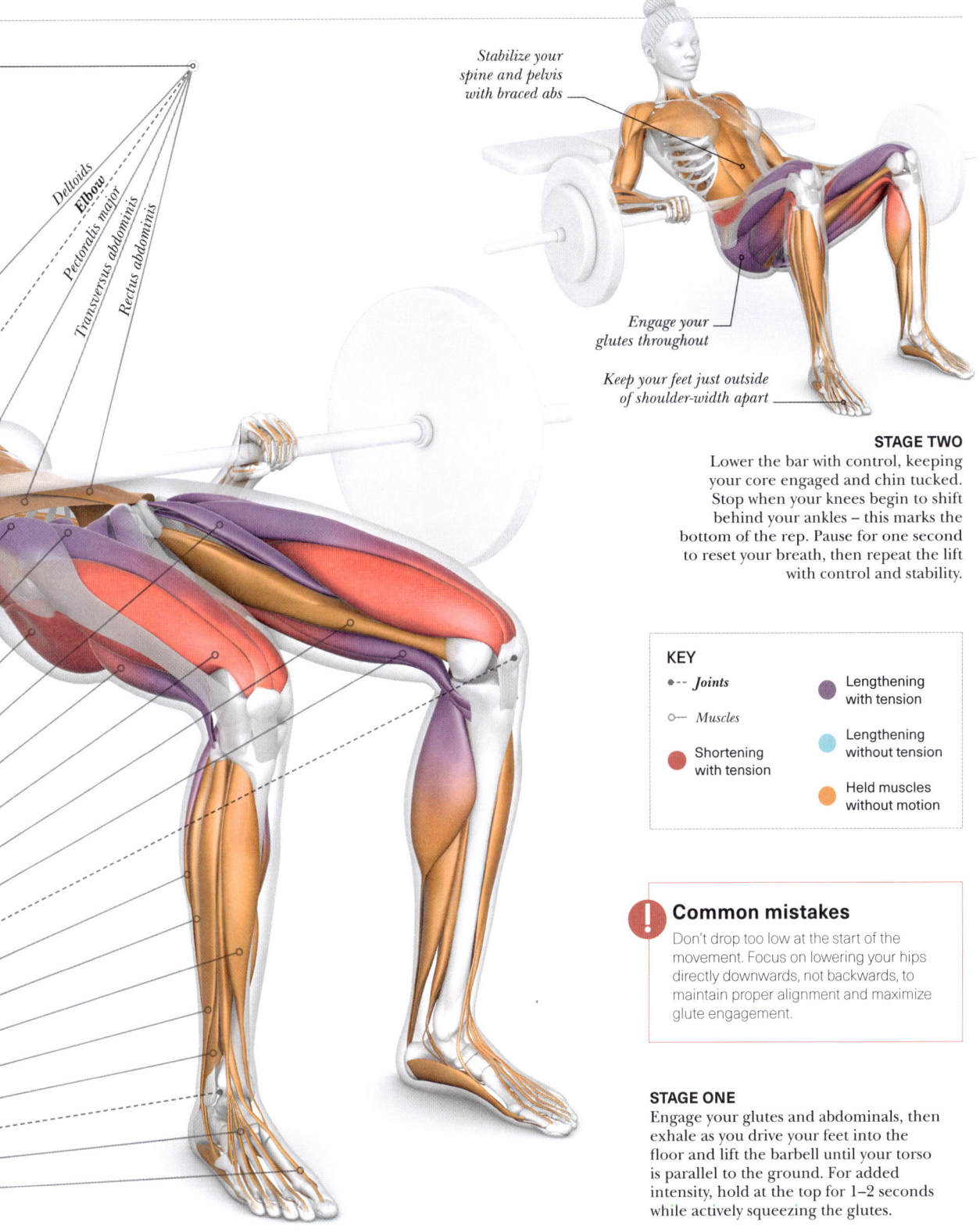

Stabilize your spine and pelvis with braced abs

Engage your glutes throughout

Keep your feet just outside of shoulder-width apart

Deltoids
Elbow
Pectoralis major
Transversus abdominis
Rectus abdominis

STAGE TWO
Lower the bar with control, keeping your core engaged and chin tucked. Stop when your knees begin to shift behind your ankles – this marks the bottom of the rep. Pause for one second to reset your breath, then repeat the lift with control and stability.

KEY
- •-- Joints
- ○— Muscles
- 🔴 Shortening with tension
- 🟣 Lengthening with tension
- 🔵 Lengthening without tension
- 🟠 Held muscles without motion

❗ Common mistakes
Don't drop too low at the start of the movement. Focus on lowering your hips directly downwards, not backwards, to maintain proper alignment and maximize glute engagement.

STAGE ONE
Engage your glutes and abdominals, then exhale as you drive your feet into the floor and lift the barbell until your torso is parallel to the ground. For added intensity, hold at the top for 1–2 seconds while actively squeezing the glutes.

LEG PRESS

The leg press is a compound lower-body exercise performed on a machine, where you push a weighted platform away using your legs. It primarily targets the quadriceps, gluteus maximus, and hamstrings, with additional support from the calves and hip adductors. Because the torso is supported, the leg press allows for heavy loading with reduced strain on the lower back. It helps build leg strength, power, and muscular endurance.

This exercise provides a comprehensive lower-body workout. Begin by setting an appropriate weight, sitting back against the pad, and adjusting the foot platform to a comfortable position. For best results, focus on movement through the hips and knees only. Hold the handles to stabilize your torso and stay securely seated throughout the exercise. Beginners can start with 4 sets of 8–10 reps.

KEY
- Joints
- Muscles
- Shortening with tension
- Lengthening with tension
- Lengthening without tension
- Held muscles without motion

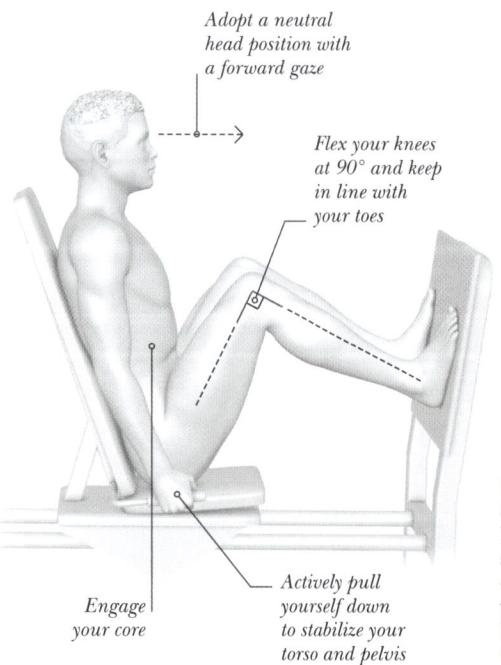

Adopt a neutral head position with a forward gaze

Flex your knees at 90° and keep in line with your toes

Engage your core

Actively pull yourself down to stabilize your torso and pelvis

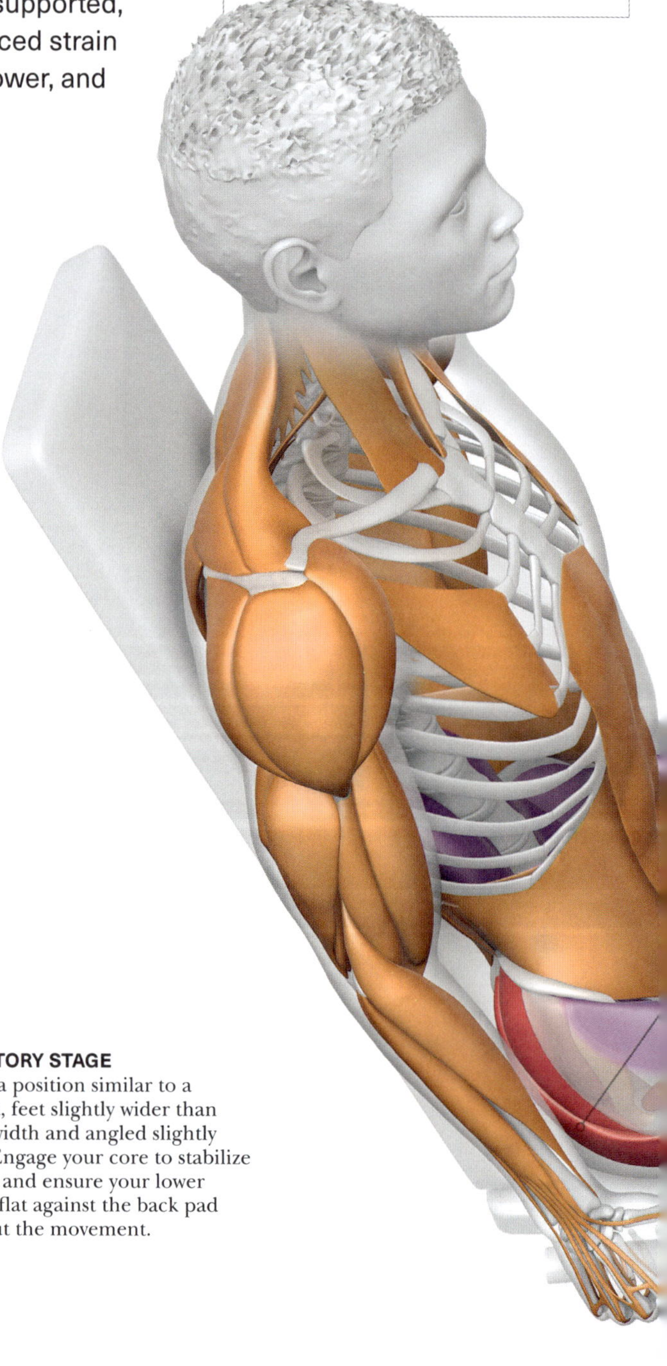

PREPARATORY STAGE
Start with a position similar to a back squat, feet slightly wider than shoulder-width and angled slightly outward. Engage your core to stabilize your torso and ensure your lower back stays flat against the back pad throughout the movement.

166

STRENGTH AND STRETCH EXERCISES | Leg Press

STAGE TWO
Inhale as you bend your knees and hips, allowing your knees to track forward in line with your first and second toes. Control the descent, slowing as you near the starting position. Reset your breath and repeat the movement from the beginning.

! Common mistakes
Avoid bending your legs too far. If your hips lift off the pad or your lower back rounds, you've exceeded a safe range of motion. Keep contact with the pad to protect your spine.

Legs
The legs do most of the work in the leg press, with the **quadriceps**, **glutes**, and **adductors** serving as the primary movers. The **hamstrings**, **psoas**, **abdominals**, and **calves** assist in stabilizing the pelvis and knees. The pressing phase represents the concentric (lifting) action.

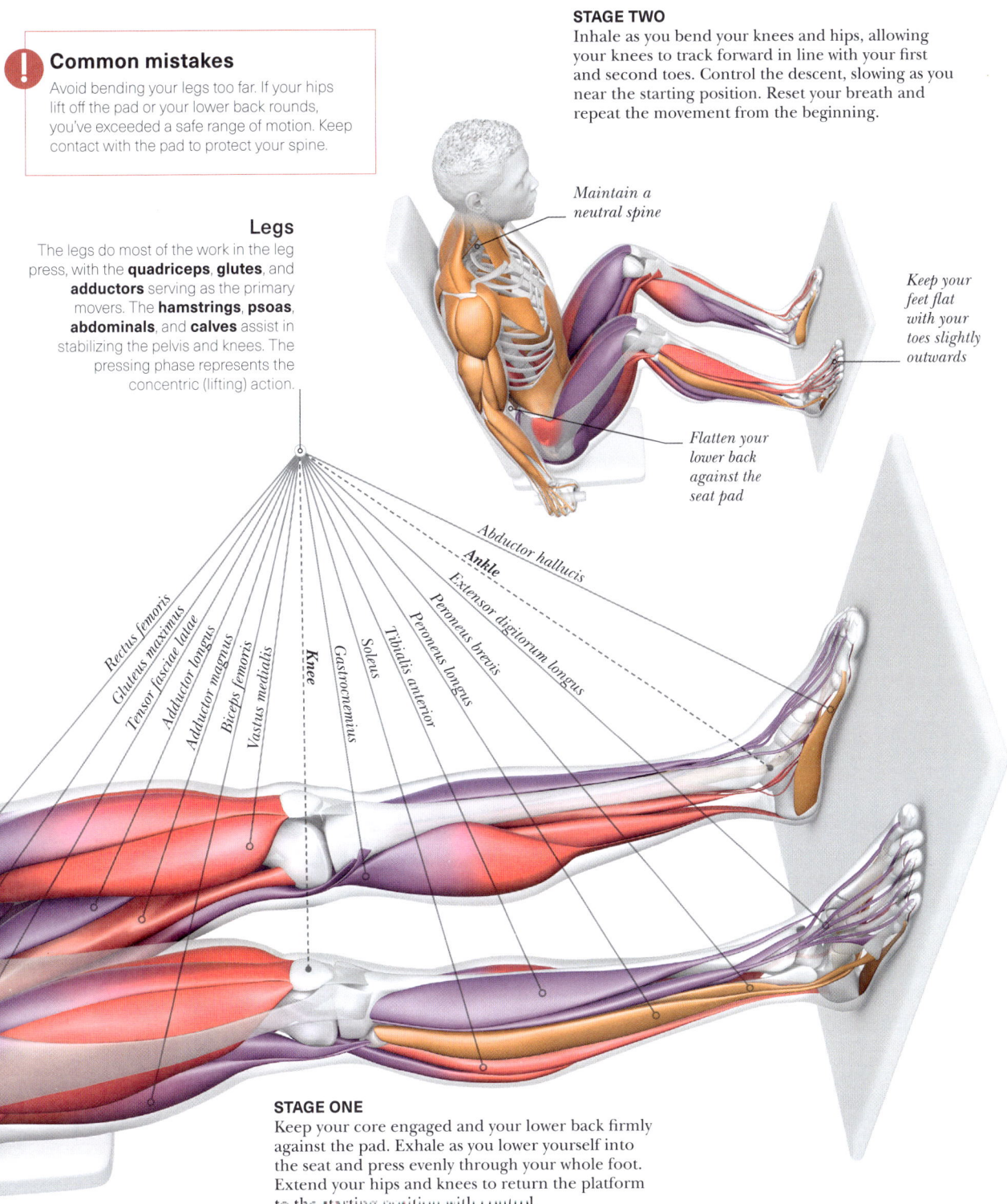

Maintain a neutral spine

Keep your feet flat with your toes slightly outwards

Flatten your lower back against the seat pad

Rectus femoris · Gluteus maximus · Tensor fasciae latae · Adductor longus · Adductor magnus · Biceps femoris · Vastus medialis · **Knee** · Gastrocnemius · Soleus · Tibialis anterior · Peroneus longus · Peroneus brevis · Extensor digitorum longus · **Ankle** · Abductor hallucis

STAGE ONE
Keep your core engaged and your lower back firmly against the pad. Exhale as you lower yourself into the seat and press evenly through your whole foot. Extend your hips and knees to return the platform to the starting position with control.

167

STRENGTH EXERCISES | Calf Raise with Dumbbells

CALF RAISE WITH DUMBBELLS

The dumbbell calf raise is a simple yet effective exercise for strengthening the lower leg. It primarily targets the gastrocnemius and soleus muscles of the calf. Secondary engagement includes the tibialis posterior and foot stabilizers, helping improve balance, power output, and injury resilience in the lower leg.

By targeting the gastrocnemius and soleus muscles, this exercise enhances plantarflexion strength and endurance, crucial for effective force transfer during the pedal stroke, particularly in the ankle's concentric and isometric phases. Improved strength in these muscles supports joint stability, optimizes neuromuscular coordination, and reduces the risk of overuse injuries such as Achilles tendinopathy. For cyclists, this contributes to greater pedalling efficiency, lower limb resilience, and improved energy economy during sustained or high-intensity efforts.

! **Caution**

Common mistakes in dumbbell calf raises include bouncing, using momentum, or letting the ankles roll inward. To avoid these, move slowly and with control, keep your feet aligned, and focus on a full range of motion. Maintain balance and avoid using excessive weight.

Upper body

The upper body muscles act mainly as stabilizers. The **forearms** and **grip muscles** are engaged to hold the dumbbells securely, while the **trapezius**, **deltoids**, and **core muscles** help maintain posture and balance throughout the movement.

- Deltoids
- Trapezius
- Iliocostalis
- Biceps
- Triceps
- Brachioradialis
- Transversus abdominis
- Extensor digitorum
- Gluteus medius

KEY
- ●--- Joints
- ○— Muscles
- ● Shortening with tension
- ● Lengthening with tension
- ● Lengthening without tension
- ● Held muscles without motion

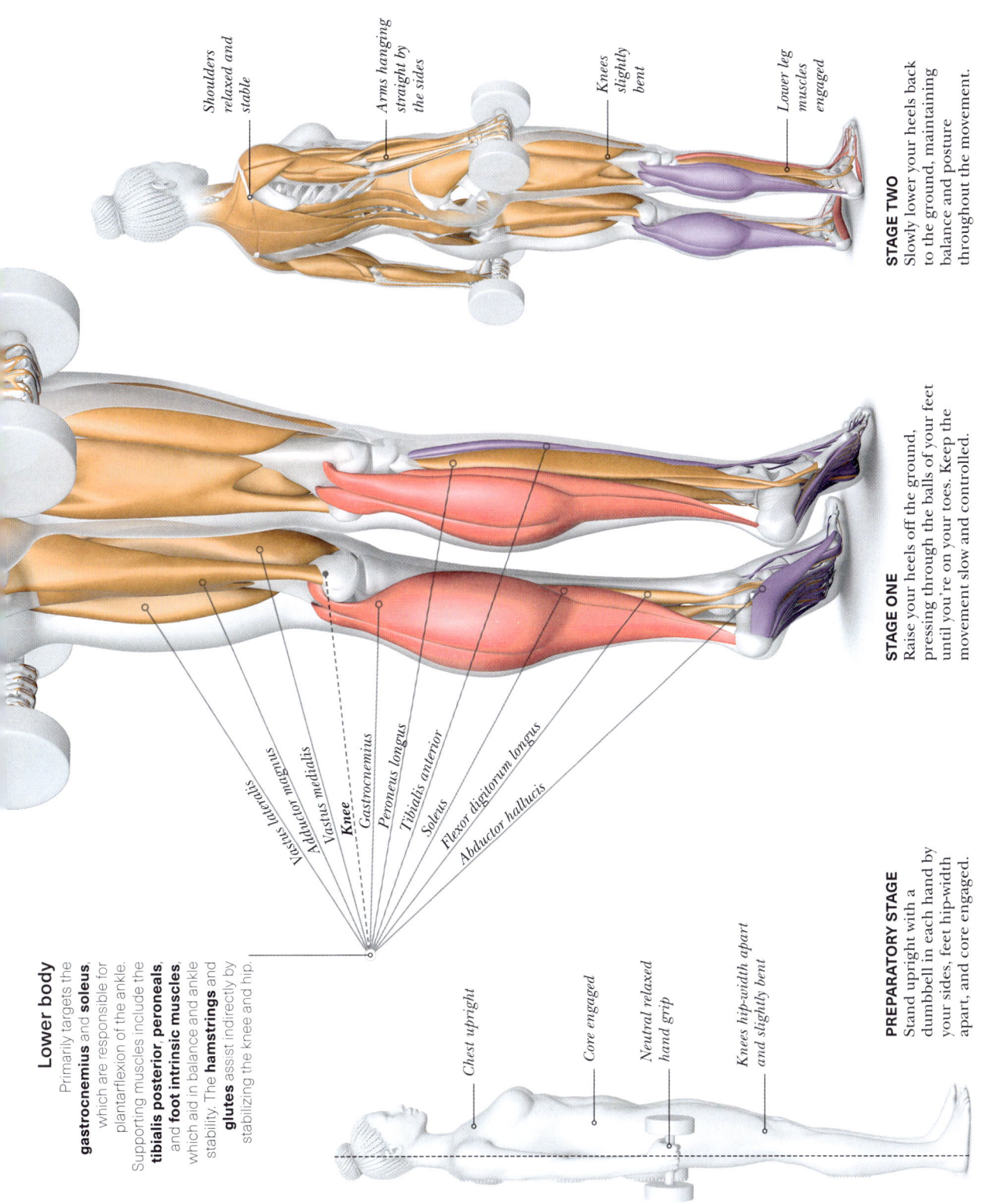

Lower body

Primarily targets the **gastrocnemius** and **soleus**, which are responsible for plantarflexion of the ankle. Supporting muscles include the **tibialis posterior**, **peroneals**, and **foot intrinsic muscles**, which aid in balance and ankle stability. The **hamstrings** and **glutes** assist indirectly by stabilizing the knee and hip.

- Vastus lateralis
- Adductor magnus
- Vastus medialis
- **Knee**
- Gastrocnemius
- Peroneus longus
- Tibialis anterior
- Soleus
- Flexor digitorum longus
- Abductor hallucis

PREPARATORY STAGE
Stand upright with a dumbbell in each hand by your sides, feet hip-width apart, and core engaged.

- Chest upright
- Core engaged
- Neutral relaxed hand grip
- Knees hip-width apart and slightly bent

STAGE ONE
Raise your heels off the ground, pressing through the balls of your feet until you're on your toes. Keep the movement slow and controlled.

STAGE TWO
Slowly lower your heels back to the ground, maintaining balance and posture throughout the movement.

- Shoulders relaxed and stable
- Arms hanging straight by the sides
- Knees slightly bent
- Lower leg muscles engaged

»VARIATIONS

Calf raise variations can correct imbalances, improve muscle endurance, and reduce the risk of overuse injuries such as Achilles tendinopathy. They also build control and resilience in the lower leg.

KEY
● Primary target muscle

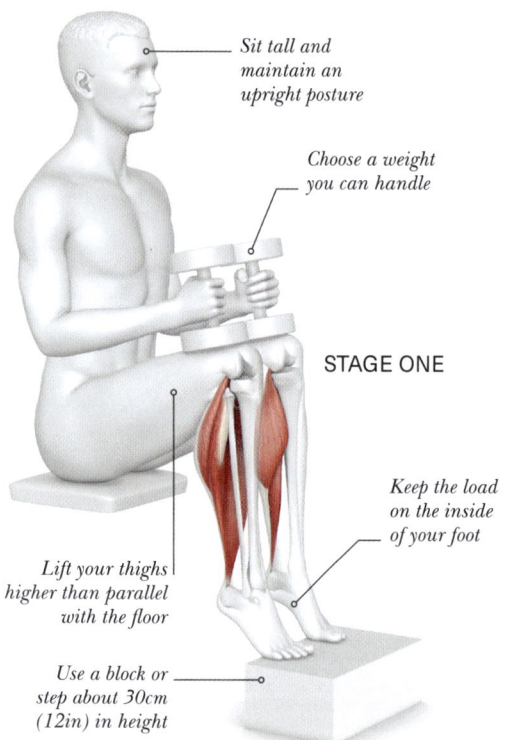

Sit tall and maintain an upright posture

Choose a weight you can handle

STAGE ONE

Keep the load on the inside of your foot

Lift your thighs higher than parallel with the floor

Use a block or step about 30cm (12in) in height

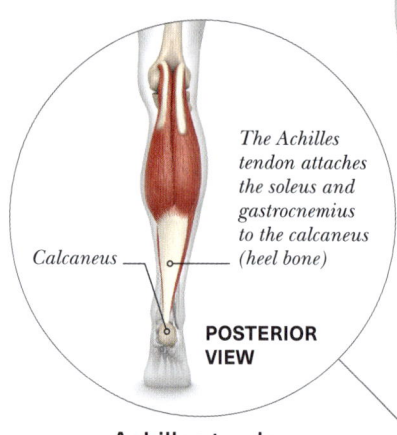

The Achilles tendon attaches the soleus and gastrocnemius to the calcaneus (heel bone)

Calcaneus

POSTERIOR VIEW

Achilles tendon
As a primary contributor to the foot's elasticity and shock-absorbing capacity, the Achilles tendon is a key mover in plantarflexion. This robust tendon, highly active during cycling, possesses the remarkable ability to withstand tensile forces equivalent to ten times the body's weight.

Stand tall, with your abs engaged and a support nearby

STAGE ONE

SEATED CALF RAISE

This seated version targets the soleus muscle more effectively than the gastrocnemius. This is because flexing your knees in a seated position shortens the gastrocnemius, putting more emphasis on the soleus.

PREPARATORY STAGE
Sit tall with your feet hip-width apart and resting on the balls of your feet on a block or step in front of you. Rest the dumbbells on your knees.

STAGE ONE
Engage your core as you inhale. As you exhale, contract your calves to raise your heels, smoothly driving your feet up and ankles forwards.

STAGE TWO
Inhale as you slowly lower your heels, ensuring your ankles remain aligned with your knees. Pause briefly at the bottom of each repetition. Then, repeat the movement.

SINGLE-LEG CALF RAISE

This unilateral calf raise loads your calves using only your bodyweight, so you don't need any additional equipment. Ensure you give equal attention to both legs.

PREPARATORY STAGE
Stand tall. Place the ball of one foot on a step and wrap your non-working leg around the back of the other. Lower your heel into the starting position.

STAGE ONE
Breathe in and engage your core. As you exhale, powerfully raise your heel, squeezing your calf muscle. Don't hesitate to use a nearby support if you need help with balance.

STAGE TWO
Inhale and slowly lower your heel, maintaining the load on the inside of your foot. Pause at the bottom before repeating the movement.

STRENGTH AND STRETCH EXERCISES | Calf Raise Variations

> " "
> *Calf exercises enhance ankle stability, pedal efficiency, and lower-leg endurance, helping cyclists generate smoother power and reduce injury risk.*

LEG PRESS CALF RAISE

This variation mimics a calf raise while providing a more stable body position and eliminating spinal load. It's an excellent alternative for those who find standing calf raise machines uncomfortable or unstable.

Ankle dorsiflexion and plantarflexion

The **ankle muscles** are crucial for both dorsiflexion and plantarflexion. Alongside the **foot muscles**, they stabilize gait and ensure the healthy function of both the foot and ankle. Properly training the **gastrocnemius**, **soleus**, and other **lower leg muscles** is essential for developing healthy movement patterns and long-term injury prevention.

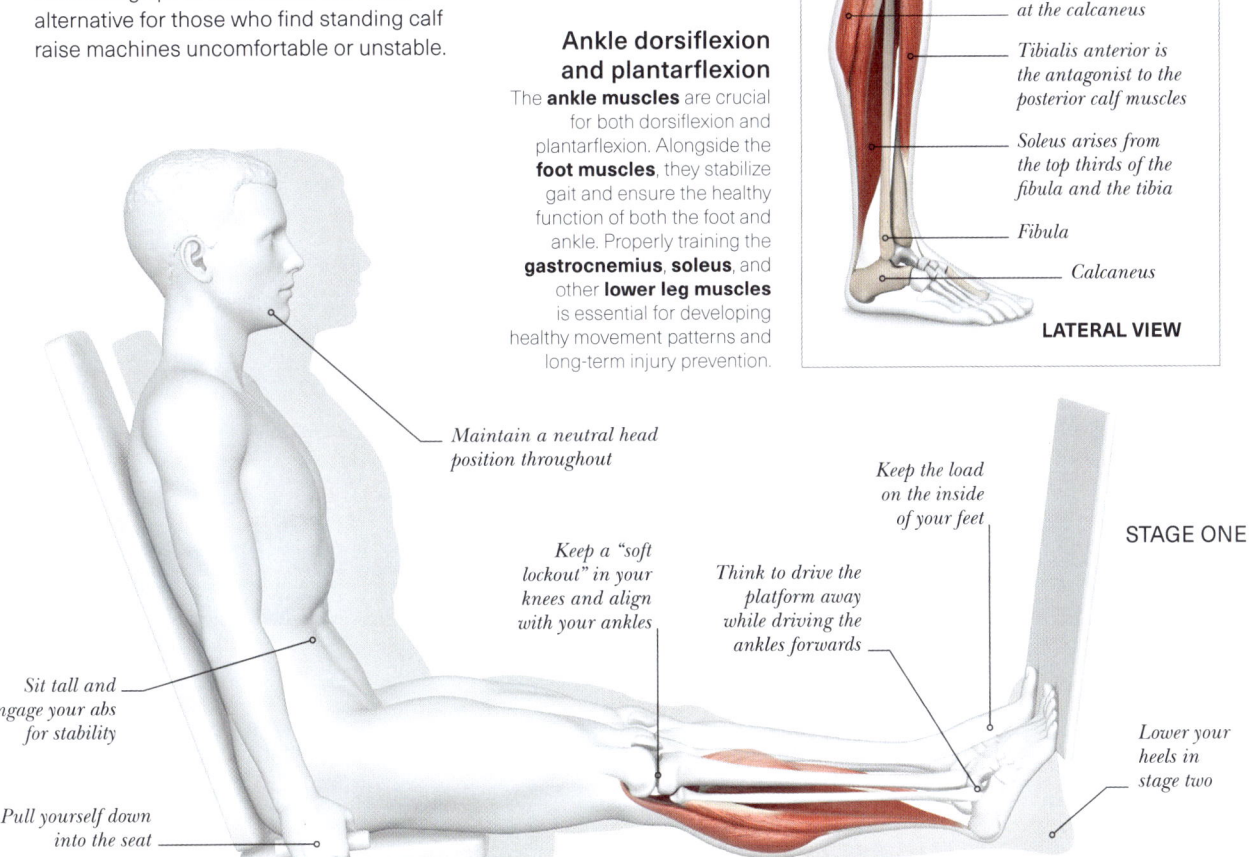

Labels:
- Femur
- Tibia
- Gastrocnemius arises from the femur and attaches at the calcaneus
- Tibialis anterior is the antagonist to the posterior calf muscles
- Soleus arises from the top thirds of the fibula and the tibia
- Fibula
- Calcaneus

LATERAL VIEW

Annotations on figure:
- Maintain a neutral head position throughout
- Keep a "soft lockout" in your knees and align with your ankles
- Think to drive the platform away while driving the ankles forwards
- Keep the load on the inside of your feet
- Sit tall and engage your abs for stability
- Pull yourself down into the seat
- Lower your heels in stage two

STAGE ONE

PREPARATORY STAGE
Set the desired weight. Sit on the machine's seat with the balls of your feet hip-width apart on the platform, and your heels lowered.

STAGE ONE
Inhale to engage your core as you pull yourself firmly into the seat. As you exhale, push through the balls of your feet to raise your heels.

STAGE TWO
Inhale and slowly lower your heels back to the starting position, maintaining control. Pause briefly at the bottom before repeating the movement.

171

BOX JUMP

The box jump develops lower-body power, speed, and coordination. It primarily targets the glutes, quadriceps, hamstrings, and calves. For cyclists, it improves sprint performance, pedal stroke explosiveness, and overall muscular reactivity, useful in accelerations, attacks, and race starts.

This exercise is designed to improve the resilience of the leg's natural elastic components, enhancing the energy storage and release capacity of the glutes, quads, calves, and hip abductors. A knee flexion angle of around 45 degrees should be maintained during the takeoff and landing phases. Using a 30-cm (12-inch) box, perform 3 sets of 10–12 repetitions. Progression involves increasing the box height and reducing the repetition count to 6–8 as the box height increases.

STAGE TWO
Explosively push through your legs, fully extending your ankles, knees, and hips to launch yourself up and forwards onto the box. Simultaneously, thrust your arms forwards and upwards to boost momentum.

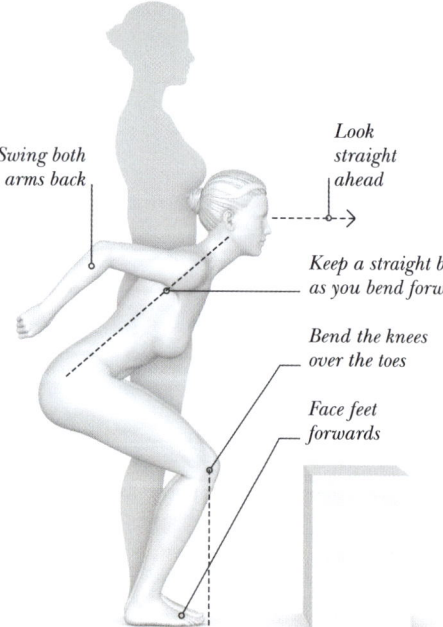

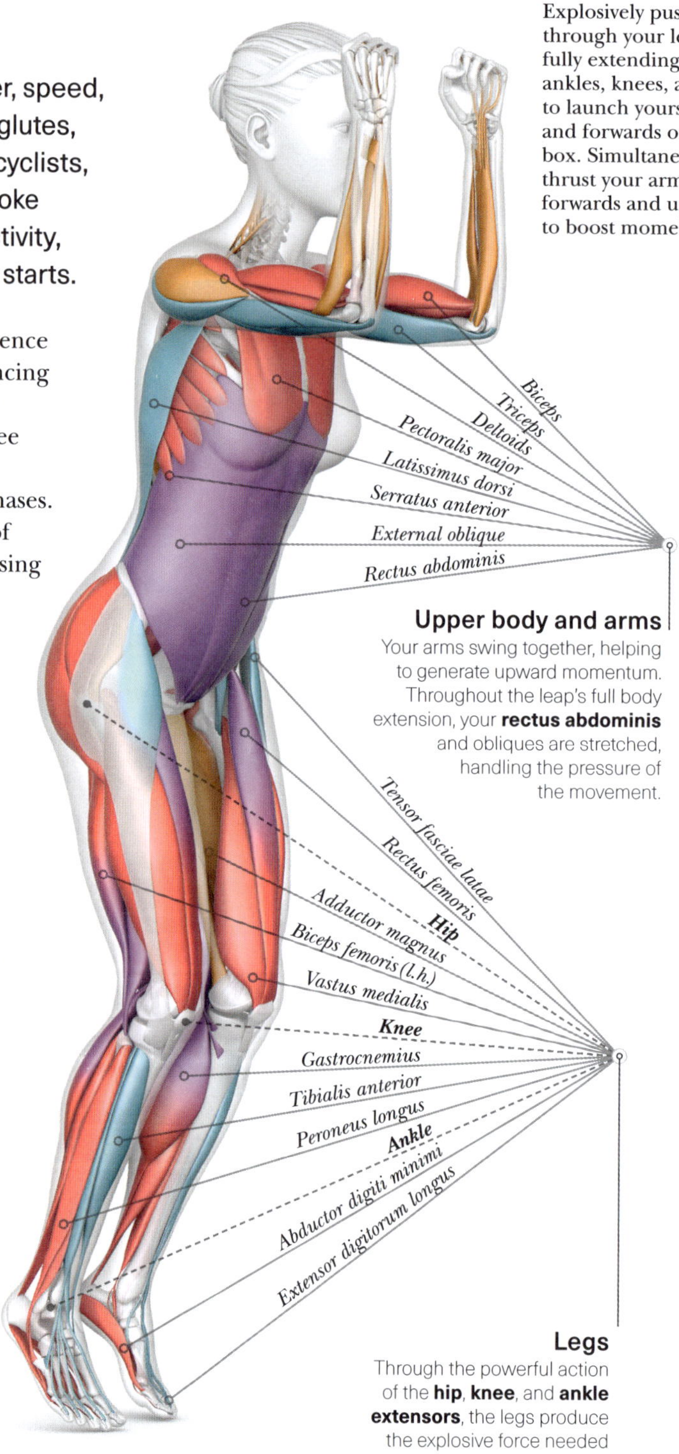

PREPARATORY STAGE/STAGE ONE
Stand facing the box with your feet hip-width apart and arms relaxed at your sides. Bend your knees to prepare for the jump, drawing your elbows back to propel your arms forwards.

Swing both arms back
Look straight ahead
Keep a straight back as you bend forward
Bend the knees over the toes
Face feet forwards

Upper body and arms
Your arms swing together, helping to generate upward momentum. Throughout the leap's full body extension, your **rectus abdominis** and obliques are stretched, handling the pressure of the movement.

Biceps
Triceps
Deltoids
Pectoralis major
Latissimus dorsi
Serratus anterior
External oblique
Rectus abdominis

Tensor fasciae latae
Rectus femoris
Hip
Adductor magnus
Biceps femoris (l.h.)
Vastus medialis
Knee
Gastrocnemius
Tibialis anterior
Peroneus longus
Ankle
Abductor digiti minimi
Extensor digitorum longus

Legs
Through the powerful action of the **hip**, **knee**, and **ankle extensors**, the legs produce the explosive force needed to lift the body off the ground.

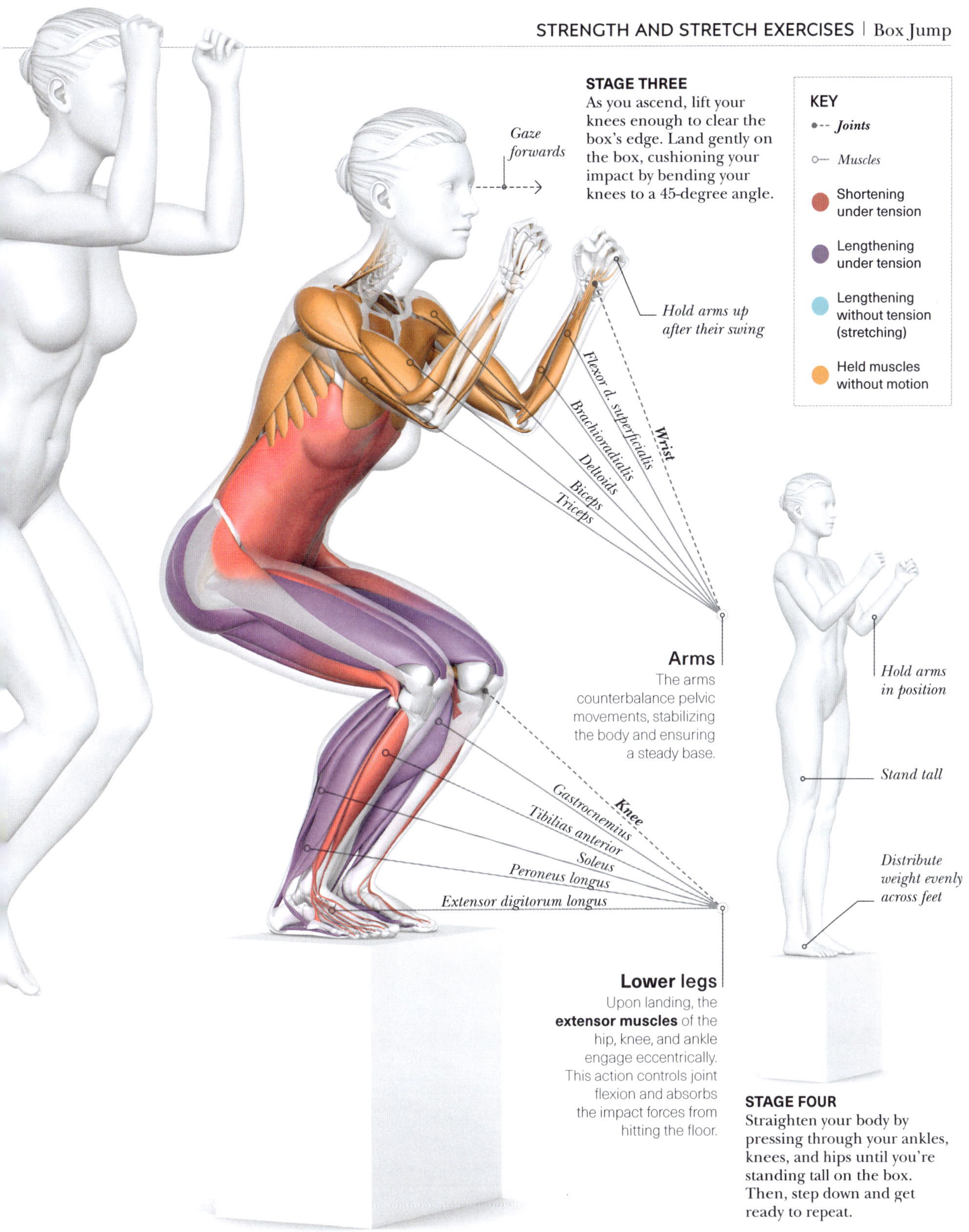

STRENGTH AND STRETCH EXERCISES | Box Jump

STAGE THREE
As you ascend, lift your knees enough to clear the box's edge. Land gently on the box, cushioning your impact by bending your knees to a 45-degree angle.

Gaze forwards

KEY
- **•--** *Joints*
- **o—** *Muscles*
- 🔴 Shortening under tension
- 🟣 Lengthening under tension
- 🔵 Lengthening without tension (stretching)
- 🟠 Held muscles without motion

Hold arms up after their swing

Flexor d. superficialis
Brachioradialis
Deltoids
Biceps
Triceps

Wrist

Arms
The arms counterbalance pelvic movements, stabilizing the body and ensuring a steady base.

Knee

Gastrocnemius
Tibialis anterior
Soleus
Peroneus longus
Extensor digitorum longus

Lower legs
Upon landing, the **extensor muscles** of the hip, knee, and ankle engage eccentrically. This action controls joint flexion and absorbs the impact forces from hitting the floor.

Hold arms in position

Stand tall

Distribute weight evenly across feet

STAGE FOUR
Straighten your body by pressing through your ankles, knees, and hips until you're standing tall on the box. Then, step down and get ready to repeat.

» VARIATIONS

Variations of the box jump benefit cyclists by developing explosive leg power, neuromuscular coordination, and reactive strength. Different jump styles can target specific movement patterns, improve balance and symmetry, and simulate race demands like sprinting or sudden accelerations. These variations also help reduce strength imbalances, enhance joint stability, and build resilience against fatigue, contributing to more efficient and powerful pedalling.

Single-leg box jump

The single-leg box jump is a plyometric exercise that builds explosive power, balance, and coordination on one leg. It targets the glutes, quads, hamstrings, and calves, while engaging the core for stability.

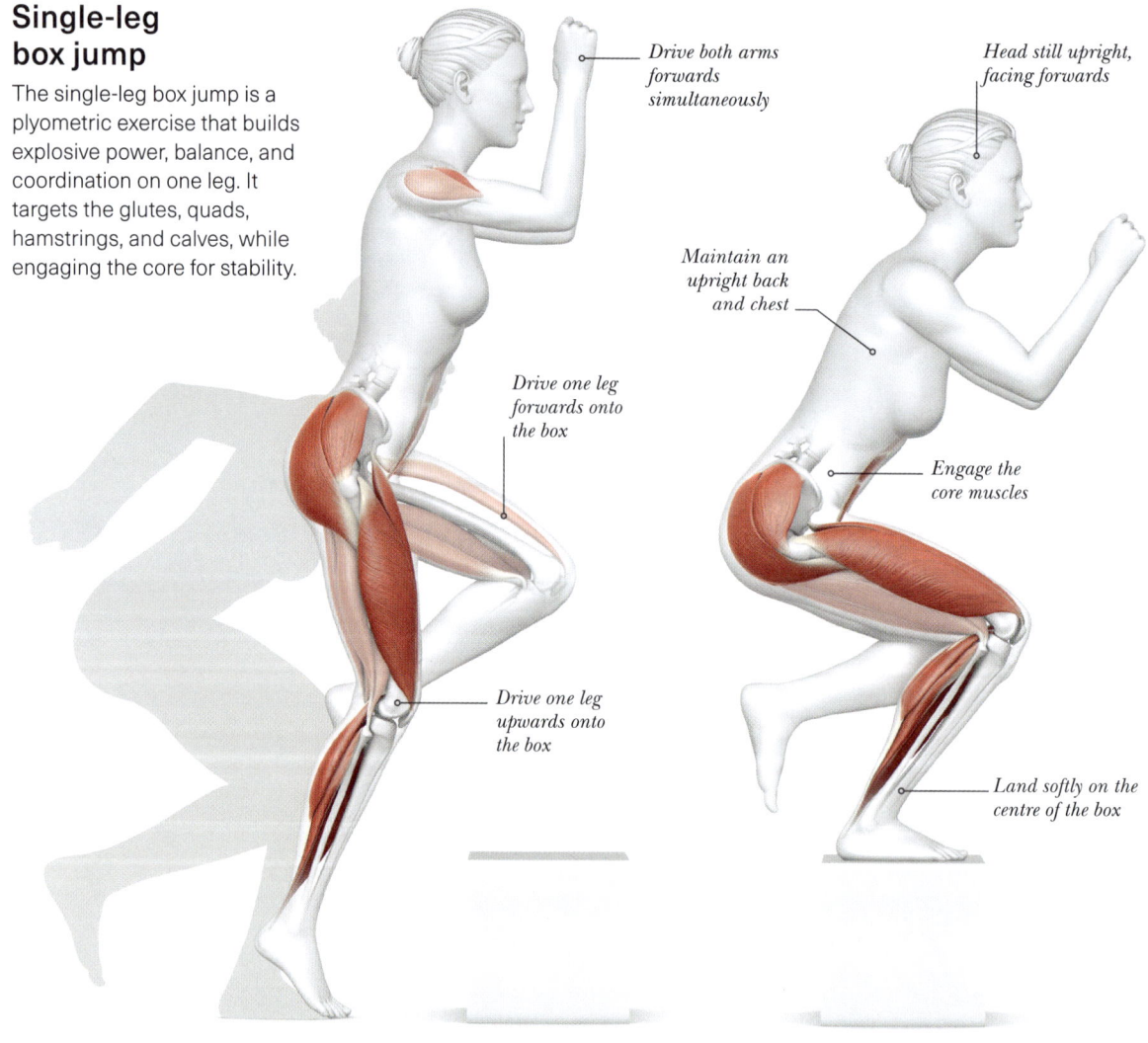

PREPARATORY STAGE
Stand on one leg facing a sturdy box or platform, with your knee slightly bent and arms ready to swing.

STAGE ONE
Swing your arms and drive through your standing leg to jump explosively onto the box, keeping your balance.

STAGE TWO
Land softly on the same leg with a slight knee bend, maintaining control and stability, then step down safely.

STRENGTH AND STRETCH EXERCISES | Box Jump Variations

Jumping off the box

The jump-off-the-box exercise, also known as a depth jump, is a plyometric drill that improves reactive strength, landing control, and explosive power. The movement involves stepping off a box, landing softly, then immediately jumping vertically or forward upon ground contact. It primarily targets the glutes, quadriceps, hamstrings, and calves, with strong activation of the core for stability.

Box jump variations build explosive leg power, balance, and neuromuscular control.

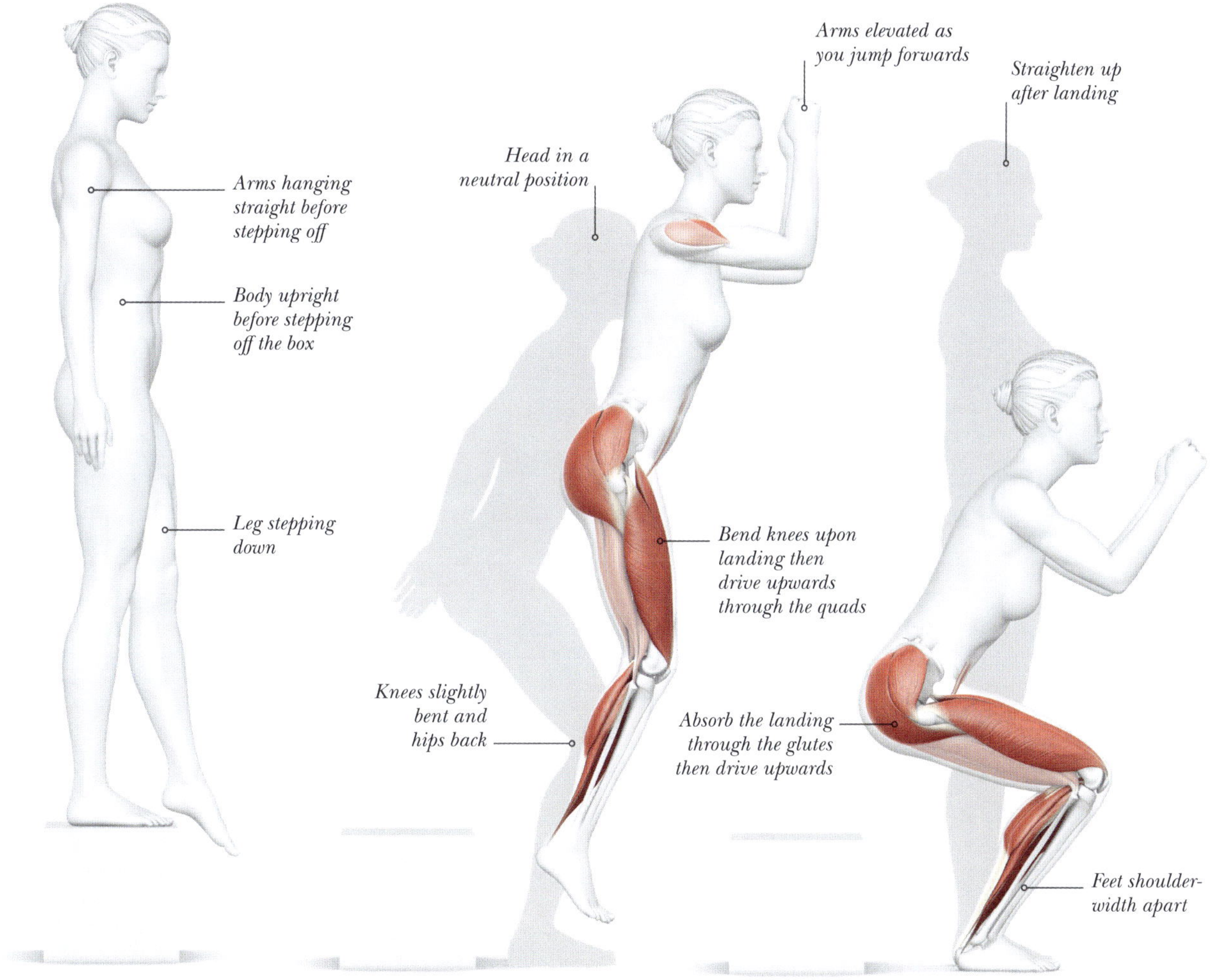

PREPARATORY STAGE
Stand on the box with your arms at your sides. Put one leg forward as you prepare to step off the box.

- Arms hanging straight before stepping off
- Body upright before stepping off the box
- Leg stepping down

STAGE ONE
Step down, landing softly, immediately exploding upwards or forwards into a jump, minimizing ground contact time.

- Head in a neutral position
- Arms elevated as you jump forwards
- Bend knees upon landing then drive upwards through the quads
- Absorb the landing through the glutes then drive upwards
- Knees slightly bent and hips back

STAGE TWO
Land with feet shoulder-width apart close to the edge of the box then stand straight.

- Straighten up after landing
- Feet shoulder-width apart

PULL UP

The pull up strengthens the upper back, biceps, and shoulder muscle, which help maintain posture, control the handlebars, and support upper-body stability during long rides and technical terrain. Regular practice enhances muscular endurance and balanced strength off the bike.

Pull-ups are essential for building strength relative to body weight, a key factor for cyclists, especially in climbing and sprinting. They target upper-body muscles that support posture, bike handling, and stability under load. Consistent pull-up training improves overall strength-to-weight ratio, contributing to more efficient riding and reduced fatigue on long or demanding efforts. Beginners can start by doing 4 sets of 6–8 reps.

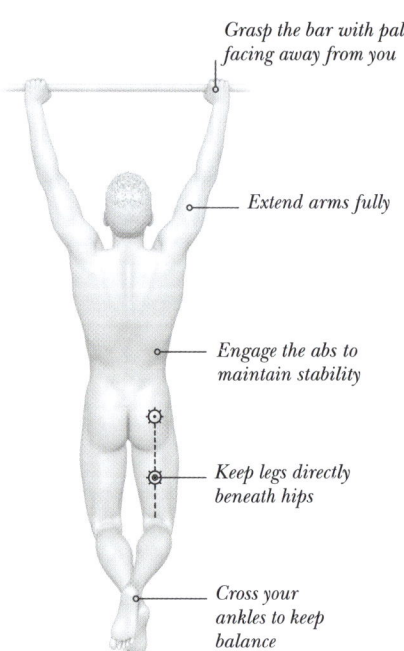

- Grasp the bar with palms facing away from you
- Extend arms fully
- Engage the abs to maintain stability
- Keep legs directly beneath hips
- Cross your ankles to keep balance

PREPARATORY STAGE
Start in a hanging position, with hands placed wider than shoulder-width apart. Engage your core, lean back slightly, and puff out the chest, ready to lift.

Extensor digitorum, Biceps, Deltoids, Infraspinatus, Trapezius, Latissimus dorsi, Teres major, Brachioradialis, Triceps brachii, External oblique

Upper body and arms
Muscles of the upper back – including the **deltoids**, **latissimus dorsi**, and **trapezius**, work together to produce a strong pulling motion, which supports cyclists in maintaining posture and control, especially during climbs and sprints. The **biceps** and **triceps** also engage to assist the lift and enhance upper-body strength.

STAGE ONE
Pull up until the bar meets your chest, focussing on driving the movement with your back and keeping your elbows pointed down.

STRENGTH AND STRETCH EXERCISES | Pull Up

KEY
- Joints
- Muscles
- Shortening with tension
- Lengthening with tension
- Lengthening without tension
- Held muscles without motion

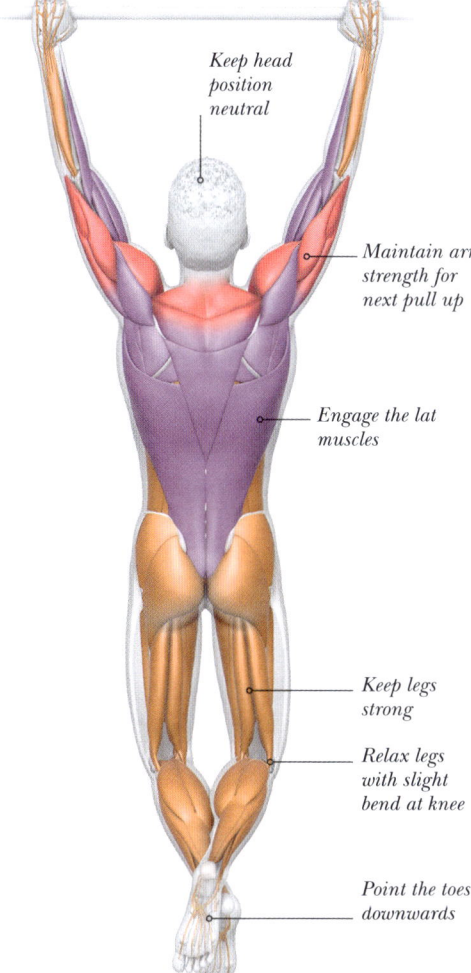

- Keep head position neutral
- Maintain arm strength for next pull up
- Engage the lat muscles
- Keep legs strong
- Relax legs with slight bend at knee
- Point the toes downwards

VARIATION: CHIN UP

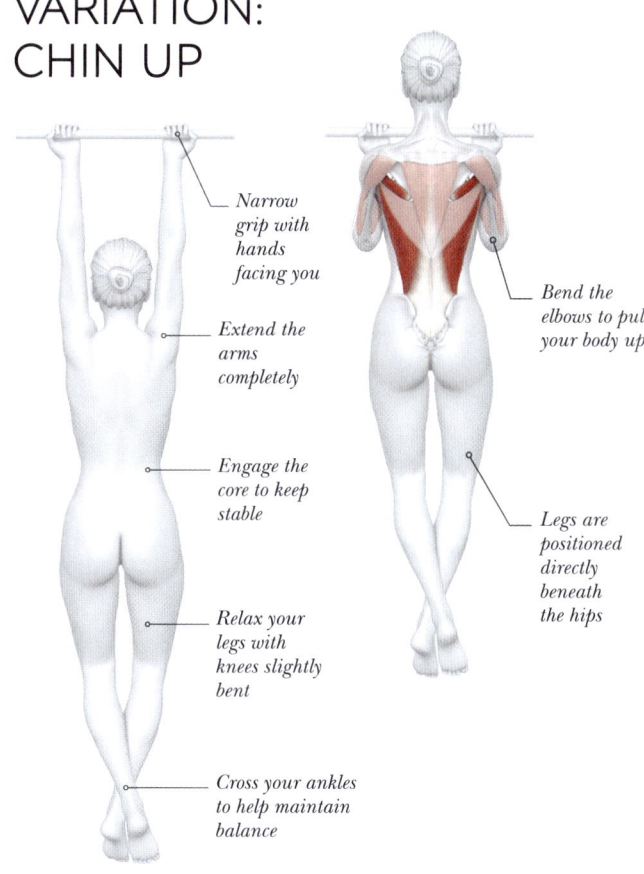

- Narrow grip with hands facing you
- Extend the arms completely
- Engage the core to keep stable
- Relax your legs with knees slightly bent
- Cross your ankles to help maintain balance
- Bend the elbows to pull your body up
- Legs are positioned directly beneath the hips

PREPARATORY STAGE
Start hanging from the bar, hands shoulder-width apart and feet crossed over each other. Make sure that your core is stabilized.

STAGE ONE
Exhale and flex your elbows to pull yourself up to the bar, clearing it with your chin. For an extra challenge, hold this position for 1–2 seconds before descending.

STAGE TWO
Inhale and extend your elbows to lower your body, being careful not to swing as you perform this movement. Repeat stages 1 and 2.

STAGE TWO
Control your descent to full arm extension, feeling the lat muscles stretch, preparing them for the next powerful upward pull.

PUSH UP

This exercise strengthens key upper-body muscles important for cycling performance, including the pectorals, deltoids, and triceps, which support handlebar control and posture. The serratus anterior and abdominals enhance core stability, while the legs contribute by stabilizing the body during the movement, promoting overall control and balance on the bike.

The push up involves a full-body engagement, and correct form and control are crucial when executing the movement. Regulate your descent, avoiding a sudden drop to the floor. Always keep the abdominal muscles activated throughout the movement. For those new to the exercise, a good starting point is to perform 4 sets, each consisting of 5–6 repetitions.

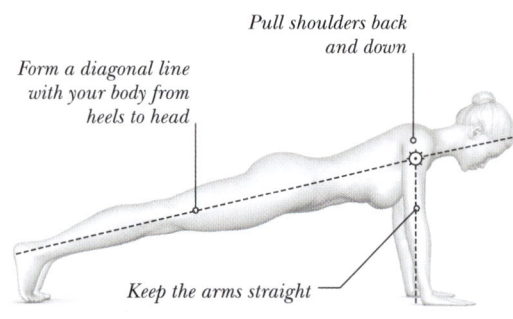

PREPARATORY STAGE
Initiate by positioning yourself in a high plank, ensuring your pelvis is aligned, your neck is neutral, and your hands are placed just beyond shoulder-width apart. Your shoulders should be pulled back and down, and your core fully engaged. Flex your toes, extending the heels backwards.

> **! Caution**
> Make sure you engage your abdominals throughout the exercise, to prevent the spine from caving in and putting pressure on the lower back and joints.

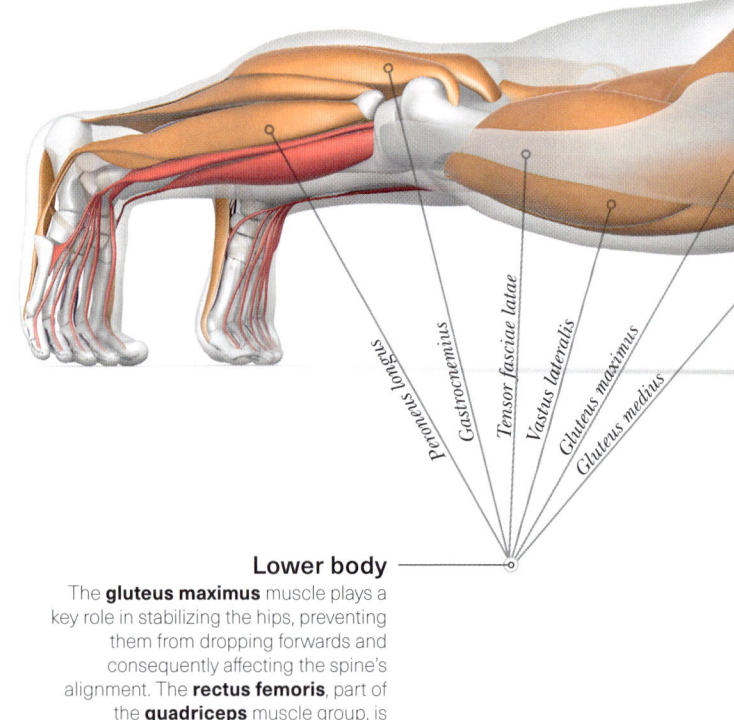

Lower body
The **gluteus maximus** muscle plays a key role in stabilizing the hips, preventing them from dropping forwards and consequently affecting the spine's alignment. The **rectus femoris**, part of the **quadriceps** muscle group, is engaged in an isometric hold.

STRENGTH AND STRETCH EXERCISES | Push Up

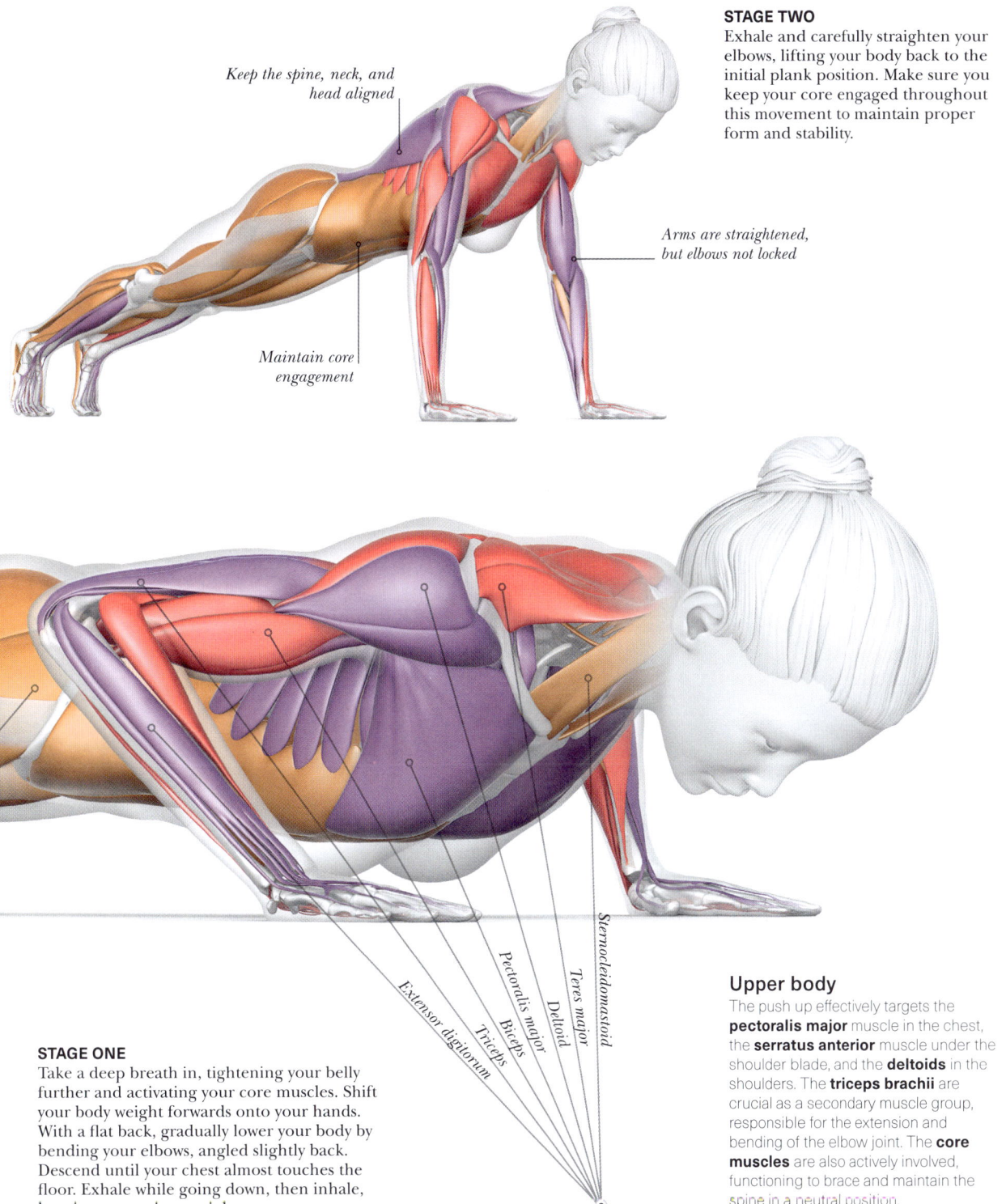

STAGE TWO
Exhale and carefully straighten your elbows, lifting your body back to the initial plank position. Make sure you keep your core engaged throughout this movement to maintain proper form and stability.

Keep the spine, neck, and head aligned

Arms are straightened, but elbows not locked

Maintain core engagement

STAGE ONE
Take a deep breath in, tightening your belly further and activating your core muscles. Shift your body weight forwards onto your hands. With a flat back, gradually lower your body by bending your elbows, angled slightly back. Descend until your chest almost touches the floor. Exhale while going down, then inhale, keeping your spine straight.

Extensor digitorum
Triceps
Biceps
Pectoralis major
Deltoid
Teres major
Sternocleidomastoid

Upper body
The push up effectively targets the **pectoralis major** muscle in the chest, the **serratus anterior** muscle under the shoulder blade, and the **deltoids** in the shoulders. The **triceps brachii** are crucial as a secondary muscle group, responsible for the extension and bending of the elbow joint. The **core muscles** are also actively involved, functioning to brace and maintain the spine in a neutral position.

179

BENT-OVER ROW WITH DUMBBELLS

This compound upper-body exercise targets the latissimus dorsi, rhomboids, trapezius, and posterior deltoids, with support from the biceps and core. Performed by hinging at the hips with a flat back and pulling the dumbbells towards the torso, it builds upper-back strength, posture, and scapular stability.

Using dumbbells offers flexibility in how you perform the bent-over row. You can work with one hand rowing while the opposite knee and hand support you on a bench, or with both feet on the ground, knees slightly bent, and hips flexed at approximately 90 degrees, maintaining a flat back. This variation increases core demand and postural control. For added difficulty, pause at the top of the row for 1–2 seconds, squeezing the shoulder blades together. Perform 3 sets of 8–10 reps on each side.

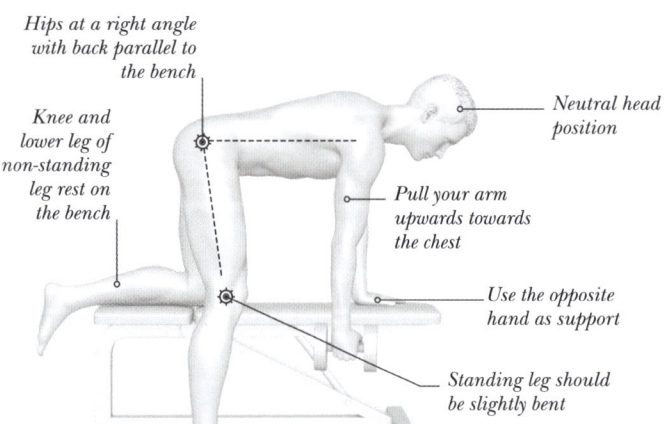

Hips at a right angle with back parallel to the bench

Knee and lower leg of non-standing leg rest on the bench

Neutral head position

Pull your arm upwards towards the chest

Use the opposite hand as support

Standing leg should be slightly bent

PREPARATORY STAGE
With one knee on the bench, position your other leg beneath your hip. Lean forwards until your back is parallel to the ground, then inhale to engage and brace your core.

STAGE ONE
Exhale as you draw your shoulder blade back and powerfully drive your arm upwards. As you do so, flex your elbow to an angle between 30–75 degrees – adjusting this angle will alter the muscle bias.

STRENGTH AND STRETCH EXERCISES | Bent-over Row with Dumbbells

> " "
>
> *The bent-over row with dumbbells builds a strong, stable upper back and core, essential for maintaining control, posture, and power on the bike.*

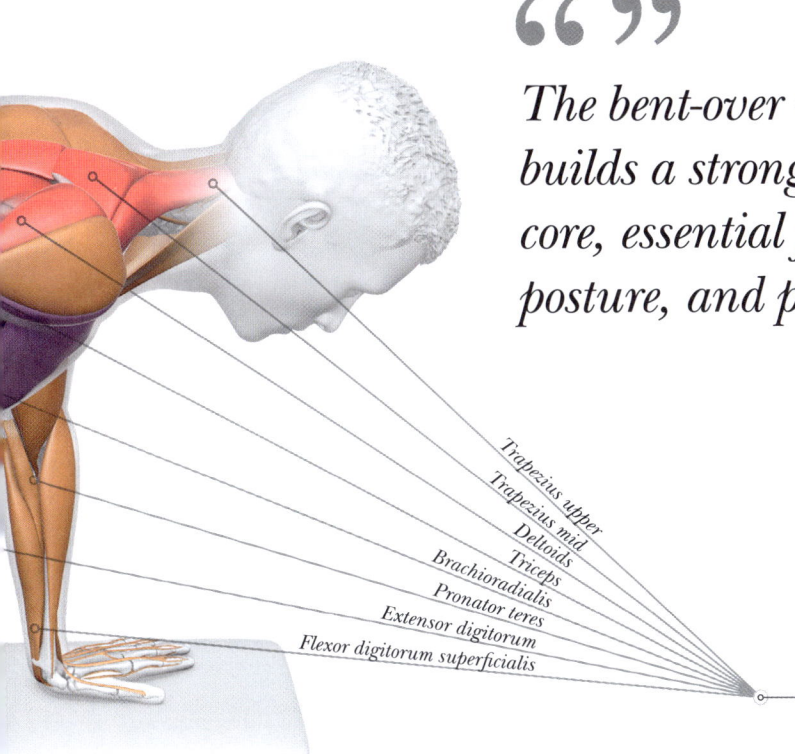

Upper body
The **latissimus dorsi** is the primary mover, responsible for pulling the arms back through shoulder extension. The **rhomboids** and **trapezius** work to retract and stabilize the shoulder blades during the movement, while the **posterior deltoids** assist in drawing the arms backwards. The **biceps** and **brachialis** contribute by flexing the elbows to lift the dumbbells. To maintain a stable position, the **erector spinae** keep the spine extended, and the core muscles – **rectus abdominis**, **obliques**, and **transversus abdominis** – stabilize the trunk.

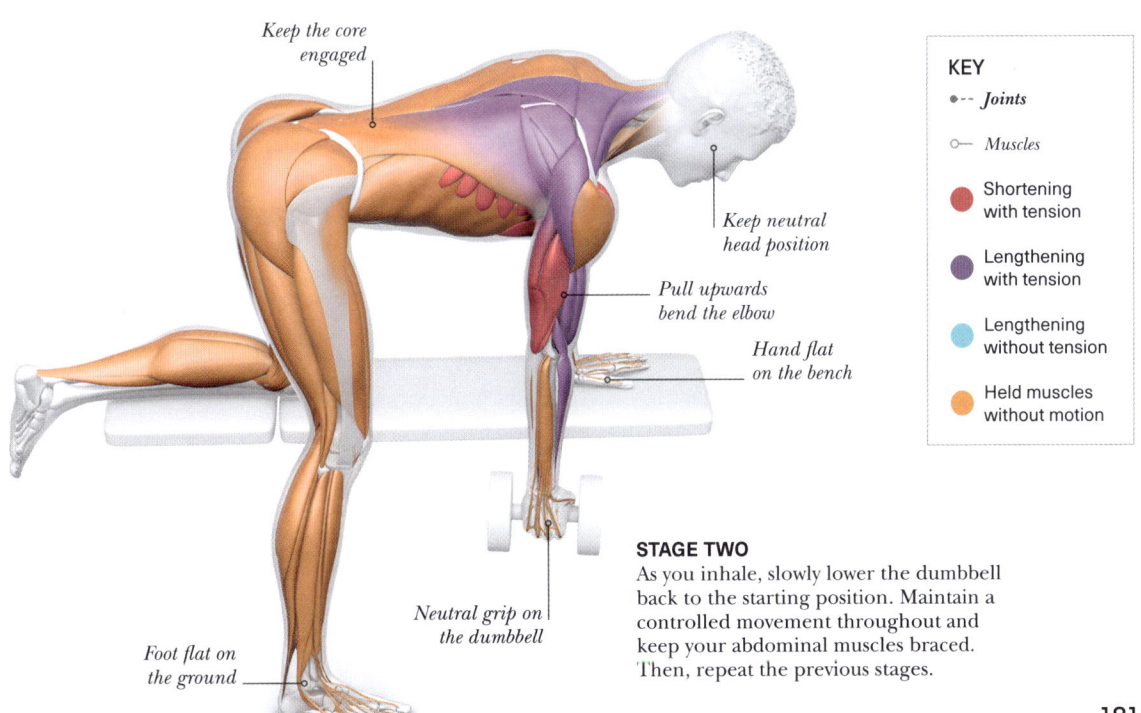

KEY
- •-- *Joints*
- o— *Muscles*
- ● Shortening with tension
- ● Lengthening with tension
- ● Lengthening without tension
- ● Held muscles without motion

STAGE TWO
As you inhale, slowly lower the dumbbell back to the starting position. Maintain a controlled movement throughout and keep your abdominal muscles braced. Then, repeat the previous stages.

DUMBBELL SHOULDER PRESS

The shoulder press is valuable for cyclists, strengthening the deltoids and triceps to support upper-body stability, handlebar control, and sustained posture during long rides and technical efforts.

This exercise builds upper-body strength, particularly in the shoulders and arms. Strengthening the deltoids and triceps enhances handlebar control, posture, and fatigue resistance, helping cyclists maintain stability, efficiency, and endurance across varied terrain and long rides. Perform 4 sets of 8–10 reps.

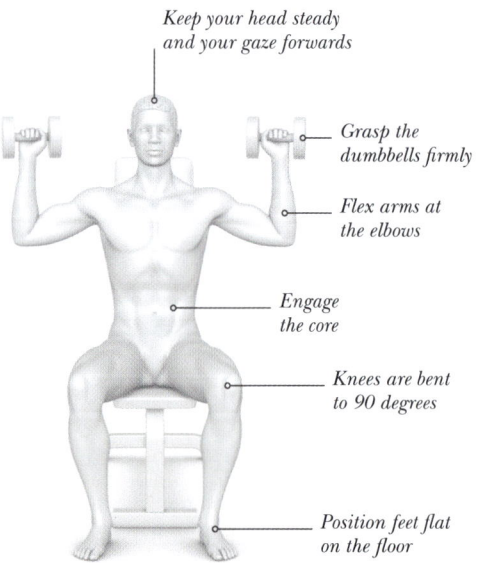

Keep your head steady and your gaze forwards

Grasp the dumbbells firmly

Flex arms at the elbows

Engage the core

Knees are bent to 90 degrees

Position feet flat on the floor

PREPARATORY STAGE
Sit on the edge of a bench with dumbbells at shoulder height, fingers facing outwards. Plant your feet firmly on the floor, with feet a bit wider than shoulder-width apart, knees slightly bent. Inhale as you prepare to lift.

KEY
- •-- Joints
- ○— Muscles
- ● Shortening with tension
- ● Lengthening with tension
- ● Lengthening without tension
- ● Held muscles without motion

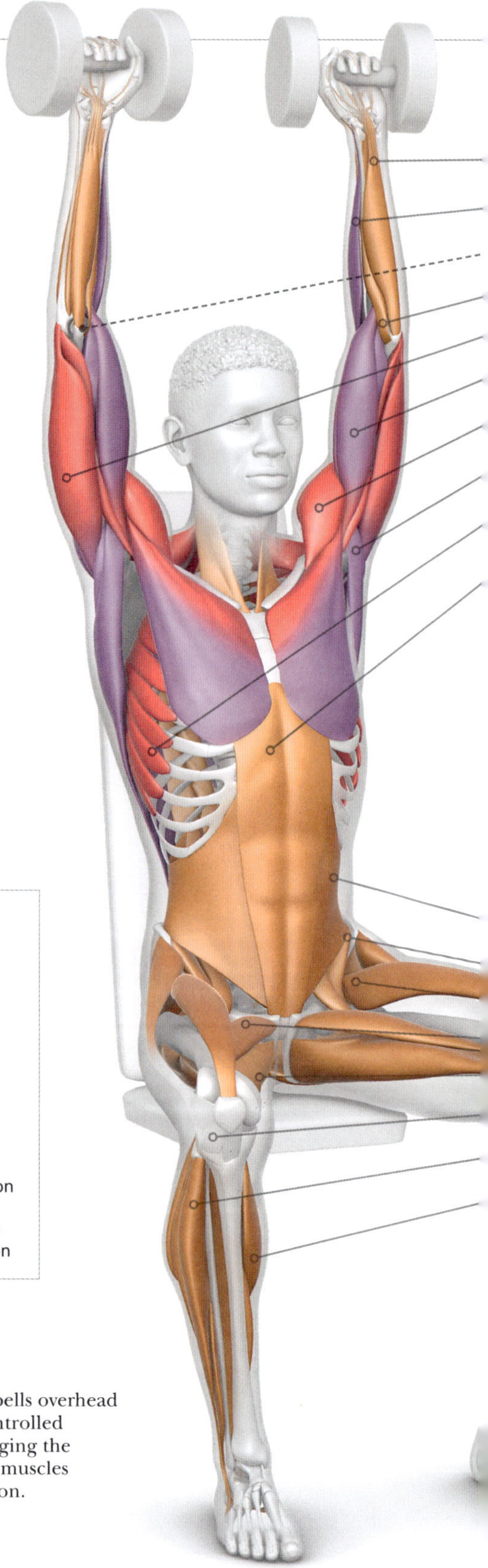

STAGE ONE
Elevate the dumbbells overhead with a smooth, controlled motion, fully engaging the shoulder and arm muscles at full arm extension.

STRENGTH AND STRETCH EXERCISES | Dumbbell Shoulder Press

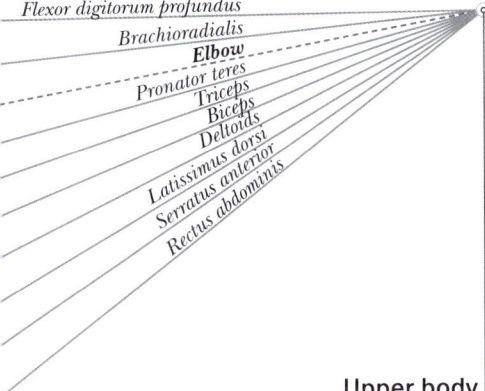

Upper body
The **deltoids** are the main drivers of the press. The **triceps** extend the elbows, aiding the lift. The **muscles of the forearms** maintain the grip and wrist position as you hold the weights overhead.

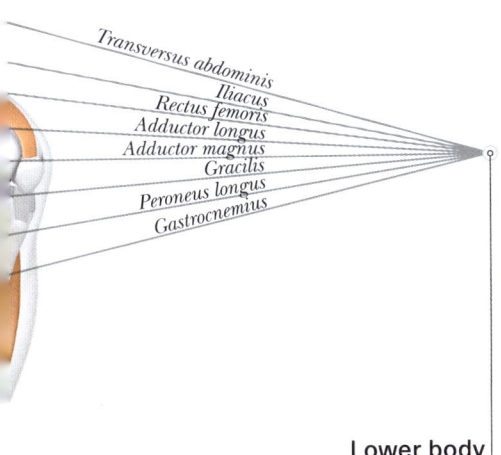

Lower body
The **muscles of the core** engage to keep the torso and pelvis stable. The **muscles of the lower body**, including the **quadriceps**, support the upper body's lift. The **calf muscles** engage to stabilize the pose throughout the lift.

> **! Caution**
> Be cautious of shoulder injuries during the press: incorrect form or excessive weight can lead to strain. Keep the movement smooth and controlled.

Strength above, stability below: shoulder presses build power for every reach and lift.

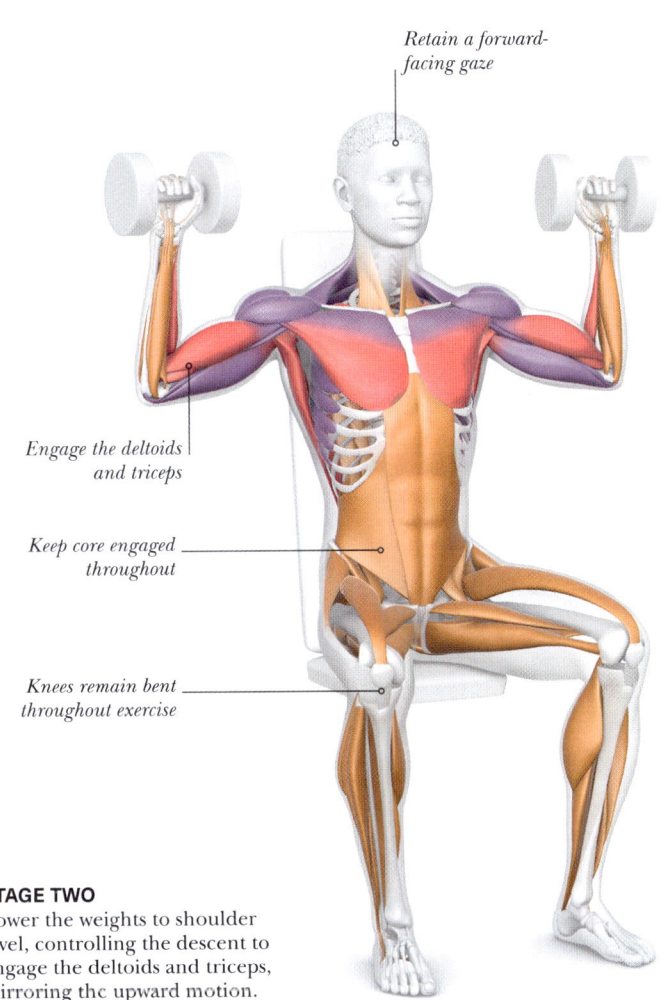

Retain a forward-facing gaze

Engage the deltoids and triceps

Keep core engaged throughout

Knees remain bent throughout exercise

STAGE TWO
Lower the weights to shoulder level, controlling the descent to engage the deltoids and triceps, mirroring the upward motion.

PLANK WITH SHOULDER TAPS

This is a core stability exercise that targets the core and lower back and engages the shoulders, chest, and glutes. For cyclists, it improves posture, enhances core endurance, and reinforces control over bike handling.

Perform from a high plank position with arms extended and hands under the shoulders, keeping your body straight and core tight. Lift one hand to tap the opposite shoulder, then alternate sides, aiming to keep the hips stable and avoid rotation. The tap develops anti-rotation strength by challenging the body to stay balanced and aligned during unilateral movement.

This exercise works the rectus abdominis, obliques, and transversus abdominis for core stability, and the deltoids, triceps, and pectorals to support the shoulders and arms. The glutes and quadriceps help stabilize the lower body, making it a full-body movement that builds strength, control, and endurance. Perform 3 sets of 6–8 reps on each side.

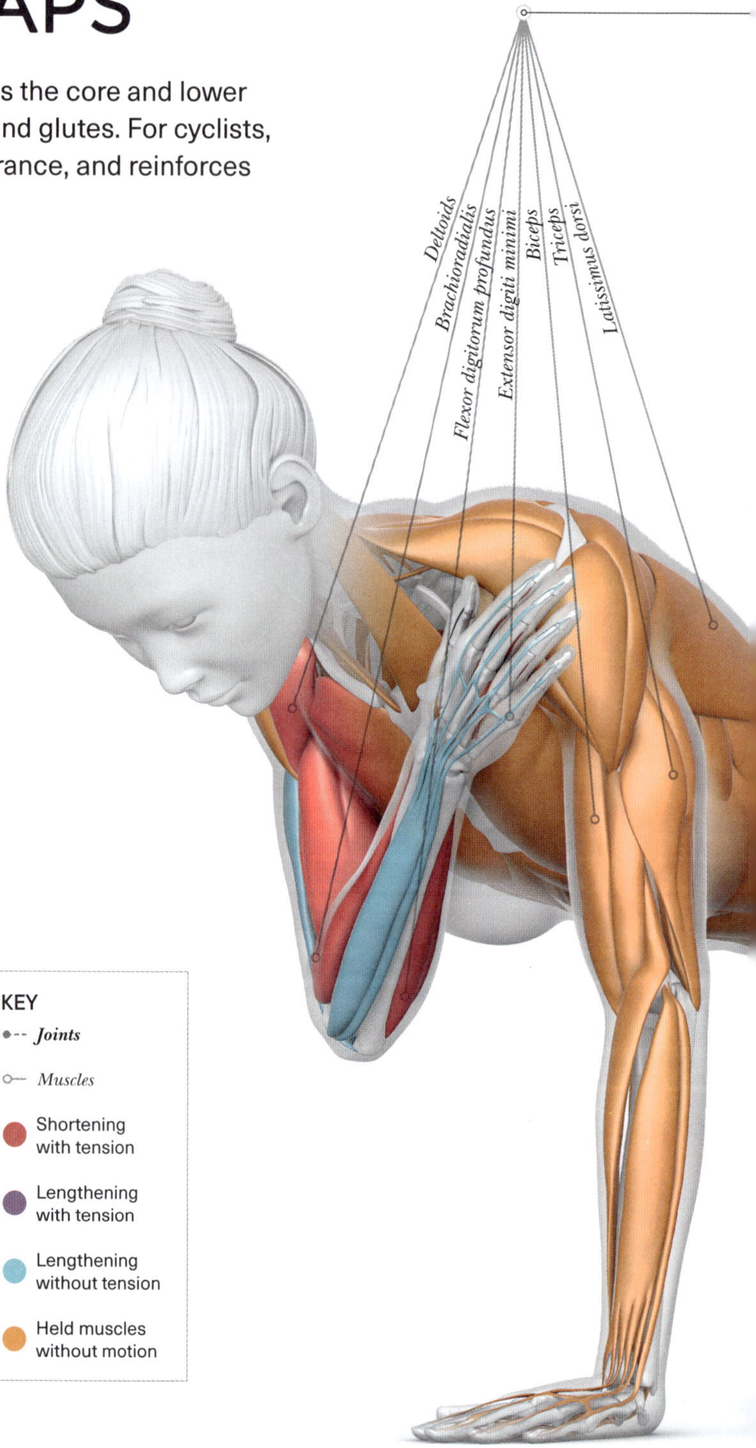

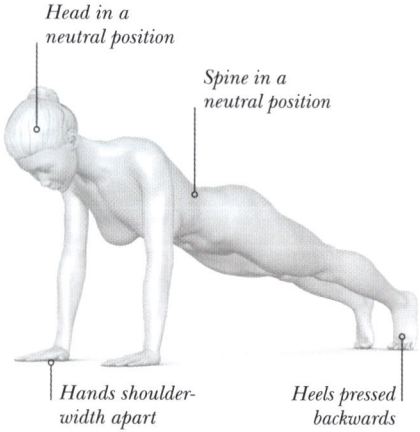

Head in a neutral position

Spine in a neutral position

Hands shoulder-width apart

Heels pressed backwards

PREPARATORY STAGE
Begin in a high plank position with hands under shoulders, feet slightly wider than hip-width, and body in a straight line.

KEY
- Joints
- Muscles
- ● Shortening with tension
- ● Lengthening with tension
- ● Lengthening without tension
- ● Held muscles without motion

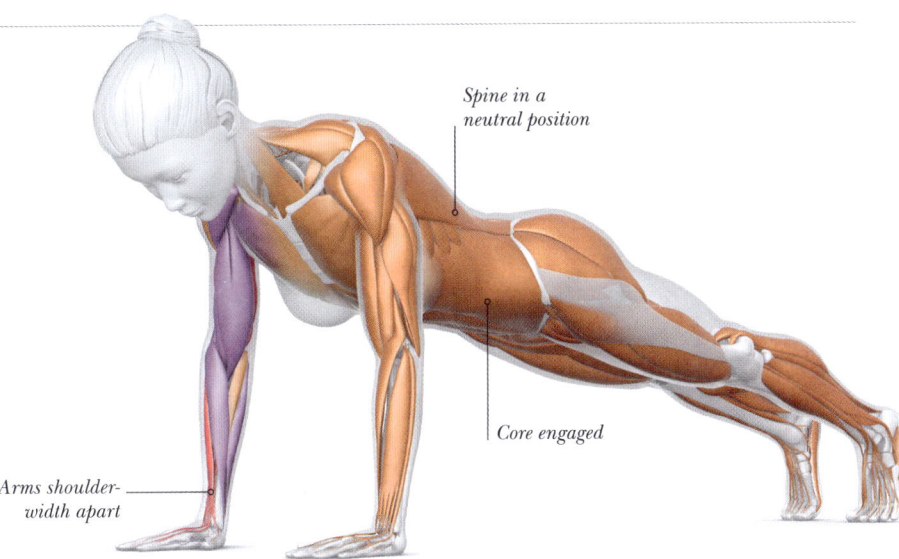

Upper body
The **upper body muscles** work primarily to stabilize and control movement. The **deltoids** and **triceps** support the arm as it lifts and lowers, while the **pectorals** assist in maintaining shoulder stability. The **trapezius** and **rhomboids** help keep the shoulders and upper back engaged, and the **forearms** contribute to wrist and grip stability.

Spine in a neutral position

Core engaged

Arms shoulder-width apart

STAGE TWO
Return the hand to the floor and repeat on the other side, alternating taps while maintaining control and alignment.

 Caution
To avoid the hips rotating, placing the feet too close together, arching the back, and misaligning the shoulders over the wrists, keep your hips stable, core engaged, and spine neutral.

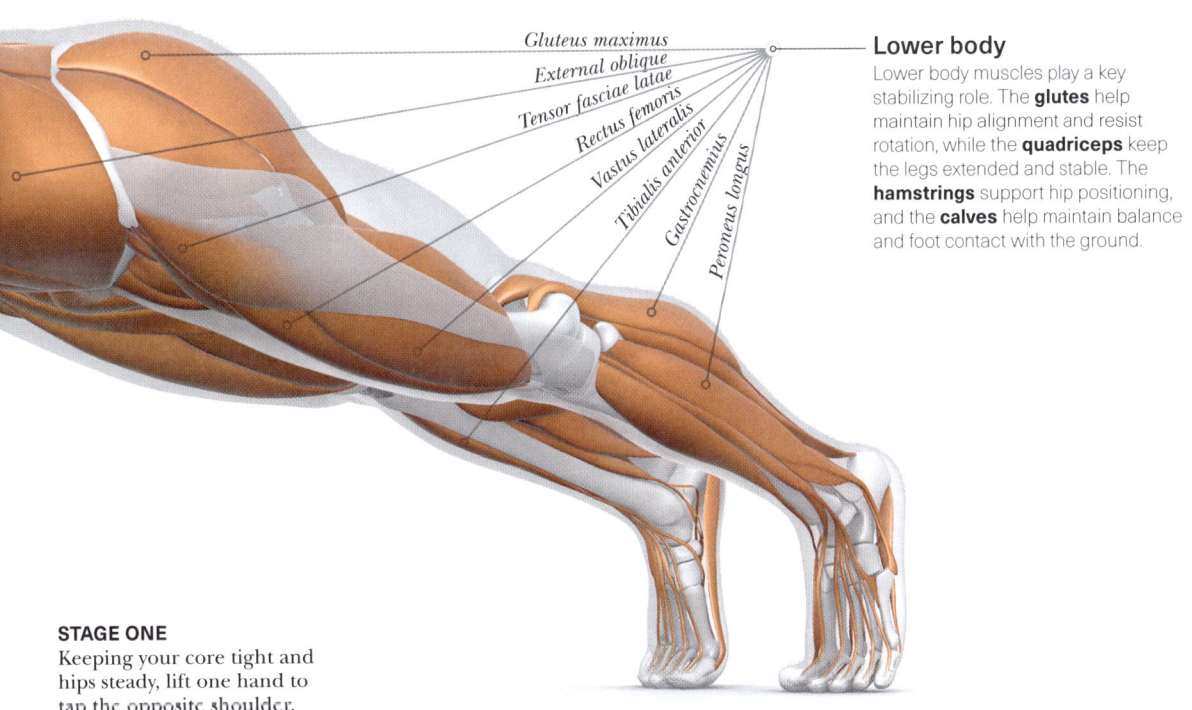

Gluteus maximus
External oblique
Tensor fasciae latae
Rectus femoris
Vastus lateralis
Tibialis anterior
Gastrocnemius
Peroneus longus

Lower body
Lower body muscles play a key stabilizing role. The **glutes** help maintain hip alignment and resist rotation, while the **quadriceps** keep the legs extended and stable. The **hamstrings** support hip positioning, and the **calves** help maintain balance and foot contact with the ground.

STAGE ONE
Keeping your core tight and hips steady, lift one hand to tap the opposite shoulder.

» VARIATIONS

The plank variations include modified versions: low plank hold and low-impact plank, and a more advanced version, the alternating foot switch and bear plank. All of them target the abdominal muscles of the transversus abdominis and the rectus abdominis.

KEY
● Primary target muscle
● Secondary target muscle

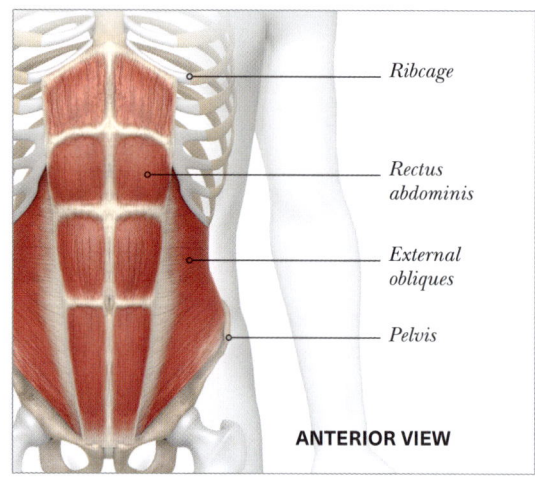

Muscles of the core
Sit-ups involve two main muscle actions: When you sit up, your abs shorten (concentric contraction), pulling your ribs and pelvis together. But when you lower yourself back down, your abs are still working hard, lengthening under control (eccentric contraction) to resist gravity.

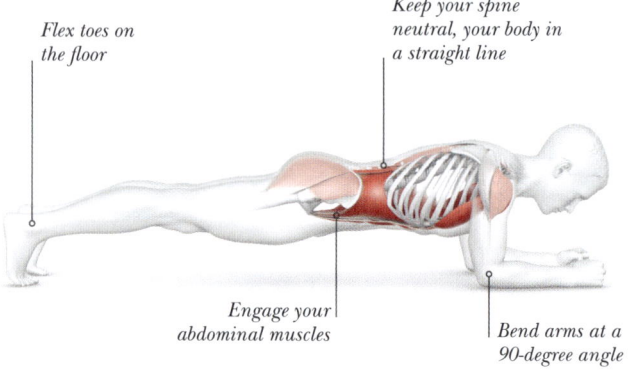

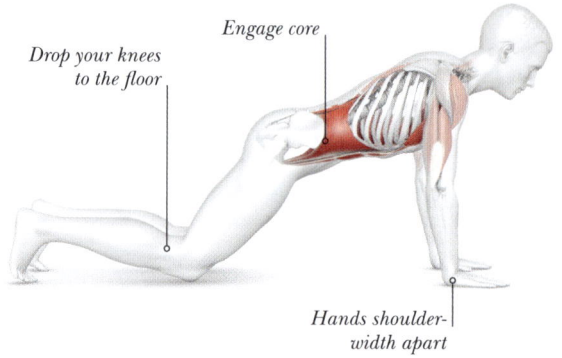

LOW PLANK HOLD

This variation involves holding a low plank. To avoid injuries to your lower back, shoulders, neck, or hips, ensure you fully engage your abs, legs, and shoulders.

PREPARATORY STAGE/STAGE ONE
To assume the low plank position, face the floor and place your forearms and toes on the ground. Extend your legs straight back and ensure your back remains flat. Align your elbows directly beneath your shoulders and point your forearms forwards. Keep your head relaxed, with your gaze directed towards the floor. Finish by tucking your hips under and squeezing your glutes. Hold this position for 30 seconds.

LOW-IMPACT PLANK

The low-impact plank is ideal if you have lower back issues or are new to HIIT workouts. It allows you to achieve the benefits of a core workout without added spinal pressure.

PREPARATORY STAGE
To begin, assume a high plank position. Your hands should be directly under your shoulders, shoulder-width apart, and your feet should be hip-width apart. Maintain a straight line from your head through your neck and spine.

STAGE ONE
With your back flat and hips tucked, slowly drop down to your knees, ensuring your back does not sag. Hold this position for 30 seconds.

STRENGTH AND STRETCH EXERCISES | Plank Variations

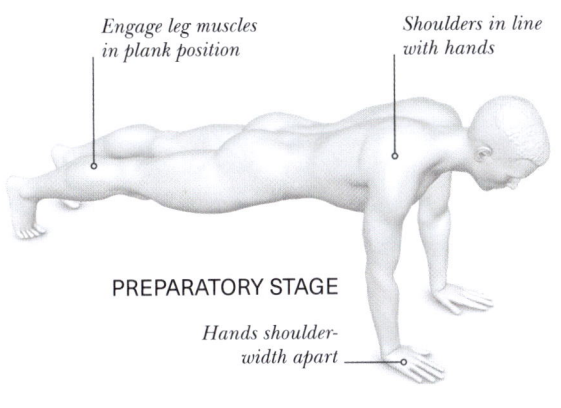

PREPARATORY STAGE
- Engage leg muscles in plank position
- Shoulders in line with hands
- Hands shoulder-width apart

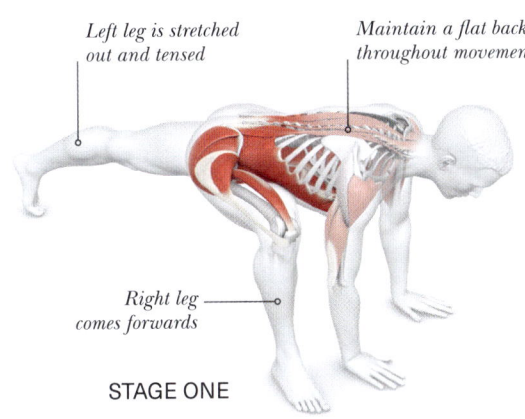

STAGE ONE
- Left leg is stretched out and tensed
- Maintain a flat back throughout movement
- Right leg comes forwards

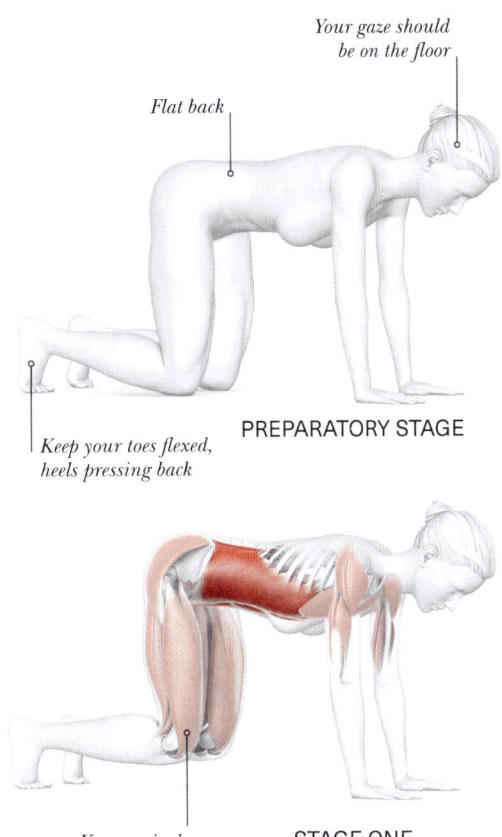

PREPARATORY STAGE
- Flat back
- Your gaze should be on the floor
- Keep your toes flexed, heels pressing back

STAGE ONE
- Knees raised

ALTERNATING FOOT SWITCH

The explosive alternating foot switch demands both fitness and coordination. Maintain a straight, flat back, aligned with your neck and head. Keep your hips tucked and avoid raising your pelvis as you perform the switches.

PREPARATORY STAGE
Get into a high plank. Make sure your wrists are right under your shoulders, your feet are hip-width apart, and your head, neck, and spine form a straight line.

STAGE ONE
Engage your core and explosively swing or jump your right foot to the right side of your body, landing with your knee bent and foot flat on the floor.

STAGE TWO
Return your right foot to the centre, nearly grazing your left foot, and seamlessly transition by swinging your left foot out to the left side. Maintain this rapid, continuous switching motion to build a steady rhythm.

BEAR PLANK

This exercise effectively strengthens your abdominals and core, which helps reduce lower back pain and injury risk. It also improves balance. Beyond the abs, the bear plank targets your gluteus medius and maximus, psoas, quadriceps, shoulders, and arms.

PREPARATORY STAGE
Start on all fours in a quadruped (tabletop) position, ensuring your back is flat. Position your hands shoulder-width apart with wrists directly under your shoulders. Your knees should be hip-width apart, and your feet flexed with toes on the floor.

STAGE ONE
Engage your core to maintain a flat back. As you exhale, push your palms into the floor, lifting your knees 8–15cm (3–6in) off the floor. Keep your toes flexed and resting on the floor, ensuring your hips remain level with your shoulders. Hold for 30–60 seconds, based on your fitness level.

STAGE TWO
Lower your knees back to the floor with control, bringing you back to the preparatory stage of the exercise.

RUSSIAN TWIST WITH MEDICINE BALL

This core-focussed exercise is valuable for cyclists to develop rotational stability and control, which is essential for maintaining posture and efficiency on the bike. The alternating motion enhances oblique engagement, helping resist unwanted torso rotation during different cycling demands.

A strong, stable core is essential for cyclists to maintain an efficient riding position, support sustained power output, and improve bike handling. The Russian twist specifically targets the obliques, which help resist excessive torso movement and maintain control during sprints, climbs, and technical terrain. Strengthening these muscles enhances riding efficiency and overall performance. Perform 3 sets of 10–12 reps each side.

STAGE ONE
Rotate your torso to the right, bringing the medicine ball towards the floor beside you. Rest on your heels and keep your knees pointed upwards as you twist from the waist, engaging your oblique muscles as you do so.

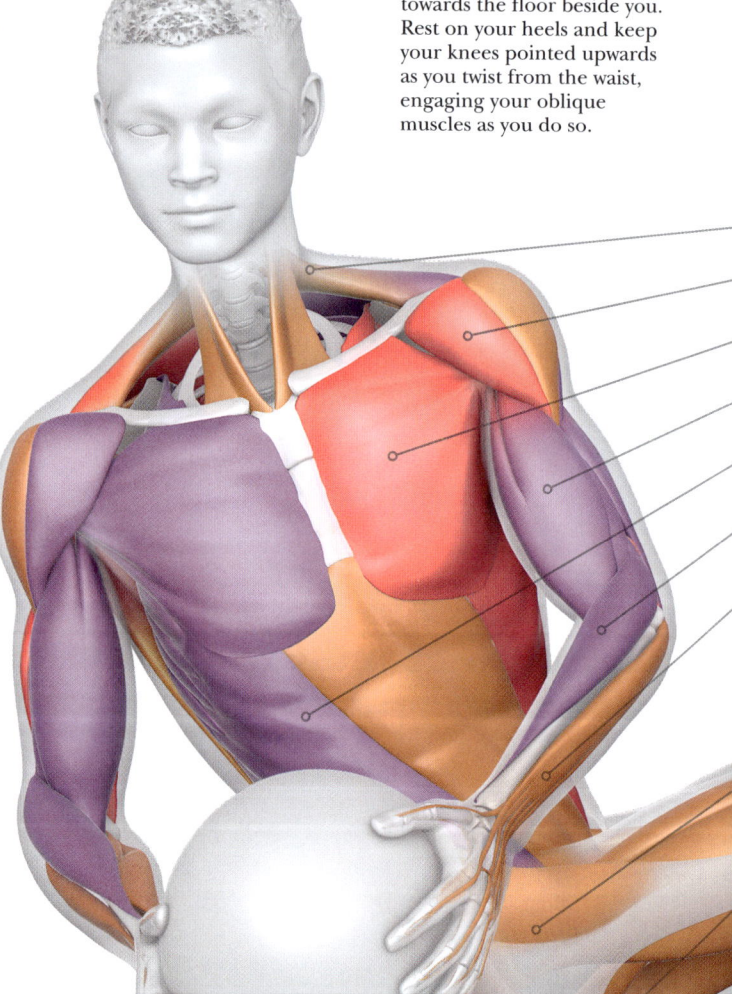

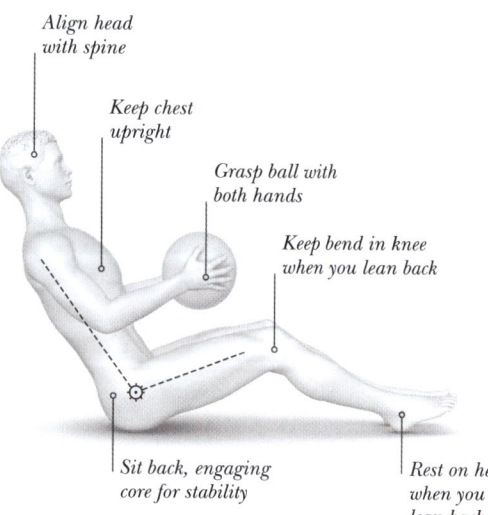

Align head with spine
Keep chest upright
Grasp ball with both hands
Keep bend in knee when you lean back
Sit back, engaging core for stability
Rest on heels when you lean back

PREPARATORY STAGE
Begin seated with knees bent and feet planted on the floor. Hold the medicine ball with both hands in front of your chest, then lean back slightly to engage the core.

STRENGTH AND STRETCH EXERCISES | Russian Twist with Medicine Ball

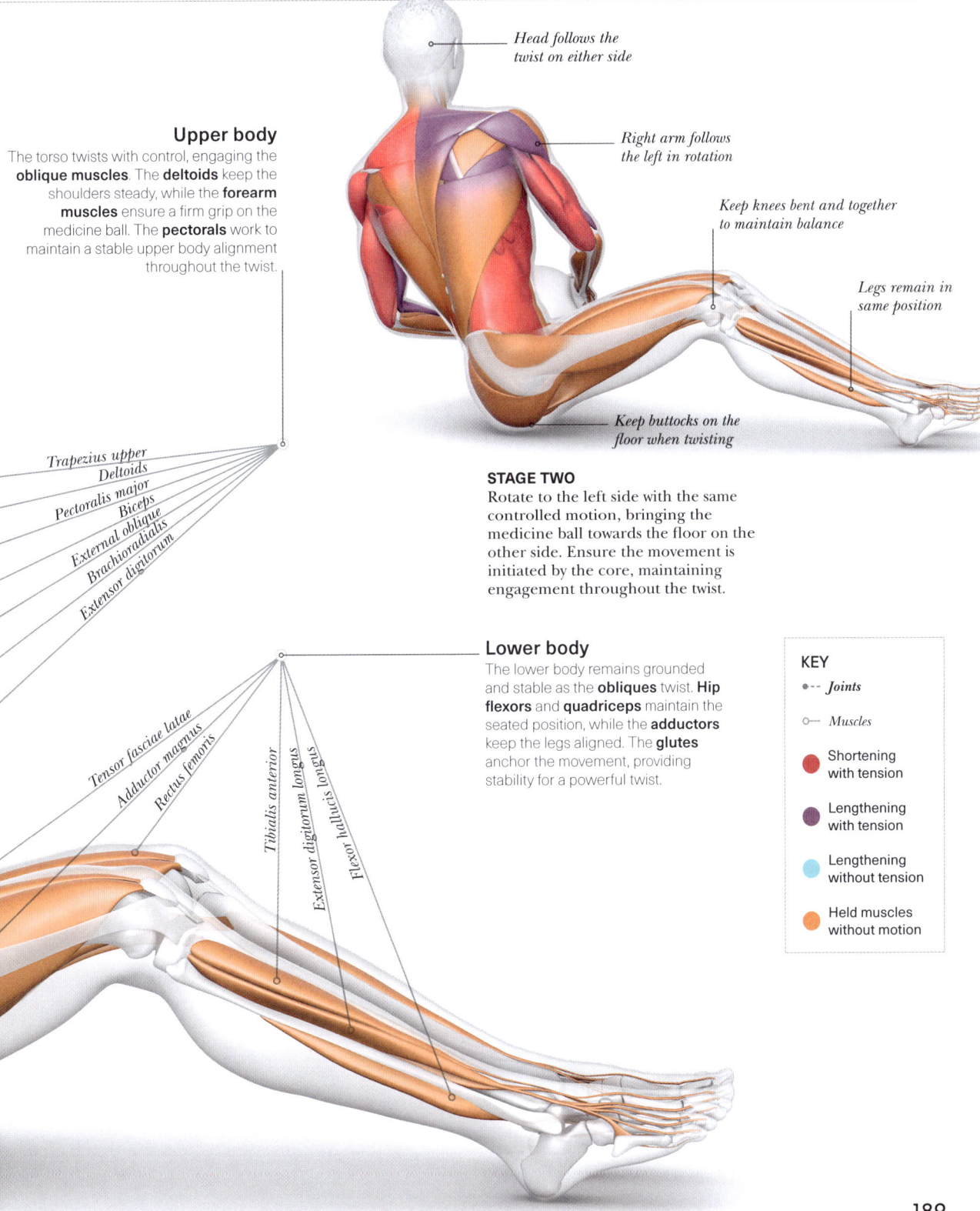

Head follows the twist on either side

Right arm follows the left in rotation

Keep knees bent and together to maintain balance

Legs remain in same position

Keep buttocks on the floor when twisting

Upper body
The torso twists with control, engaging the **oblique muscles**. The **deltoids** keep the shoulders steady, while the **forearm muscles** ensure a firm grip on the medicine ball. The **pectorals** work to maintain a stable upper body alignment throughout the twist.

Trapezius upper
Deltoids
Pectoralis major
Biceps
External oblique
Brachioradialis
Extensor digitorum

STAGE TWO
Rotate to the left side with the same controlled motion, bringing the medicine ball towards the floor on the other side. Ensure the movement is initiated by the core, maintaining engagement throughout the twist.

Lower body
The lower body remains grounded and stable as the **obliques** twist. **Hip flexors** and **quadriceps** maintain the seated position, while the **adductors** keep the legs aligned. The **glutes** anchor the movement, providing stability for a powerful twist.

Tensor fasciae latae
Adductor magnus
Rectus femoris
Tibialis anterior
Extensor digitorum longus
Flexor hallucis longus

KEY
- •-- Joints
- o— Muscles
- ● Shortening with tension
- ● Lengthening with tension
- ● Lengthening without tension
- ● Held muscles without motion

DEADBUG

The deadbug is a core stability exercise performed lying facing upwards and alternately extending opposite arms and legs while keeping your lower back flat. It can help cyclists develop trunk control, pelvic stability, and cross-body coordination essential for efficient pedalling.

This exercise reinforces lumbar and pelvic stability, key for maintaining a strong, efficient posture on the bike, especially under fatigue. It trains the deep core muscles (like the transversus abdominis and multifidus), which support force transfer and protect against common cycling issues like lower back pain. The contralateral limb movement also mirrors the pedalling pattern, enhancing neuromuscular control without adding spinal load.

Used regularly, deadbugs can improve trunk control in aero positions, help manage imbalances, and serve as a safe yet effective way to activate the core pre-ride or in strength sessions. Perform 3 sets of 6–8 reps per side.

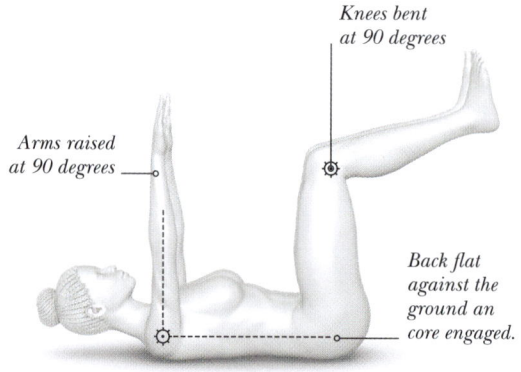

PREPARATORY STAGE
Lie on your back with your arms flexed at the shoulders, legs flexed at the hips and knees, and head lifted off the ground, in a neutral position.

Upper body
Muscles targeted include the **transversus abdominis**, which acts like a corset to stabilize the spine and pelvis, and the **rectus abdominis** and **obliques** for trunk control, while the **multifidus** and **erector spinae** provide spinal support.

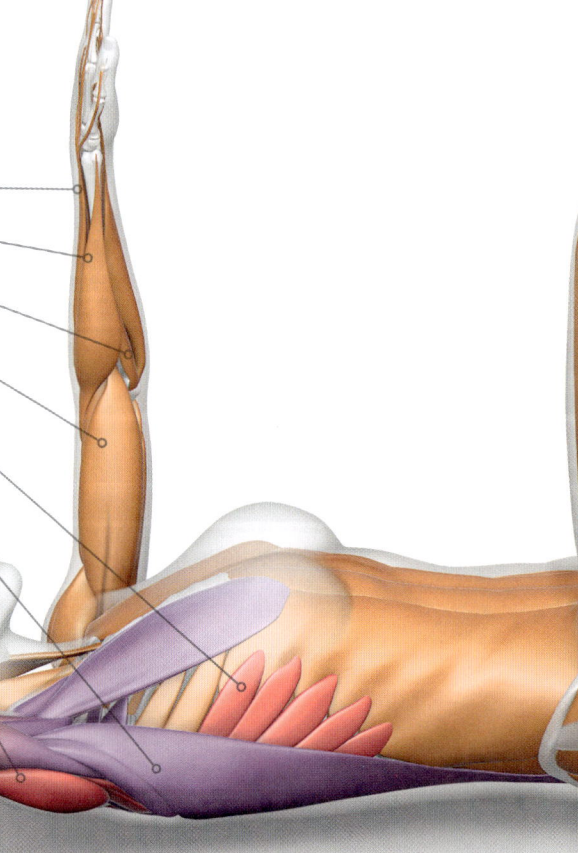

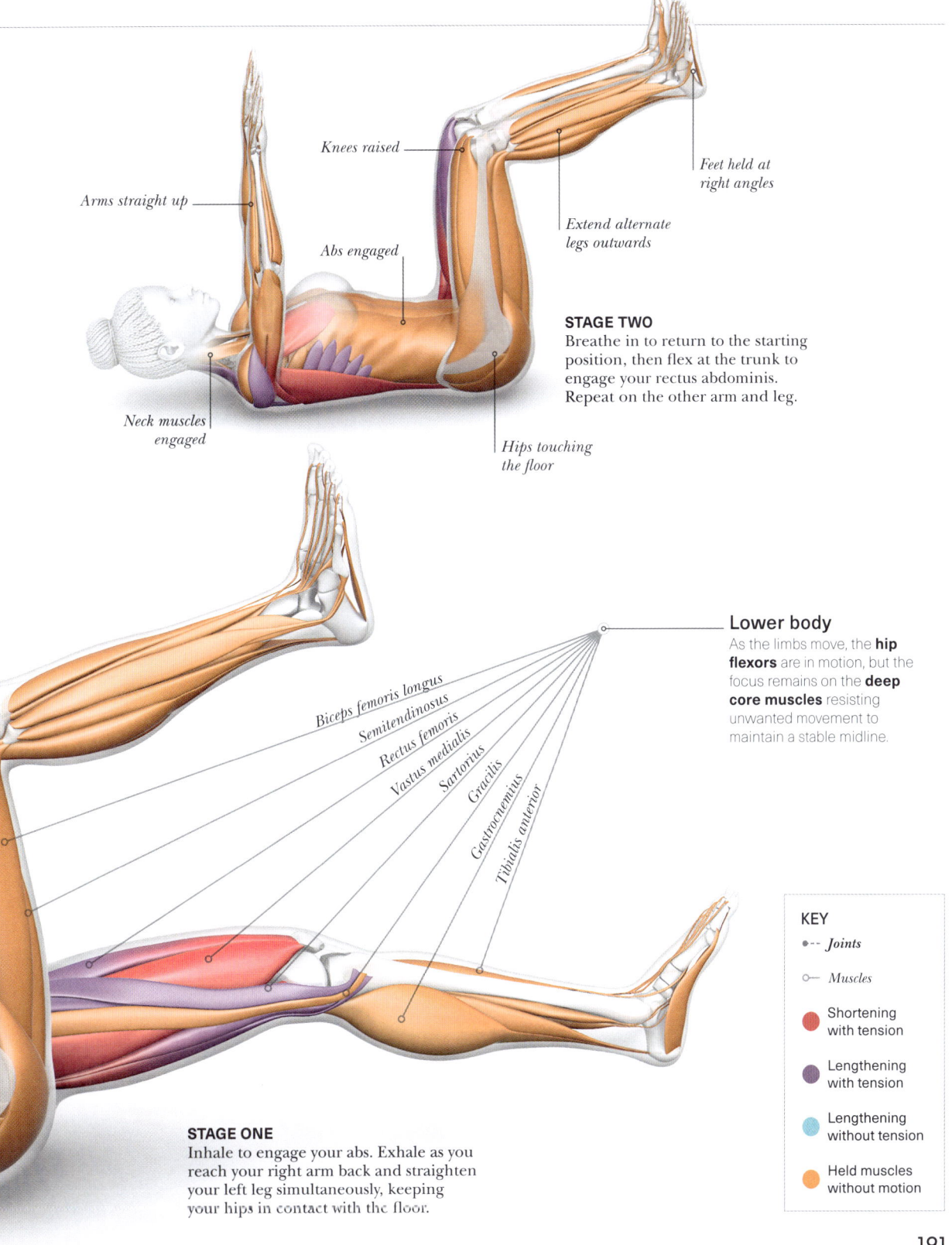

GLUTE BRIDGE

This floor exercise strengthens the glutes and hamstrings, key muscle groups for delivering power during accelerations and climbing efforts. It also enhances core stability, which is essential for maintaining an efficient riding position and maximizing power transfer on the bike.

The glute bridge activates the gluteal muscles while strengthening the rectus abdominis, obliques, quadriceps, and erector spinae – muscles essential for cyclists to maintain a stable pelvis and powerful lower-body engagement during pedalling. To perform it correctly, avoid lifting your hips too high, as this can strain the lower back. Keeping the core engaged helps prevent excessive arching. If your hips begin to sag, lower your pelvis to the floor and reset. Beginners can start with short holds for 1 set of 8–12 reps, gradually increasing hold time and sets as control and strength improve.

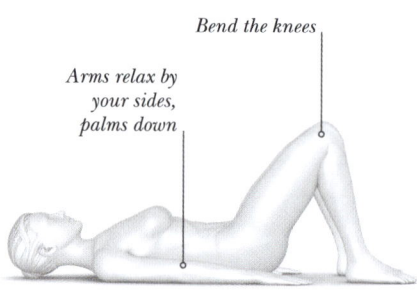

PREPARATORY STAGE
Begin by positioning yourself on your back, arms by your sides with palms facing down, knees bent, and feet planted firmly on the floor. Activate your core muscles by pressing your lower back against the floor and engaging your glutes in preparation for the lift.

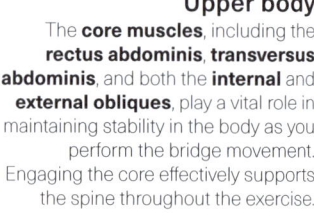

Upper body
The **core muscles**, including the **rectus abdominis**, **transversus abdominis**, and both the **internal** and **external obliques**, play a vital role in maintaining stability in the body as you perform the bridge movement. Engaging the core effectively supports the spine throughout the exercise.

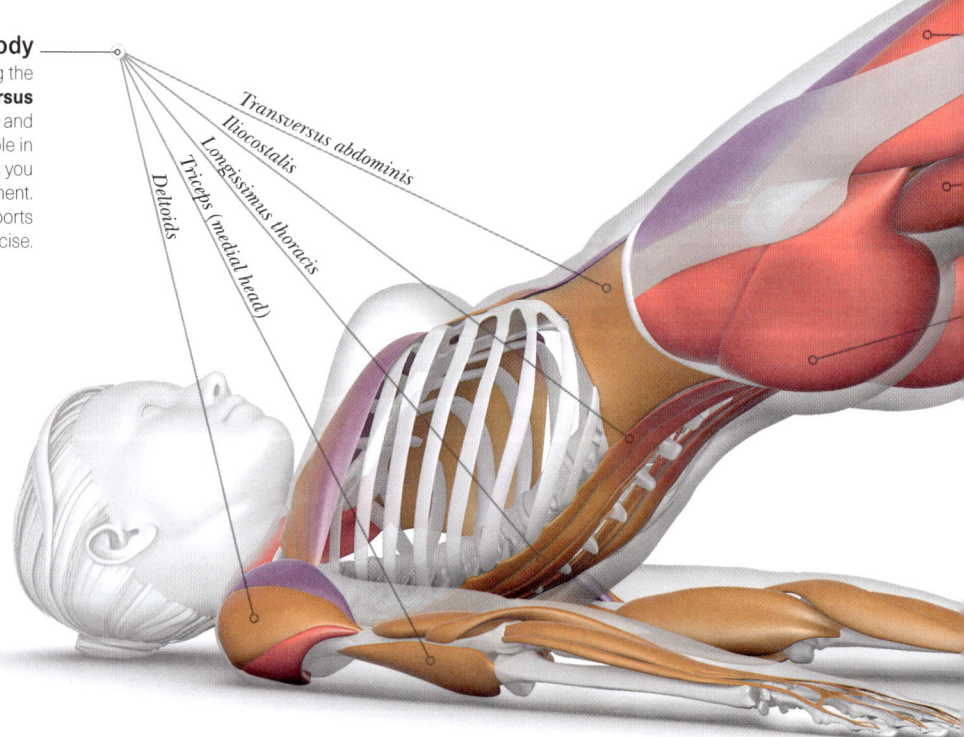

STAGE ONE
Inhale, and as you exhale, press through your heels to gradually elevate your hips, forming a straight line from your knees to your shoulders. Keep your arms flat on the ground as you lift your hips into the bridge position, ensuring your core remains tight.

STRENGTH AND STRETCH EXERCISES | Glute Bridge

KEY
- •-- *Joints*
- ○— *Muscles*
- 🔴 Shortening with tension
- 🟣 Lengthening with tension
- 🔵 Lengthening without tension
- 🟠 Held muscles without motion

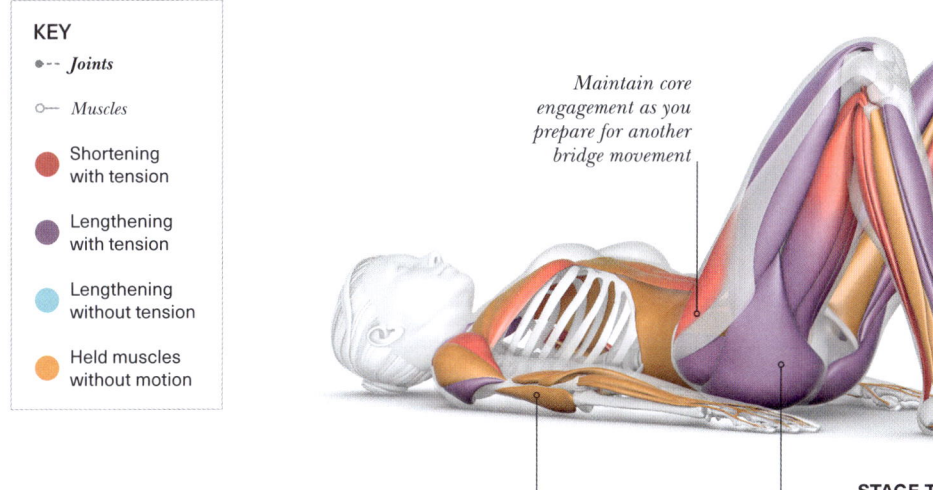

Maintain core engagement as you prepare for another bridge movement

Plant feet on floor, with toes pointed forwards

Keep arms relaxed by your sides throughout

Squeeze your glutes as you descend from bridge

STAGE TWO
Hold this raised position for 20–30 seconds, continuously engaging your gluteal muscles, then carefully lower back to your starting posture, managing the descent. Avoid collapsing back to the floor abruptly. Then, prepare for the next rep.

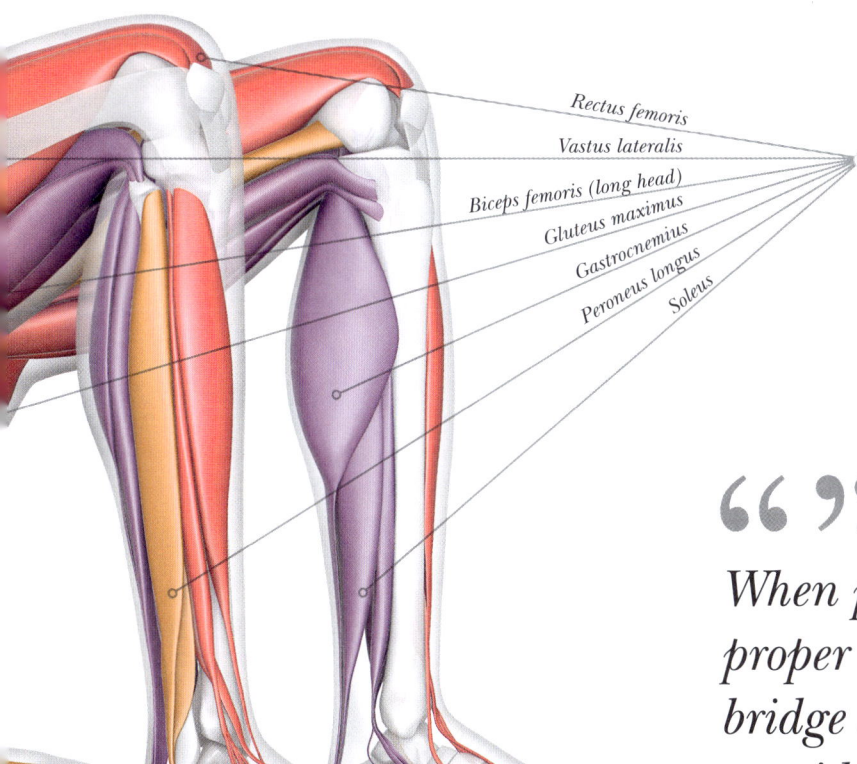

Rectus femoris
Vastus lateralis
Biceps femoris (long head)
Gluteus maximus
Gastrocnemius
Peroneus longus
Soleus

Lower body
This exercise specifically targets the **muscles in the posterior chain**, such as the **gluteus maximus**, **medius**, and **minimus**. Additionally, it engages the **hamstrings** and **hip abductors**. The **quadriceps** contribute to stabilizing the lower body during the exercise, and the **calf muscles** are also involved.

> 66 99
>
> *When performed with proper technique, glute bridge exercises are considered safe for those with persistent back issues.*

HANGING KNEE RAISE

This exercise develops control and coordination between the hips and core, enhancing body awareness and movement efficiency, both crucial for cyclists. It specifically targets the hip flexors and rectus abdominis, with your own body weight providing sufficient resistance to build strength and stability.

The hanging knee raise may appear simple, but it requires practice and control to master. While hanging from a pull-up bar, you engage the hip flexors and abdominal muscles to lift your knees and flex your spine, aiming for maximum height with each rep. Bracing your core beforehand helps stabilize the spine and maintain proper form. Using arm straps can reduce grip fatigue, allowing you to focus more on core activation. Beginners can start with 4 sets of 8–10 controlled repetitions.

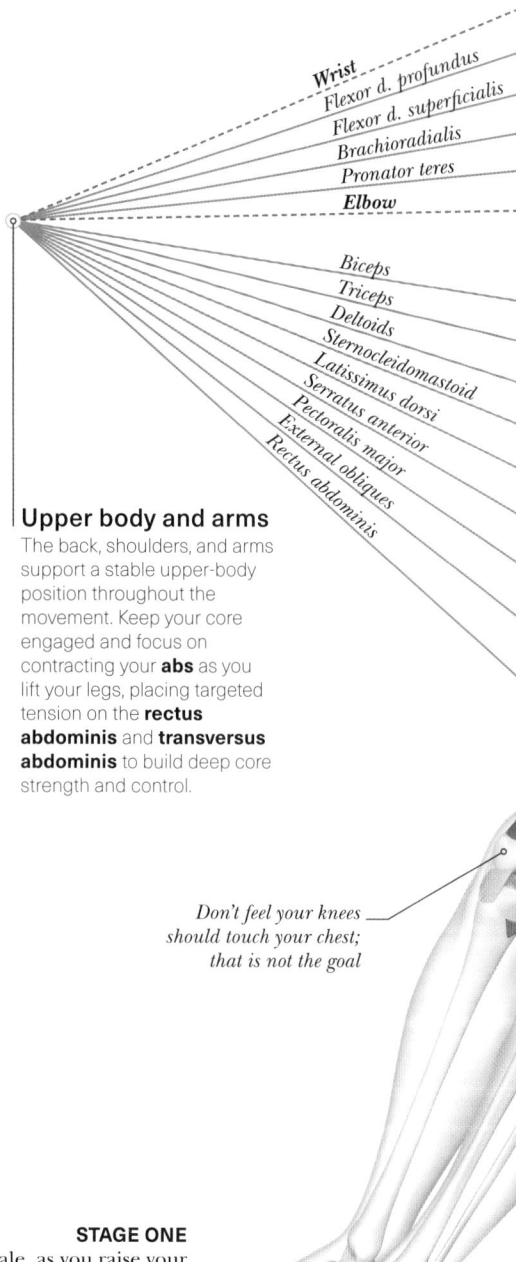

Upper body and arms
The back, shoulders, and arms support a stable upper-body position throughout the movement. Keep your core engaged and focus on contracting your **abs** as you lift your legs, placing targeted tension on the **rectus abdominis** and **transversus abdominis** to build deep core strength and control.

Don't feel your knees should touch your chest; that is not the goal

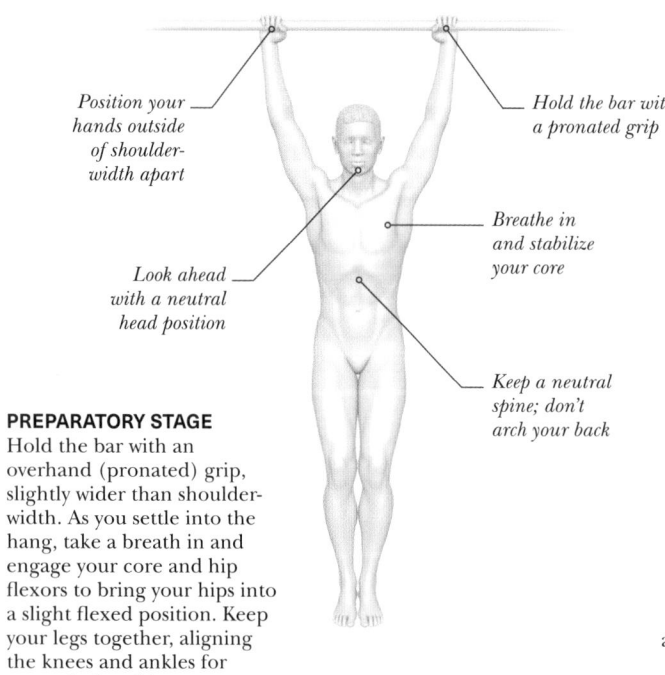

Position your hands outside of shoulder-width apart

Hold the bar with a pronated grip

Breathe in and stabilize your core

Look ahead with a neutral head position

Keep a neutral spine; don't arch your back

PREPARATORY STAGE
Hold the bar with an overhand (pronated) grip, slightly wider than shoulder-width. As you settle into the hang, take a breath in and engage your core and hip flexors to bring your hips into a slight flexed position. Keep your legs together, aligning the knees and ankles for controlled, stable movement.

STAGE ONE
Exhale as you raise your knees, allowing your abs to contract and shorten while your hips flex. Focus on tucking your pelvis under and contracting abdominals at the top to bring your knees in towards your chest.

STRENGTH AND STRETCH EXERCISES | Hanging Knee Raise

⚠️ Common mistakes
Maintaining a controlled speed of movement combined with breathing is key to avoiding unwanted swinging. Focus on flexing both your spine and hips so the pelvis tilts forwards and the knees can lift fully. Avoid using your arms to pull yourself; let the core do the work.

KEY
- •-- *Joints*
- ○— *Muscles*
- 🔴 Shortening with tension
- 🟣 Lengthening with tension
- 🔵 Lengthening without tension
- 🟠 Held muscles without motion

Knee, Rectus femoris, Vastus lateralis, Biceps femoris, Gluteus max., Tensor f. latae

Legs
Engage the muscles around the hips to stabilize your lower body and prevent swaying or swinging. The **hip flexors** coordinate with the core to integrate hip and spinal flexion as you lift your legs.

Engage your core throughout

Keep your knees together and slightly flexed

Position your ankles under your knees

Maintain your arm position

Activate the hip flexors to slightly flex your hips

STAGE TWO
From the top position, lower your knees back down with control, inhaling as you return to the starting position. Keep your core engaged to maintain stability and prevent swinging. Once back at the starting point, reset your breathing and repeat the movement with the same focus and control.

195

CAT COW

The cat pose mimics the arched back of a cat, while the cow pose reflects the gentle dip in a cow's spine. Performed as a flowing sequence, this is an effective stretch for improving spinal mobility and control. This helps relieve stiffness from prolonged time in a flexed riding position and supports better movement on the bike.

The spinal flexion and extension in the cat cow sequence mobilize the entire spine while gently activating the abdominal and chest muscles. It's an ideal exercise for reducing joint stiffness and can be used as part of a warm-up or daily mobility routine to maintain spinal health and fluid movement.

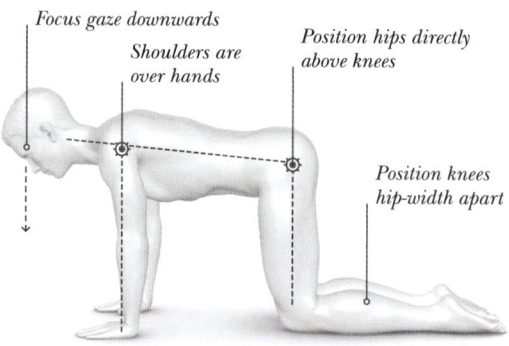

PREPARATORY STAGE
Start on your hands and knees with your shoulders over your wrists, hips aligned above your knees, and your head and neck in line with your spine. Maintain a neutral spine between full flexion and full extension.

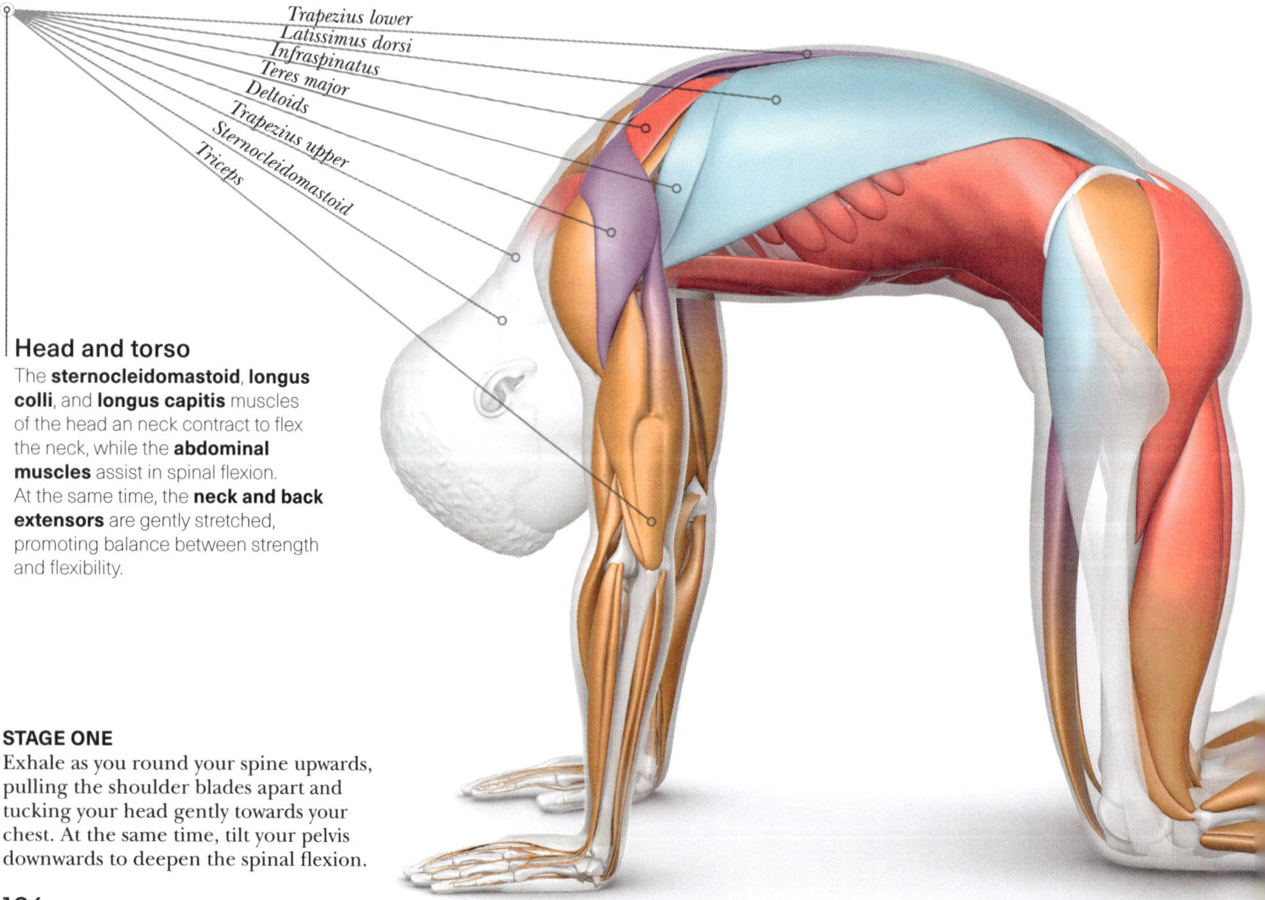

Head and torso
The **sternocleidomastoid**, **longus colli**, and **longus capitis** muscles of the head an neck contract to flex the neck, while the **abdominal muscles** assist in spinal flexion. At the same time, the **neck and back extensors** are gently stretched, promoting balance between strength and flexibility.

STAGE ONE
Exhale as you round your spine upwards, pulling the shoulder blades apart and tucking your head gently towards your chest. At the same time, tilt your pelvis downwards to deepen the spinal flexion.

STRENGTH AND STRETCH EXERCISES | Cat Cow

KEY
- Joints
- Muscles
- Shortening with tension
- Lengthening with tension
- Lengthening without tension
- Held muscles without motion

VARIATION: SEATED

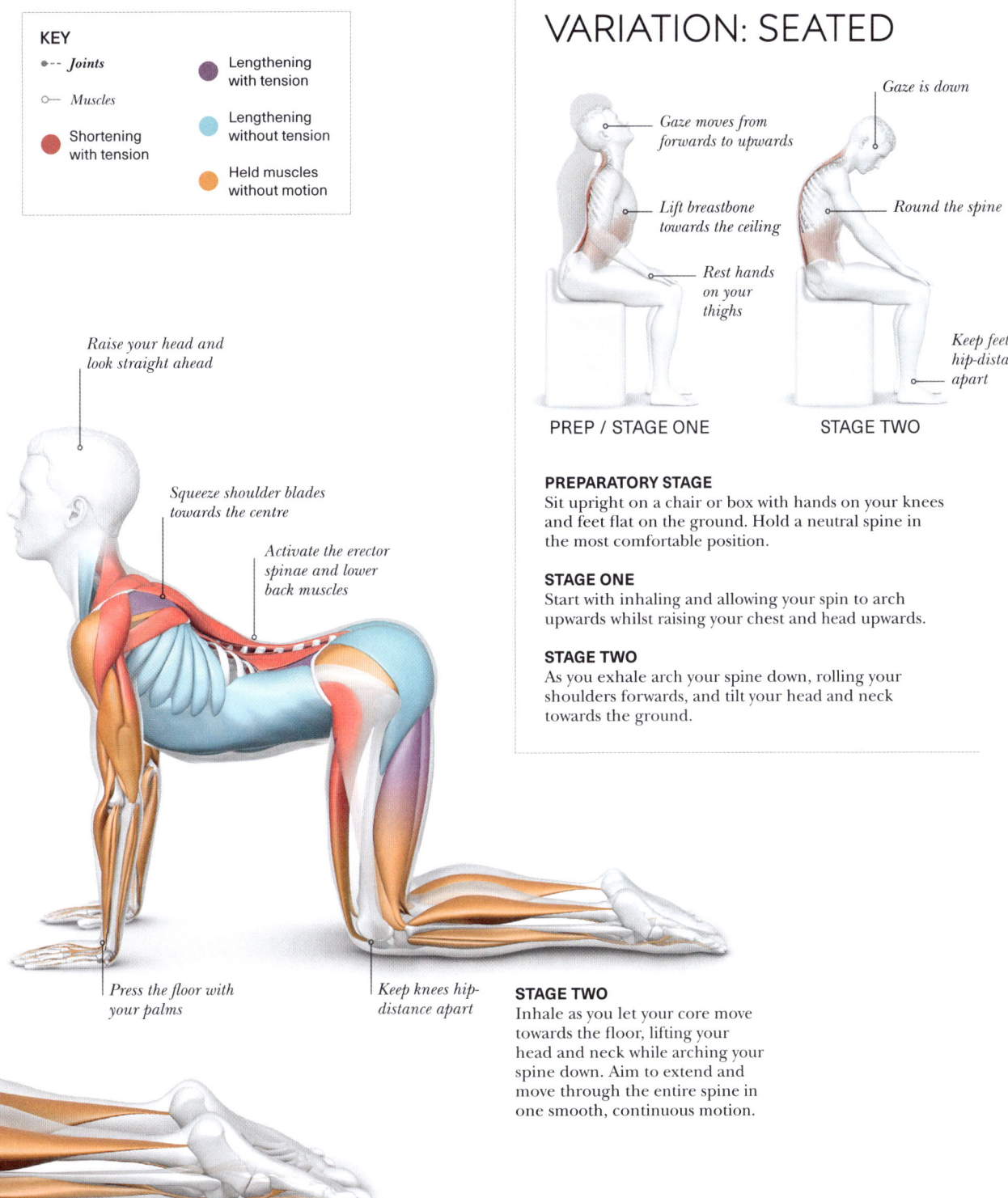

PREP / STAGE ONE — Gaze moves from forwards to upwards; Lift breastbone towards the ceiling; Rest hands on your thighs

STAGE TWO — Gaze is down; Round the spine; Keep feet hip-distance apart

PREPARATORY STAGE
Sit upright on a chair or box with hands on your knees and feet flat on the ground. Hold a neutral spine in the most comfortable position.

STAGE ONE
Start with inhaling and allowing your spin to arch upwards whilst raising your chest and head upwards.

STAGE TWO
As you exhale arch your spine down, rolling your shoulders forwards, and tilt your head and neck towards the ground.

Main pose labels: Raise your head and look straight ahead; Squeeze shoulder blades towards the centre; Activate the erector spinae and lower back muscles; Press the floor with your palms; Keep knees hip-distance apart.

STAGE TWO
Inhale as you let your core move towards the floor, lifting your head and neck while arching your spine down. Aim to extend and move through the entire spine in one smooth, continuous motion.

197

CHILD'S POSE

This restorative stretch helps release tension in the back and pelvis while gently stretching the arms and ankles. For cyclists, the mild spinal flexion is especially beneficial in easing lower back tightness and relieving joint stiffness caused by long hours in a forward-flexed riding position.

Incorporate a lateral variation (see right) to stretch the sides of the back and target the latissimus dorsi. Stay within a comfortable range, avoiding any forceful movement, and use your breath to help ease into the stretch.

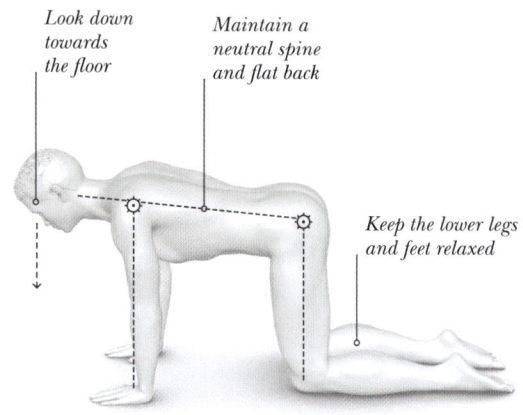

Look down towards the floor
Maintain a neutral spine and flat back
Keep the lower legs and feet relaxed

PREPARATORY STAGE
Start on your hands and knees, facing downwards with your shoulders over your wrists, hips over your knees, and your head and neck in line with your spine.

Neck and arms
This stretch targets the **splenius capitis** and **splenius cervicis muscles** in the neck, along with the **posterior deltoids**. As your arms reach overhead, they act as anchors, helping to lengthen the shoulders and upper back for a deeper, more effective stretch.

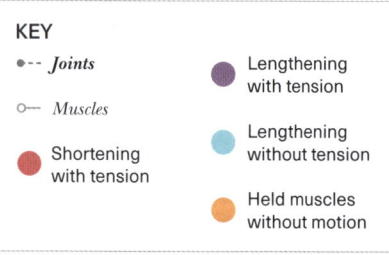

KEY
- ●--- Joints
- ○— Muscles
- ● Shortening with tension
- ● Lengthening with tension
- ● Lengthening without tension
- ● Held muscles without motion

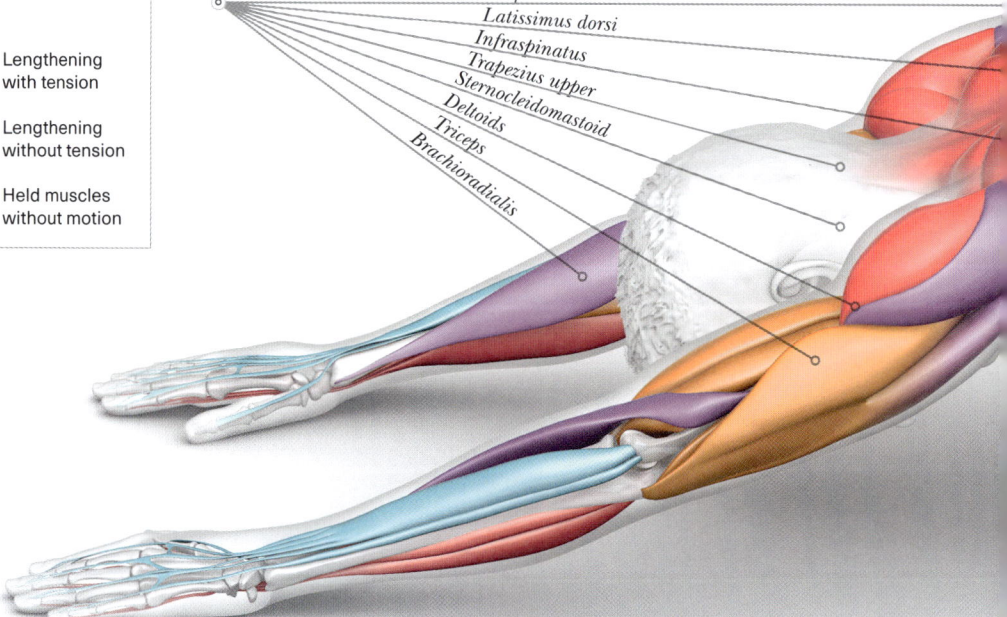

Trapezius lower
Latissimus dorsi
Infraspinatus
Trapezius upper
Sternocleidomastoid
Deltoids
Triceps
Brachioradialis

STAGE ONE
Slowly sit back onto your heels, reaching your arms forwards with your palms on the floor. Allow your torso to fold between your thighs and relax your back and shoulders as you exhale, sinking deeper into the stretch.

STRENGTH AND STRETCH EXERCISES | Child's Pose

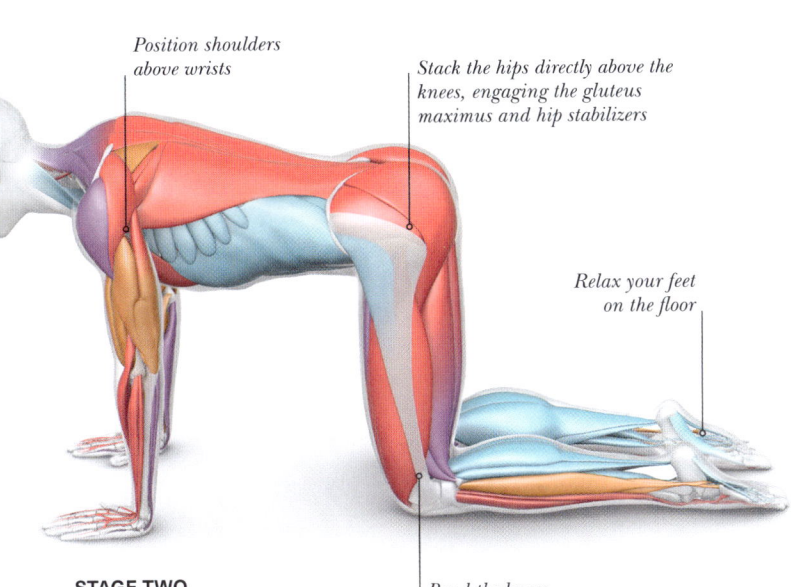

Position shoulders above wrists

Stack the hips directly above the knees, engaging the gluteus maximus and hip stabilizers

Relax your feet on the floor

Bend the knees to 90 degrees

STAGE TWO
Return to the preparatory position, ensuring your shoulders are over your wrists, your hips directly above your knees, and your head and neck are aligned with your spine.

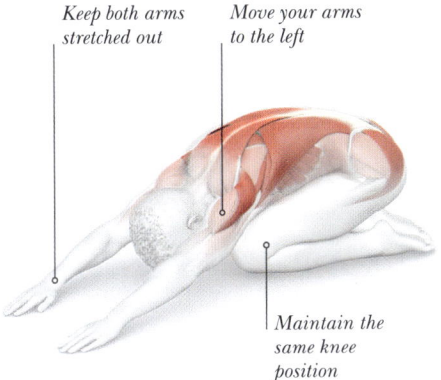

VARIATION: LATERAL POSE

Keep both arms stretched out

Move your arms to the left

Maintain the same knee position

ALTERNATIVE STAGE TWO
Walk your hands to one side while keeping them extended in front of you to create a stretch along the side of your torso. Hold briefly, then repeat on the opposite side.

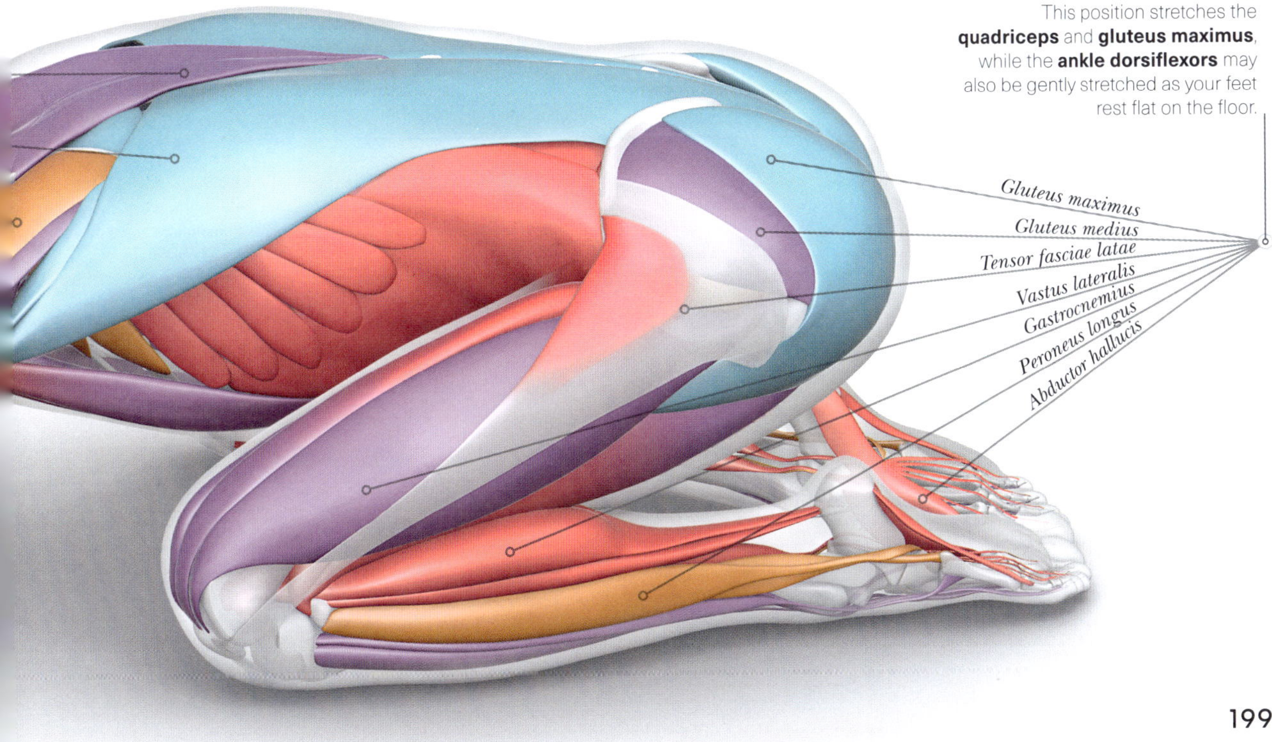

Lower body
This position stretches the **quadriceps** and **gluteus maximus**, while the **ankle dorsiflexors** may also be gently stretched as your feet rest flat on the floor.

Gluteus maximus
Gluteus medius
Tensor fasciae latae
Vastus lateralis
Gastrocnemius
Peroneus longus
Abductor hallucis

199

THREAD THE NEEDLE

This mobility exercise helps reduce stiffness and enhance rotation in the thoracic spine, an area often restricted in cyclists due to prolonged time in a flexed position.

This dynamic mobility exercise promotes thoracic spine rotation by anchoring one arm while the other threads beneath the body. It's effective as a warm-up, daily mobility drill, or integrated into an upper-body movement routine. It is highly beneficial for cyclists to counteract stiffness from prolonged static postures and improve spinal and shoulder mobility.

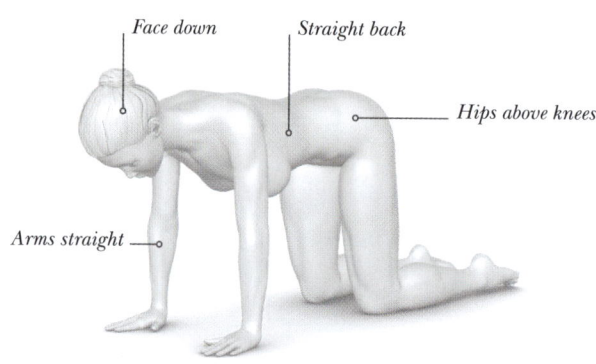

PREPARATORY STAGE
Start on your hands and knees facing downwards with your shoulders aligned over your wrists, hips directly above your knees, and your head and neck in line with your spine.

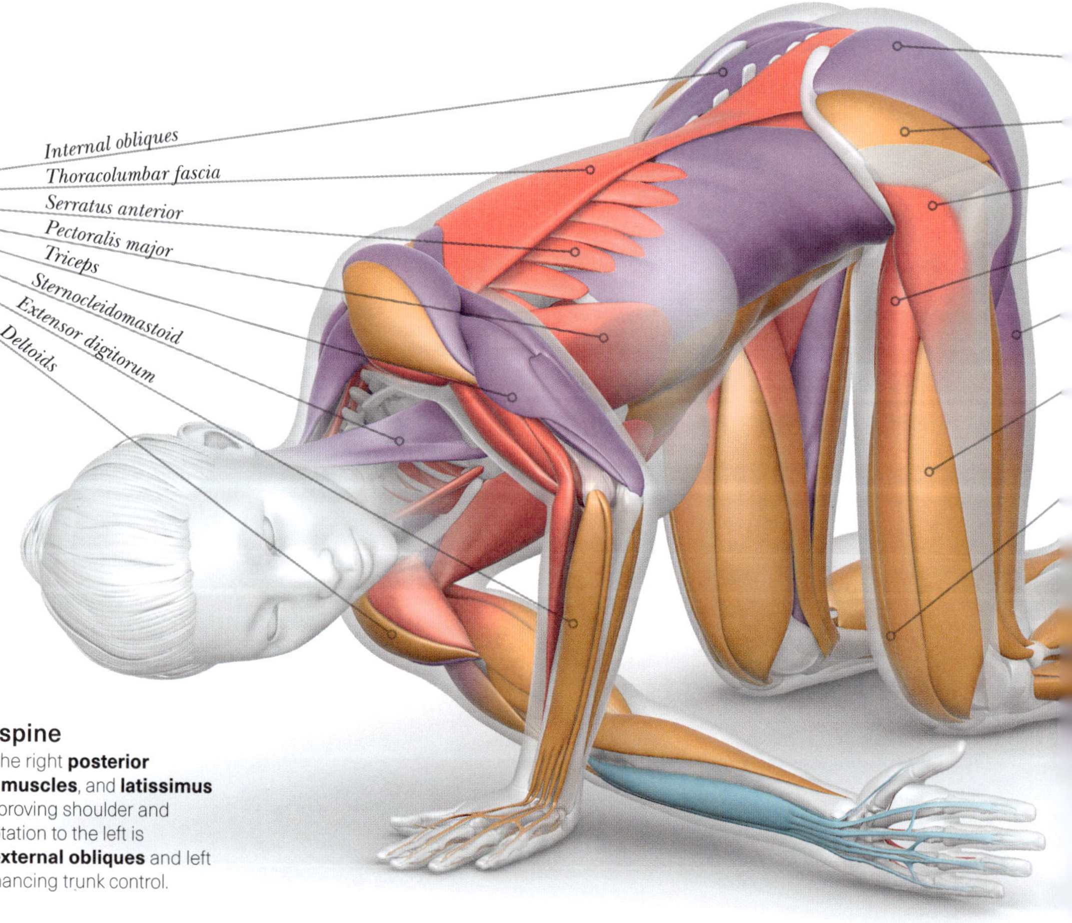

Upper body and spine
As you reach through, the right **posterior deltoid**, **periscapular muscles**, and **latissimus dorsi** are stretched, improving shoulder and upper back mobility. Rotation to the left is facilitated by the right **external obliques** and left **internal obliques**, enhancing trunk control.

STRENGTH AND STRETCH EXERCISES | Thread the Needle

⚠ Caution
To reduce wrist discomfort, you can place a rolled towel under the base of your palms. If there's any knee discomfort, use a soft pad or folded towel beneath the knees to provide cushioning and support.

Lower body
As the trunk lowers and rotates, the **gluteals** and **hip abductors** engage to stabilize the pelvis and control the movement. At the same time, the **toe extensors** and **ankle dorsiflexors** are gently stretched against the floor, supporting lower limb mobility.

Gluteus maximus
Gluteus medius
Tensor fasciae latae
Rectus femoris
Biceps femoris (long head)
Vastus lateralis
Vastus medialis

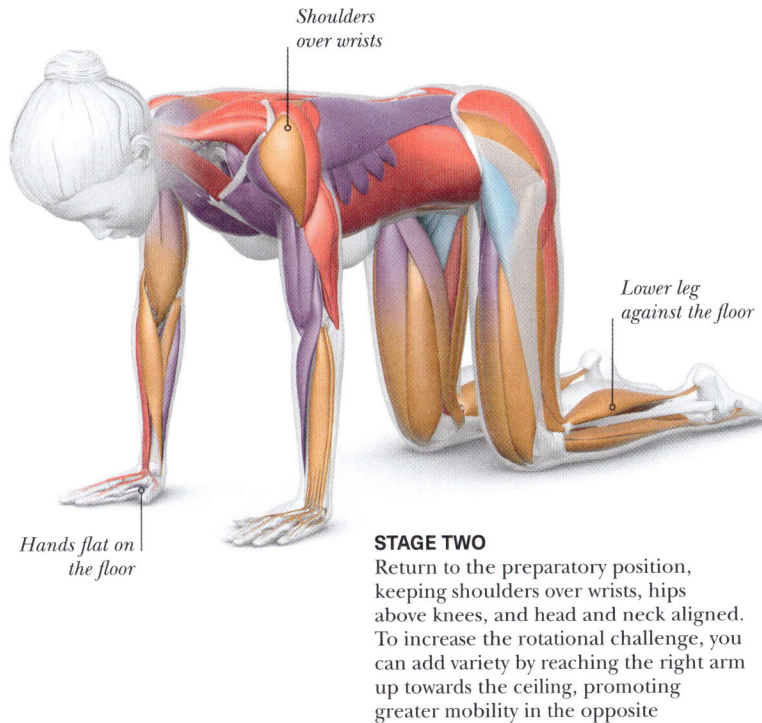

Shoulders over wrists

Lower leg against the floor

Hands flat on the floor

STAGE TWO
Return to the preparatory position, keeping shoulders over wrists, hips above knees, and head and neck aligned. To increase the rotational challenge, you can add variety by reaching the right arm up towards the ceiling, promoting greater mobility in the opposite direction. Repeat on the other side.

> 66 99
>
> *The rotational movement promotes thoracic spine mobility, a key area to target for maintaining shoulder mechanics and neck function.*

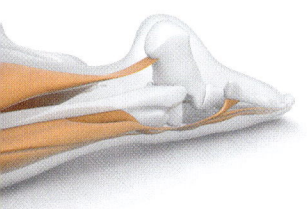

STAGE ONE
With your left hand anchored on the floor, reach your right arm through the space between your left arm and thighs, rotating through the thoracic spine to encourage controlled upper-back mobility.

KEY
- •-- Joints
- ○— Muscles
- ● Shortening with tension
- ● Lengthening with tension
- ● Lengthening without tension
- ● Held muscles without motion

STRENGTH AND STRETCH EXERCISES | Figure 4 Stretch

FIGURE 4 STRETCH

Performed lying on the floor, this stretch effectively targets the gluteal complex and hip rotators to improve external rotation. For cyclists, it helps counteract hip tightness from prolonged time in a flexed position.

This is an effective exercise for enhancing hip mobility and reducing tension in the gluteal muscles, which can contribute to referred discomfort such as sciatic pain. Performed on the ground, it provides support for the lower back, making it a suitable option for those with spinal sensitivity. The seated variation offers a convenient alternative that can be easily incorporated into daily routines. This stretch is commonly included in rehabilitation and mobility programmes aimed at improving hip function and overall lower-limb movement quality.

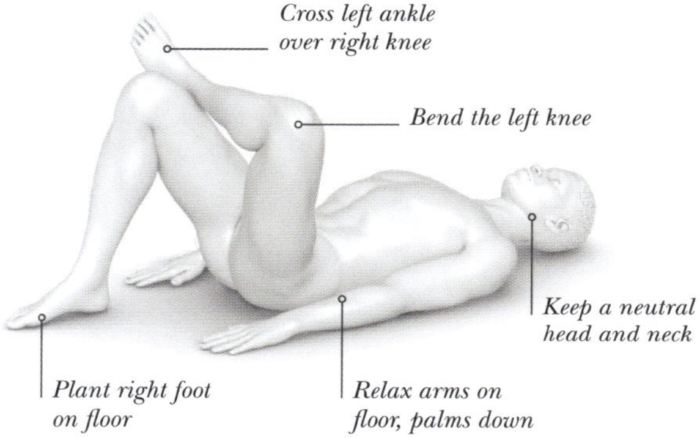

Cross left ankle over right knee

Bend the left knee

Keep a neutral head and neck

Relax arms on floor, palms down

Plant right foot on floor

PREPARATORY STAGE
Start lying on the ground with one leg crossed so that the ankle is resting on the knee of the other leg.

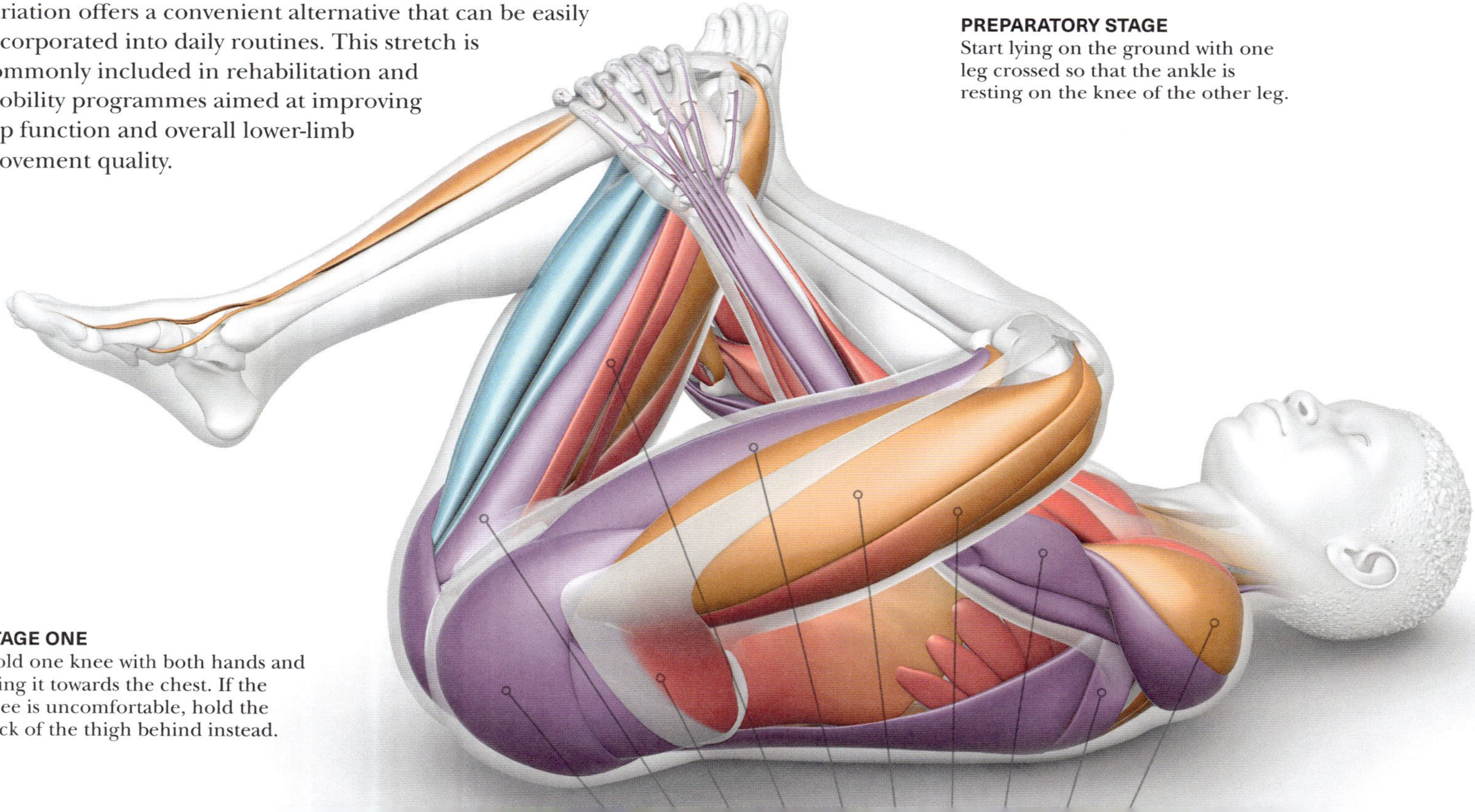

STAGE ONE
Hold one knee with both hands and bring it towards the chest. If the knee is uncomfortable, hold the back of the thigh behind instead.

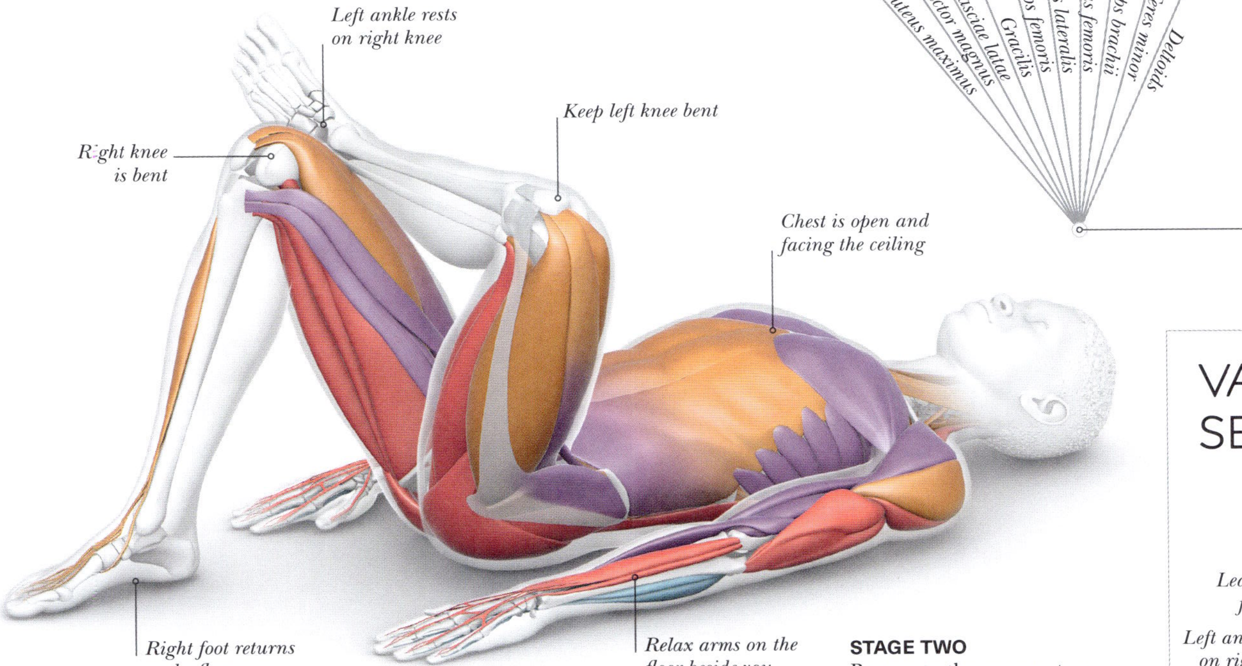

Left hip

When the hip is positioned in external rotation and flexion, the muscles of the posterior hip, such as the **glutes** and **hip rotators**, undergo lengthening. Conversely, the **hip flexors** and **adductors** in this position are shortened and remain untensioned.

- Deltoids
- Teres minor
- Triceps brachii
- Rectus femoris
- Vastus lateralis
- Biceps femoris
- Gracilis
- Tensor fasciae latae
- Adductor magnus
- Gluteus maximus

VARIATION: SEATED

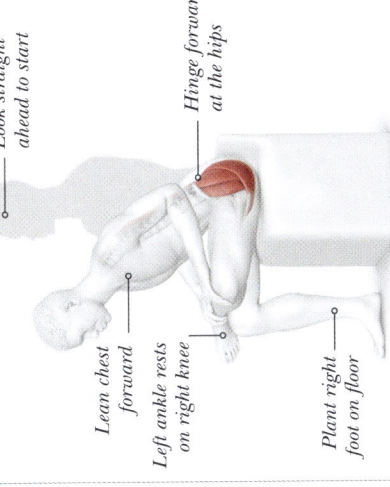

- Look straight ahead to start
- Hinge forward at the hips
- Lean chest forward
- Left ankle rests on right knee
- Plant right foot on floor

PREPARATORY STAGE
Start seated with one ankle crossed over the opposite leg, with your hands resting on top.

STAGE ONE
Sit straight backed and flex at the hip to achieve the stretch.

STAGE TWO
Return to the preparatory position.

- Chest is open and facing the ceiling
- Keep left knee bent
- Left ankle rests on right knee
- Right knee is bent
- Relax arms on the floor beside you
- Right foot returns to the floor

STAGE TWO
Return to the preparatory position, arms and right foot back on the floor.

> " "
> *A simple yet effective stretch to relieve glute tension, support hip mobility, and ease lower back or sciatic discomfort.*

KEY
- ●-- Joints
- ○— Muscles
- 🔴 Shortening with tension
- 🟣 Lengthening with tension
- 🔵 Lengthening without tension
- 🟠 Held muscles without motion

HALF-KNEEL HIP FLEXOR STRETCH

This exercise effectively targets the anterior hip, stretching the quadriceps and hip flexors often shortened in cyclists due to prolonged time in a flexed position. The half-kneeling posture, with both legs bent, creates shorter levers, making the stretch more accessible while still providing a deep, targeted release.

Through gluteal activation and controlled breathing, this stretch facilitates a reciprocal stretch of the hip flexors. For individuals with restricted range of motion, achieving this stretch with minimal joint displacement is often preferred.

KEY
- •-- *Joints*
- o— *Muscles*
- ● Shortening with tension
- ● Lengthening with tension
- ● Lengthening without tension
- ● Held muscles without motion

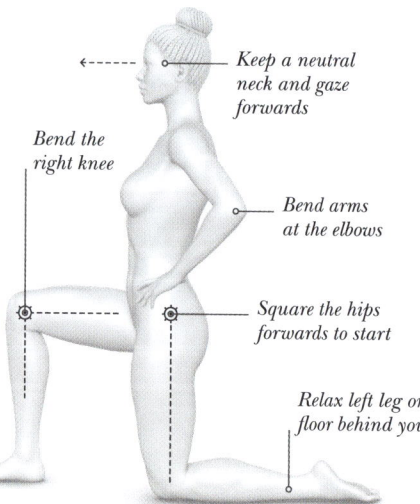

Keep a neutral neck and gaze forwards

Bend the right knee

Bend arms at the elbows

Square the hips forwards to start

Relax left leg on floor behind you

PREPARATORY STAGE
Begin in a half-kneeling position with one knee resting on the floor and your opposite foot flat in front, knee bent at 90 degrees. Maintain a neutral spine and look forwards to ensure proper alignment throughout the stretch.

STRENGTH AND STRETCH EXERCISES | Half-kneel Hip Flexor Stretch

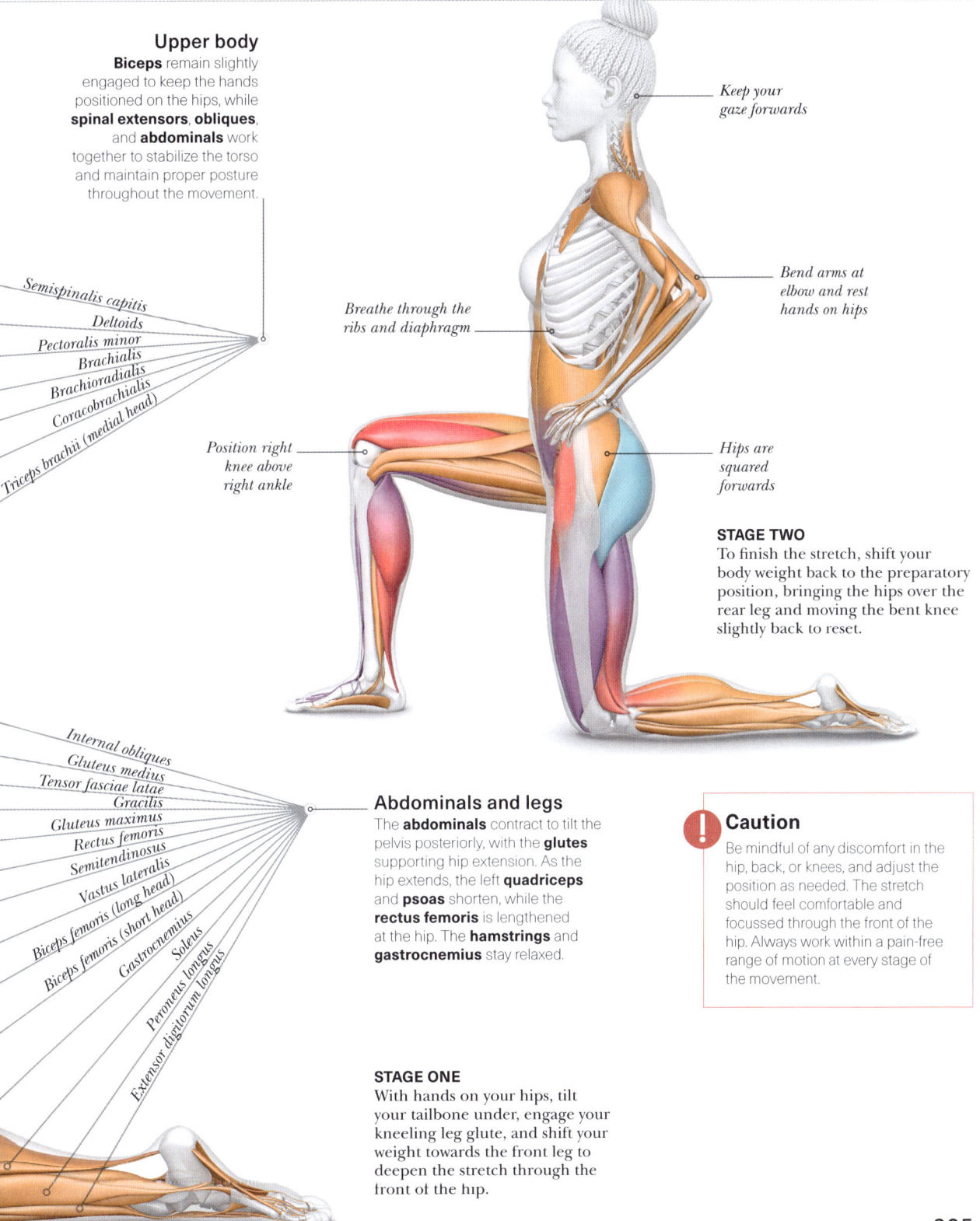

Upper body
Biceps remain slightly engaged to keep the hands positioned on the hips, while **spinal extensors**, **obliques**, and **abdominals** work together to stabilize the torso and maintain proper posture throughout the movement.

Semispinalis capitis
Deltoids
Pectoralis minor
Brachialis
Brachioradialis
Coracobrachialis
Triceps brachii (medial head)

Keep your gaze forwards

Breathe through the ribs and diaphragm

Bend arms at elbow and rest hands on hips

Position right knee above right ankle

Hips are squared forwards

STAGE TWO
To finish the stretch, shift your body weight back to the preparatory position, bringing the hips over the rear leg and moving the bent knee slightly back to reset.

Internal obliques
Gluteus medius
Tensor fasciae latae
Gracilis
Gluteus maximus
Rectus femoris
Semitendinosus
Vastus lateralis
Biceps femoris (long head)
Biceps femoris (short head)
Gastrocnemius
Soleus
Peroneus longus
Extensor digitorum longus

Abdominals and legs
The **abdominals** contract to tilt the pelvis posteriorly, with the **glutes** supporting hip extension. As the hip extends, the left **quadriceps** and **psoas** shorten, while the **rectus femoris** is lengthened at the hip. The **hamstrings** and **gastrocnemius** stay relaxed.

> **! Caution**
> Be mindful of any discomfort in the hip, back, or knees, and adjust the position as needed. The stretch should feel comfortable and focussed through the front of the hip. Always work within a pain-free range of motion at every stage of the movement.

STAGE ONE
With hands on your hips, tilt your tailbone under, engage your kneeling leg glute, and shift your weight towards the front leg to deepen the stretch through the front of the hip.

QUAD STRETCH WITH FOOT RAISED

This stretch targets the rectus femoris and hip flexors, helping cyclists counteract tightness from prolonged hip flexion on the bike. It promotes better hip extension, glute activation, and posture, while reducing the risk of back and knee discomfort.

The stretch involves bending one knee with the foot placed on a raised surface behind the body while the other leg supports in a standing position. This places the rectus femoris in an extended position at both the hip and knee, effectively lengthening it. The stretch also involves the quads, as these are monoarticular knee extensors. To maintain proper form and avoid lumbar extension, the gluteus maximus of the rear leg should be actively engaged, while the abdominals assist in maintaining a posterior pelvic tilt and trunk stability.

KEY
- •-- *Joints*
- ○— *Muscles*
- ● Shortening with tension
- ● Lengthening with tension
- ● Lengthening without tension
- ● Held muscles without motion

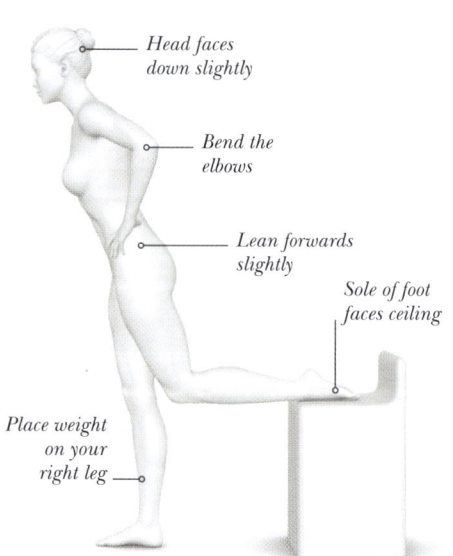

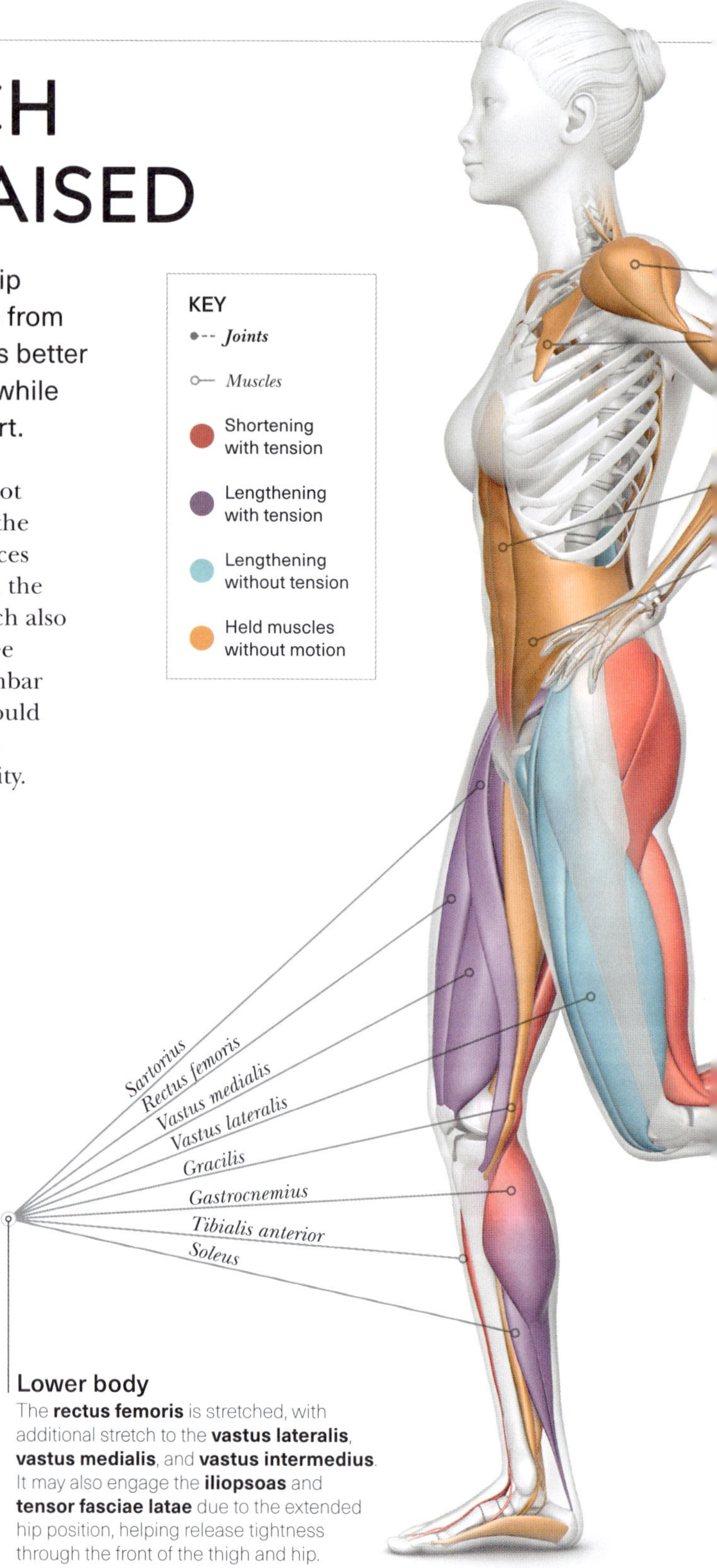

- Head faces down slightly
- Bend the elbows
- Lean forwards slightly
- Sole of foot faces ceiling
- Place weight on your right leg

PREPARATORY STAGE
Start standing with the foot of one leg resting on a box or chair behind at knee height. Hands placed on hips or, for balance, on a wall or chair to the side.

Sartorius
Rectus femoris
Vastus medialis
Vastus lateralis
Gracilis
Gastrocnemius
Tibialis anterior
Soleus

Lower body
The **rectus femoris** is stretched, with additional stretch to the **vastus lateralis**, **vastus medialis**, and **vastus intermedius**. It may also engage the **iliopsoas** and **tensor fasciae latae** due to the extended hip position, helping release tightness through the front of the thigh and hip.

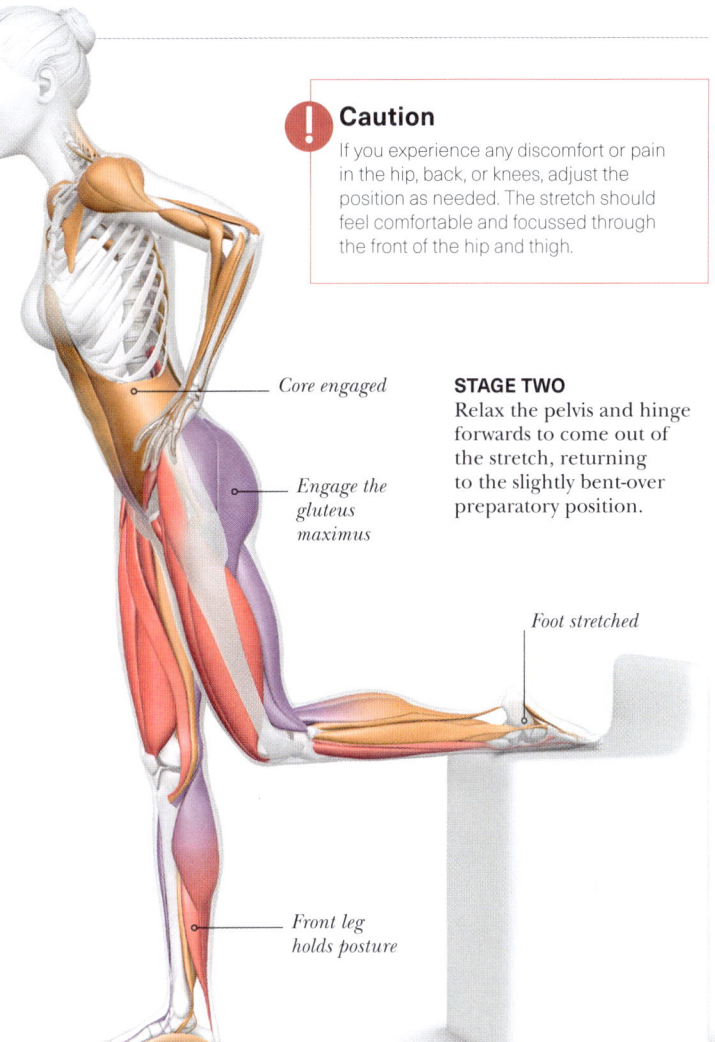

Caution
If you experience any discomfort or pain in the hip, back, or knees, adjust the position as needed. The stretch should feel comfortable and focussed through the front of the hip and thigh.

Core engaged

STAGE TWO
Relax the pelvis and hinge forwards to come out of the stretch, returning to the slightly bent-over preparatory position.

Engage the gluteus maximus

Foot stretched

Front leg holds posture

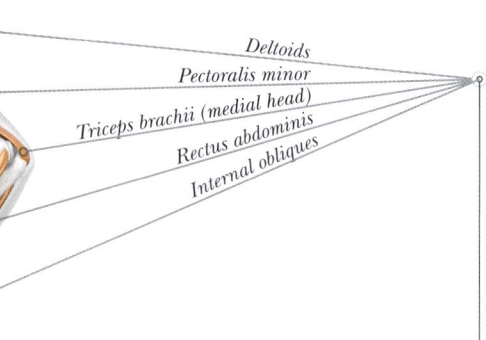

Deltoids
Pectoralis minor
Triceps brachii (medial head)
Rectus abdominis
Internal obliques

Upper body
There is a mild stretch through the **abdominals**, particularly the **rectus abdominis**, and the **spinal stabilizers** are engaged to maintain posture. The **gluteus maximus** and **core muscles** also activate to stabilize the pelvis and prevent excessive lumbar extension during the stretch.

STAGE ONE
Rotate the pelvis forwards and hold the back leg straight to stretch the anterior quad muscles of the standing leg.

> *A powerful stretch to open up the front of the hip and thigh, restoring length to tight quads and hip flexors essential for strong, efficient cycling.*

STATIC HAMSTRING STRETCH

This foundational stretch targets the hamstrings and is particularly useful for cyclists to counteract tightness from sustained hip and knee flexion on the bike. It supports overall mobility in the hips and knees and fits well into a lower-body recovery or flexibility routine.

The hamstrings originate at the pelvis, run along the back of the femur, and attach below the knee, playing a key role in hip extension and knee flexion, both critical in cycling. Tight or inflexible hamstrings can reduce knee extension efficiency and cause knee discomfort.

Limited mobility of the sciatic nerve, which travels from the lower back to the feet, can also impact hamstring function. For cyclists, combining this stretch with targeted strength training helps improve both flexibility and performance in this essential muscle group.

KEY
- Joints
- Muscles
- ● Shortening with tension
- ● Lengthening with tension
- ● Lengthening without tension
- ● Held muscles without motion

> *The hamstrings include the semitendinosus, semimembranosus, and biceps femoris – used for hip extension and knee flexion in cycling.*

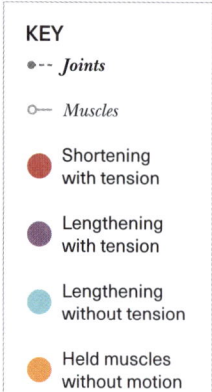

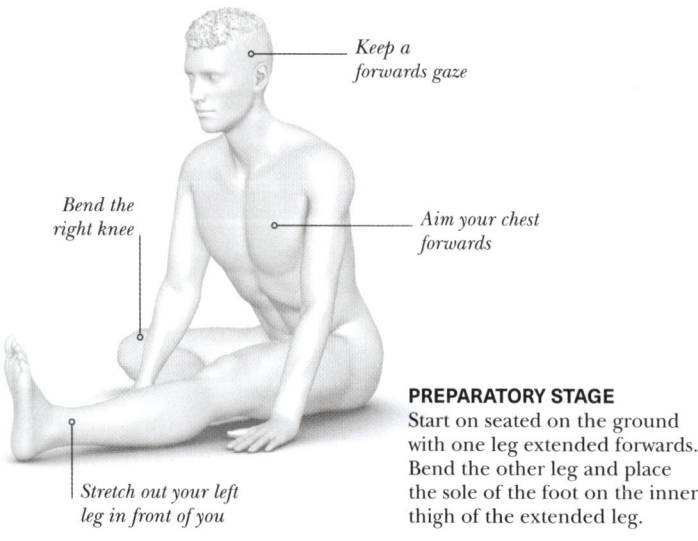

Keep a forwards gaze

Bend the right knee

Aim your chest forwards

Stretch out your left leg in front of you

PREPARATORY STAGE
Start on seated on the ground with one leg extended forwards. Bend the other leg and place the sole of the foot on the inner thigh of the extended leg.

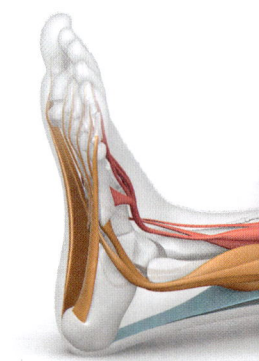

Lower body
The **hamstrings** are lengthened with tension with the **hip** flexed and **knee** extended. The **gastrocnemius** is lengthened without tension, while the **ankle** is dorsiflexed.

STRENGTH AND STRETCH EXERCISES | Static Hamstring Stretch

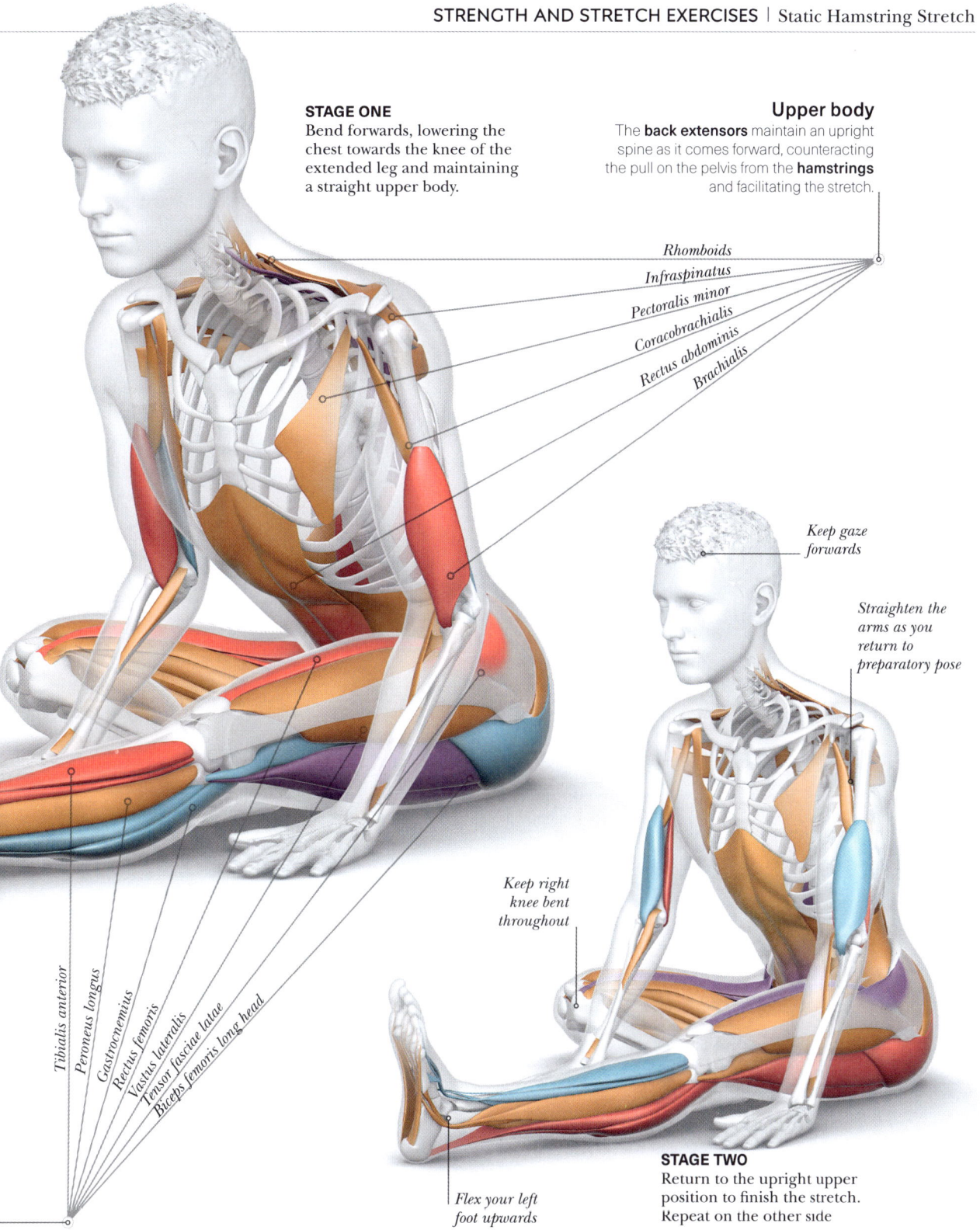

STAGE ONE
Bend forwards, lowering the chest towards the knee of the extended leg and maintaining a straight upper body.

Upper body
The **back extensors** maintain an upright spine as it comes forward, counteracting the pull on the pelvis from the **hamstrings** and facilitating the stretch.

Rhomboids
Infraspinatus
Pectoralis minor
Coracobrachialis
Rectus abdominis
Brachialis

Tibialis anterior
Peroneus longus
Gastrocnemius
Rectus femoris
Vastus lateralis
Tensor fasciae latae
Biceps femoris long head

Keep gaze forwards

Straighten the arms as you return to preparatory pose

Keep right knee bent throughout

Flex your left foot upwards

STAGE TWO
Return to the upright upper position to finish the stretch. Repeat on the other side

209

GASTROCNEMIUS STRETCH

This stretch targets the upper calf by keeping the back leg straight with the heel on the ground. For cyclists, it helps maintain ankle mobility, reduces lower leg tension, and supports efficient pedalling mechanics.

The stretch is done by stepping one foot back, keeping the heel down and knee straight, while leaning with hands flat against a wall or resting on a support. This targets the gastrocnemius, which crosses both the knee and ankle. For cyclists, it helps maintain ankle mobility, supports smooth pedal strokes, and reduces the risk of calf and Achilles tightness.

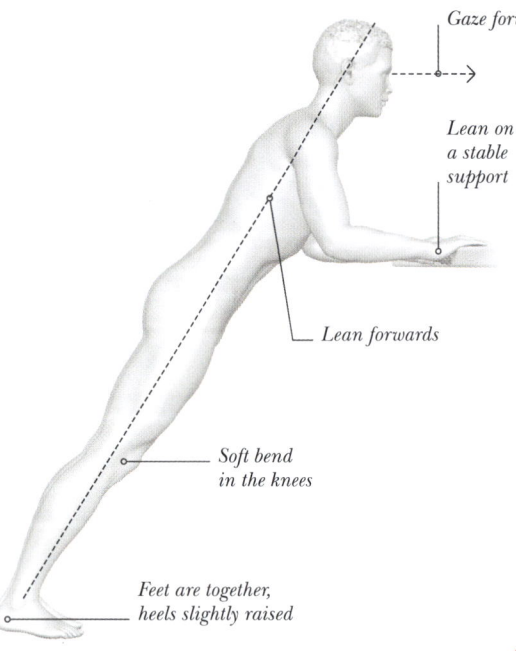

Gaze forwards

Lean on a stable support

Lean forwards

Soft bend in the knees

Feet are together, heels slightly raised

PREPARATORY STAGE
Rest your hands on the support, then step back to increase the angle of your lean to roughly 45 degrees, with your body forming a straight line from your heels to the top of your head. Your heels should be just off the ground, your knees slightly soft.

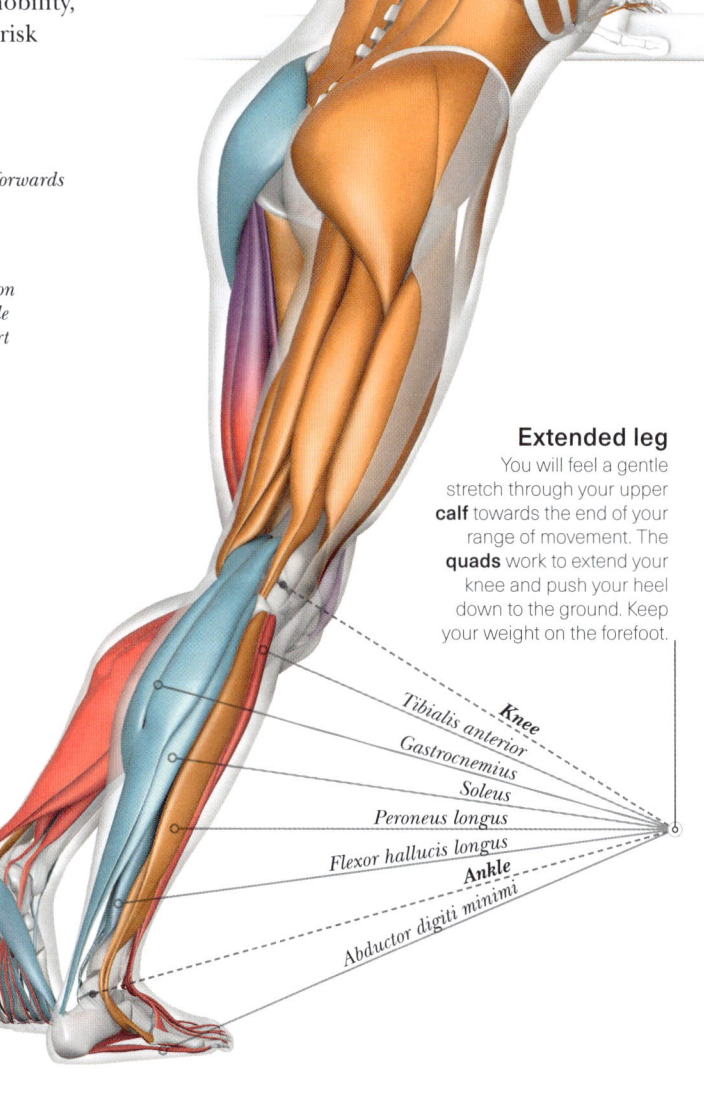

Extended leg
You will feel a gentle stretch through your upper **calf** towards the end of your range of movement. The **quads** work to extend your knee and push your heel down to the ground. Keep your weight on the forefoot.

Knee
Tibialis anterior
Gastrocnemius
Soleus
Peroneus longus
Flexor hallucis longus
Ankle
Abductor digiti minimi

STRENGTH AND STRETCH EXERCISES | Gastrocnemius Stretch

STAGE ONE
Bend your left knee and raise your left heel off the ground to bring your weight onto your forefoot. Simultaneously, push down towards the floor with your right heel and extend your right knee. You should feel a gentle stretch in your right upper calf. As soon as you feel the stretch, move fluidly into stage 2.

STAGE TWO
Reverse the movement. Unlock your right knee and raise the right heel off the floor while you push down with your left heel and extend your left knee. Repeat stages 1 and 2 in a continuous movement.

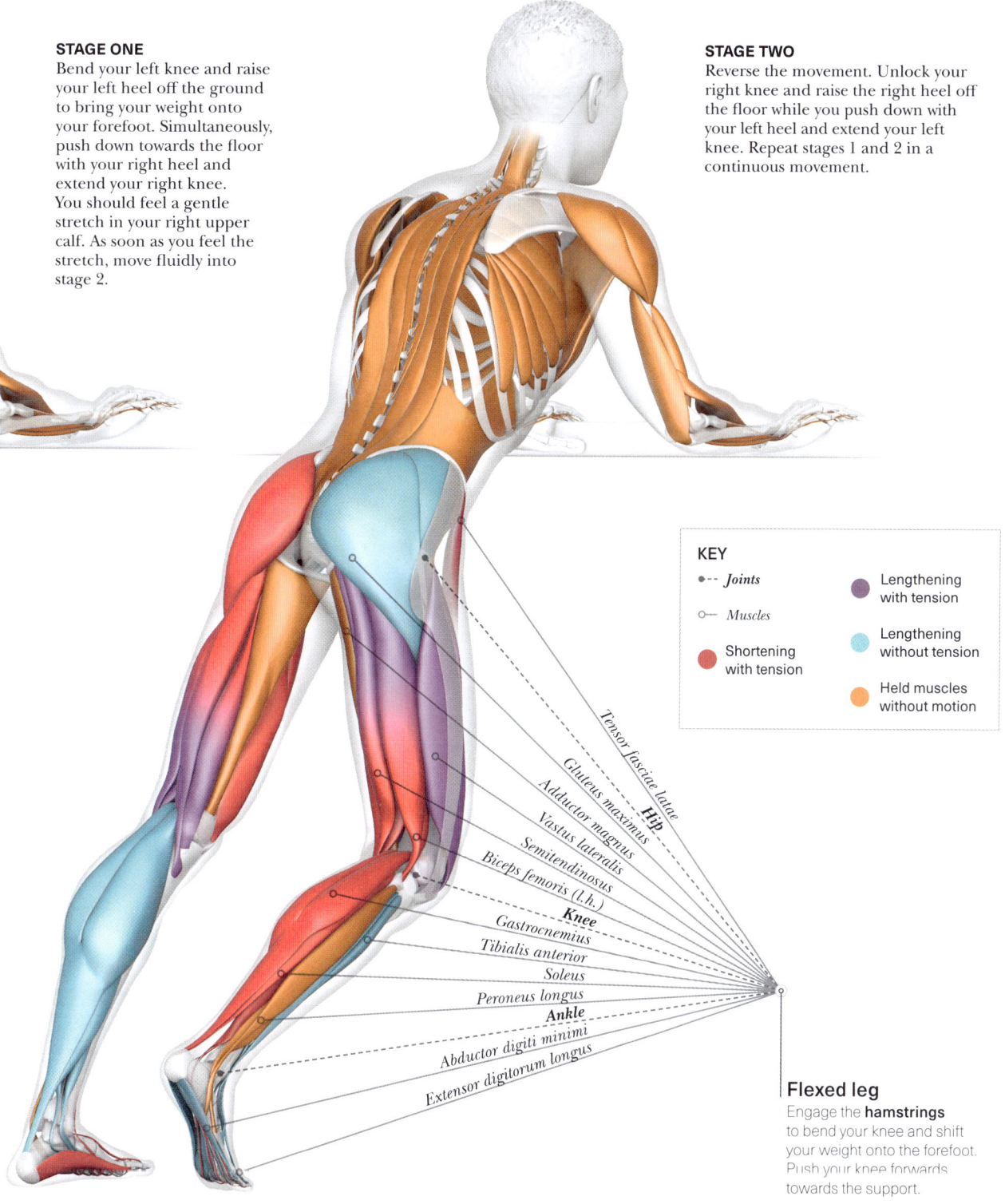

KEY
- Joints
- Muscles
- Shortening with tension
- Lengthening with tension
- Lengthening without tension
- Held muscles without motion

Tensor fasciae latae
Gluteus maximus
Adductor magnus
Vastus lateralis
Semitendinosus
Biceps femoris (l.h.)
Hip
Knee
Gastrocnemius
Tibialis anterior
Soleus
Peroneus longus
Ankle
Abductor digiti minimi
Extensor digitorum longus

Flexed leg
Engage the **hamstrings** to bend your knee and shift your weight onto the forefoot. Push your knee forwards towards the support.

211

INDEX

20th-century cycling 8
1980s cycling 8–9

A

abdominal muscles 14, 18, 21
acclimatization
 altitude 58–9
 temperature 72, 73
Achilles tendinopathy 43
acidosis 69
Acute Training Load (ATL) 121
adductors 14
adrenaline 116
aerobic energy systems
 (oxidative system) 48, 49, 51, 61, 108
aerobic fitness
 cyclocross 23
 aerobic training 53
aerodynamic positioning 9, 25, 26–27, 88
 biomechanical trade-offs 27
 coordination and postural stability 27
 kinematics 26
 muscles that aid 18, 22, 23
 and respiratory muscle function 52
 time trials 109
 training and position sustainability 27
aerodynamics
 aerodynamic drag 10, 11, 107
 aerodynamic force 108–109
 aerodynamic positioning 9, 18, 22, 23, 25, 26–27, 52, 88, 109
 aerodynamic sensors 93
 bike design 82, 83, 84, 86
 drafting and the peloton 100, 101
 and energy 100
 improvements in 8, 9
 wheel design 85, 89
AI (artificial intelligence) 94
air resistance 10, 26, 83, 100
alignment 32–33, 40
alkaline water 68

Alternative Foot Switch 187
altitude 105
 acclimatization and adaptation 58–59
 altitude sensors 93
 effects of 58–59, 104
 practical considerations 59
 training strategies 59
amygdala 114
anaerobic capacity 46, 51, 120
 interval training 51, 124, 125, 131
anaerobic energy systems 108
 anaerobic alactic energy (phosphagen) system 48, 49, 50, 51, 69, 106
 anaerobic lactic (glycolytic) system 48, 49, 50, 51, 106
anaerobic respiration 61
anaerobic threshold 56
anaerobic training 53
anatomy of cycling 12–43
ankles
 ankle dorsiflexion 14, 20
 ankle flexion 24
 ankle joint actions 20
 ankle plantarflexion 15, 16, 20
anterior cingulate cortex (ACC) 114
anxiety, managing 116–17
arm muscles 19, 23, 127
arteries 53
ATP (adenosine triphosphate) 48, 49, 50, 61, 63
attentional control 115
Australian Institute of Sport (AIS) 68

B

the back
 back muscles 18, 19, 23
 lower back pain 41
backward induction 113
Banister Training Impulse model (TRIMP) 121
Barbell Front Squat 154–55
BCAAs 68
Bear Plank 187

bearings 85, 86
 maintenance 91
bench 151
Bent-Over Row with Dumbbells 180–81
beta-alanine 68, 69
bicycles
 component selection 88–89
 design advances 84–87
 efficient setup 88–91
 evolution of 82–83
 frame design 8, 9, 83, 84–85, 87
 general inspection 91
 mechanical maintenance 91
 physical principles 82–83
 tyres 8, 11, 85, 86, 88, 89, 90, 91, 92, 109
 Union Cycliste Internationale (UCI) regulations 87
 wheels 83, 84, 85–86, 88, 89, 91, 108
bike computers 93, 94
bike fits 25, 27, 35, 36–37, 95
 common fit challenges 39, 40–43
 fit adjustments 38–39
biomechanics 95
 alignment 32–33
 biomechanical efficiency 27
 cadence and 32–33
biometrics
 interpreting 35
 key variables in biometrics monitoring 34
 monitoring 34–35
 reasons for tracking 34
blood buffering 69
BMX cycling
 cadence 33
 energy use 47
 strength training 127
body composition, managing 66–67
body position see positioning
bone shaker 82
box 151
Box Jump 172–73
 Jump-off-the Box 175

 Single Leg Box Jump 174
the brain
 hormones and 116
 mental performance 114–15
braking systems 83, 85, 86
 maintenance 91
branched-chain amino acids (BCAAs) 63
breakaways 102–103, 112, 113
 factors for success 103
breathing 52, 116
 breathing rates 52
 importance of 151
 ventilatory thresholds 56–57
bronchoconstriction 73
Bulgarian Split Squats 158

C

cable maintenance 91
cadence 30–31, 83, 89, 92
 cadence and biomechanics 32–33
 cadence drills 122
 cadence ranges and context 30
 efficiency and economy 31
 high and low cadence 30
 and muscles 32
 pedal stroke and fatigue 30
 and performance strategy 33
caffeine 68, 69
calf muscles 14, 16, 17, 22
 Calf Raise with Dumbbells 168–69
 pedalling mechanics 29
 Seated Calf Raise 170
 Single-Leg Calf Raise 170
 use of in different disciplines 23
capillaries 53
carbohydrates 8, 9, 60, 61
 fear of carbohydrates 71
 fuelling recovery with 71
 fuelling with 62–63, 64, 65, 77
 high-carbohydrate availability 67
 metabolism of 48, 49, 51
 supplements 68
carbon fibre 83, 84, 85, 89
cardiorespiratory system 52–53

212

cardiovascular system 52
  assessing cardio fitness 52
Cat Cow 196–7
  Seated Cat Cow 197
catecholamines 116
cervical extensors 15
chains 86, 91
cherry juice 68
chest muscles 19
Child's Pose 127, 198–99
  Lateral Pose 199
Chin Up 177
chin-up bar 151
Chronic Training Load (CTL) 121
circulation 53
cleats
  cleat adjustment 43
  cleat alignment 25
  cleat positioning 38, 39, 40
climbing
  cadence 33
  drafting 101
  how to race climbs 104–105
  muscles involved 21, 22
  performance in climbing 104
  psychological and tactical
    readiness 105
  tactical climbing 105
clothing
  aerodynamic 84
  body temperature regulation
    72
  time trials 109
coaching, AI for 94
coefficient of drag area (CdA)
  109, 113
cognitive-behavioural model of
  performance anxiety 116
cognitive restructuring 115, 117
cold conditions 73
cold-induced diuresis 73
cold water immersion 9, 79, 131
collagen fibres 15
collagen with vitamin C
  supplements 68
commercialization 99
competitive cycling 96–117
compression 9, 79, 131

computational fluid dynamics
  (CFD) 83, 84, 85
conditioning work 43
conduction 72
continuous glucose monitors
  (CGMs) 92, 93, 94
convection 72
cooling down 125, 131
cooperation, strategic 112
coordination, aerodynamic
  positioning 27
core body temperature 94
core muscles 14, 18, 21, 126
  core stability 22
  use of in different disciplines
    23
core temperature sensors 93
cortisol 116
cranks
  across the disciplines 89
  crank angle 28
  crank length 25, 31, 37, 39, 88
  crank torque profiles 25
  power cranks 34, 86, 92
  Q-factor 88, 89
creatine monohydrate 68, 69
creatine phosphate 48
criterium races
  8-week training plan
    132–33
  fuelling strategies 64, 65
Critical Power (CP) 51, 105,
  108
crosswinds 102, 107
cycling
  anatomy of 12–43
  competitive cycling 96–117
  evolution of 8–9
  physics of 10–11
  physiology of 44–79
  technology 80–95
cyclist's palsy 42
cyclocross
  8-week training plan 136–37
  crank length and Q-factor 89
  energy use 47
  fuelling strategies 64, 65
  gearing systems 90

muscle recruitment 23
  strength training 127
  tyres and pressure 90

D

data 99
  data analytics 83, 92–93, 94
  future of data 94–95
  real-time data processing 94
  using data in pacing regulation
    105
Deadbug 190–91
deadlifts 126
  Trap Bar Deadlift 160–61
dehydration 71, 73, 74, 75
deltoids 21
derailleur systems 82, 89
descending, and drafting 101
diet 60
  see also fuelling; nutrition
dimensional limitations 87
disc brakes 9, 83, 85, 86, 91
disc wheels 84, 85, 108
DMAA 68
domestiques 99, 101, 105
doping 9
  anti-doping 99
dorsiflexors 29
downstroke
  force application and torque
    production 29
  joint actions 20, 21
  muscles that aid 23, 29
drafting 10, 49, 100–101, 103,
  104, 112
drag 83, 84, 93, 100
  aerodynamic drag 10, 11, 107
Drais, Baron Karl von 82
Drasine 82
drinks 74–75
  electrolyte balance 75, 77
  electrolyte drinks 64, 68, 71,
    75, 77
drivetrain 82, 86, 89, 108
  efficiency 11
  maintenance 91
  resistance 10

dropped riders 102
dumbbells 151
  Bent-Over Row with Dumbbells
    180–81
  Dumbbell Goblet Squat 159
  Dumbbell Shoulder Press
    182–83
  Step Up with Dumbbells
    162–63

E

echelons 102–103
effective force ratio (EFR) 29, 34
efforts 131
elbow flexors 14
electrolytes 93
  drinks 64, 68, 71, 75, 77
  electrolyte balance 75
  fatigue and recovery 77
  neglecting 71
electromyography (EMG) studies
  29, 34
electronic shifting systems 83,
  86, 89
electronics 83
EMG sensors 34
emotional flexibility 115
endurance 8
  aerodynamic positioning 27
  building endurance 47
  fuelling 62, 63, 64, 65, 66
  general endurance fitness 130
  muscle fibres suited for 54
  muscles that aid 23
  oxygen uptake 52
  training implications 51
energy
  and aerodynamics 100
  energy deficiency 67
  energy duration 50–51
  energy efficiency 11
  energy systems 48–49, 50,
    106
  energy use across different
    disciplines 47
  fuelling energy systems 60–61
  how energy is released 48

requirements in a crosswind 102
training implications 51
equipment
    training equipment 151, 152
    UCI rules 87
    see also bicycles
erector spinae 21
erythropoietin (EPO) 58–59
evaporation 72, 73
events, nutrition planning 67
evolution of cycling 8–9
Exercise Associated Hyponatraemia (EAH) 75
Exercise-Induced Arterial Hypoxemia (EIAH) 52
exposure-based training 117

F

fascicles 55
fat 60, 61
    fat oxidation 47, 51, 56, 63
    fuelling with 63, 65
    metabolism of 48, 49, 51
fatigue 76–77
    acute fatigue 76
    altitude and 58
    and build-up of hydrogen ions 69
    central fatigue 76, 77
    cumulative fatigue 76
    dehydration and 74
    delaying with correct fuelling 62, 63, 64
    importance of thermoregulation and hydration 47
    indicators of 34
    lactate acid 48
    managing 77
    pedal stroke and 30
feet
    Achilles tendinopathy 43
    feet issues 43
    foot angle 34
fibre, dietary 60, 67
Figure 4 Stretch 202–203
    Seated Figure 4 Stretch 203
finite element analysis (FEA) 84
forces 83
    pedalling mechanics 29, 34

forearm muscles 19, 21
frame design 8, 9, 83, 84–85, 87
freewheeling 49
fuelling 60–65
    inadequate in-ride fuelling 70
    for moderate to high intensity exercise 62
    strategies for different events 64–65
    timing and availability of fuel sources 62–63, 70
    under-fuelling 70
    see also nutrition
Functional Threshold Power (FTP) 51, 57
    racing a climb 104, 105
    time trials 108
    training zones 120, 128, 129

G

game theory 112–13
gastrocnemius 16, 20, 21
    Gastrocnemius Stretch 210–11
gastrointestinal health 60, 65
    gastrointestinal issues 68, 69, 70
gearing systems 8, 82, 86, 88, 89, 90
gelatin 68
General Adaptation Syndrome (GAS) 120
general classification (GC) contenders 99, 103, 105, 111
glucose 49, 50, 60, 61
    breakdown of 48
    continuous glucose monitors (CGMs) 92, 93, 94
    maintaining levels 62, 63
glutamine 68
gluteal muscles 14, 15, 17, 22, 126
    Glute Bridge 192–93
    gluteus maximus 17, 20
    gluteus medius 17
    gluteus minimus 17
    pedalling mechanics 29
    use of in different disciplines 22, 23
glycogen 8, 48, 54, 60, 61
    availability 62, 64, 65
    glycogen depletion 47, 76, 77, 104
    glycogen sparing 51

premature depletion 104
    resynthesis and recovery 63, 64, 65, 67, 78, 79
glycolysis 49, 50
glycolytic energy systems 48, 49, 50, 51, 106
goal persistence 111, 115
Goblet Squats, Dumbbell 159
GPS devices 83, 93
gradient, and gravitational force 11
Gran Fondo, 12-week training plan 142–44
Grand Tours 98
gravel cycling
    crank length and Q-factor 89
    energy use 47
    gearing systems 90
    muscle activation 23
    strength training 127
    tyres and pressure 90
gravity 10, 11, 104, 108–109
grip strength 19
gutter riders 102

H

Half-Kneel Hip Flexor Stretch 204–205
hamstrings 14, 17, 20, 21
    pedalling mechanics 29
    Static Hamstring Stretch 208–209
    strength training 126
    use of in different disciplines 22, 23
handlebars 85, 87, 109
    design 86
    reach and drop 38, 41, 42
hands, numbness 39, 42
Hanging Knee Raise 194–95
the heart, cardiorespiratory system 52, 53
heart rate 74
    anxiety and 116
    at altitude 58
    cardiovascular response 52
    data 83, 92
    heart rate variability (HRV) 79, 92, 93, 117, 120, 121
    monitoring recovery 79
heart rate monitors 8, 92, 93, 105
    pacing with 77

heat
    acclimatization 73
    heat loss 72, 73
helmets 85
hip flexors 14, 16, 20, 22
    Half-Kneel Hip Flexor Stretch 204–205
    pedalling mechanics 29
    use of in different disciplines 23
hips
    aerodynamic positioning 26, 27
    Half-Kneel Hip Flexor Stretch 204–205
    hip extension 17, 20, 24
    hip extensors 15
    hip flexion 24
    hip flexors 14, 16, 20, 22, 23, 29
    hip joint actions 20
    Hip Thrust 164–65
    preventing rocking 18
holistic monitoring platforms 95
homeostasis 72, 77
hormones 131
    hormonal changes 116
    hormone synthesis 60
hoses 81
housing 91
humidity sensors 93
hydration 8, 47, 60, 74–75
    carbohydrate-electrolyte drinks 64
    during exercise 75
    electrolyte balance 75
    electrolyte drinks 64, 68, 71, 75, 77
    fatigue and recovery 77
    how much water to drink per day 74
    neglecting hydration 71
    overhydration 75
    personalized strategies 75, 93
    post-exercise rehydration 75
    pre-race strategies 75
    race context 75
    recovery drinks 65
    rehydration 78
    road races 65
    sweat sensors 93
hydraulic brakes 91
hydrogen ions (H+) 69
hyperthermia 72
hyponatremia 71, 75

hypothalamic-pituitary-adrenal (HPA) axis 114, 116
hypothalamus 72–73

I

iliopsoas 20
iliotibial band syndrome (ITBS) 25, 40
Index of Force Effectiveness (IFE) 31
indexed shifting systems 82
individual time trials (ITT) 108–11
inertial measurement units (IMUs) 34, 92
injuries
   bike fit challenges 39
   common injuries 40–3
   injury prevention 24, 25, 127
   kinematics 24, 25
   reducing injury risk 24–25
   training to avoid 43
innovations 84
integration 85
intensity
   increasing 130
   low-intensity rides 78–79
   polarized intensity 130
intermittent hypoxic training (IHE/IHT) 59
interval training 8, 124–25
   training structure 125
iron 68

J

joints
   cycling joint actions 20–21
   joint angles 34, 35
   pedal stroke and joint kinematics 24
Jump-off-the-Box 175

K

kinematics 24–25, 107
   aerodynamic positioning 25, 26–27
   bike fit and positioning 25
   kinematic variables 34
   performance and injury prevention 25
kinetic variables 34
knees
   Hanging Knee Raise 194–95
   knee extension 21, 24
   knee flexion 17, 21, 24
   knee joint actions 21
   knee pain 24, 33, 39, 40
   knee stability 16
   knee tracking 25
   misalignment 18
   movements at 90 RPM 20
   valgus and varus alignment 33, 34

L

lactate 46, 50, 120
   blood lactate accumulation (LT1) 53, 56, 57
   blood lactate dynamics 56
   blood lactate quantification 94
   elevated blood lactate levels 104
   glycolytic system 48
   lactate clearance 53, 57, 106, 124
   lactate threshold (LT) 47, 51, 57, 108, 122, 124
   maximal lactate steady state (MLSS) 57
   mitochondria and 49
   training zones and 121
lactic acid 61, 69
Lateral Pose 199
lead-outs 99, 101, 106–107, 113
   role of the lead-out train 107
   tactical complexity 107
leg muscles 17
   Leg Press 166–67
   see also individual muscles
live high, train high (LHTH) 59
live high, train low (LHTL) 59
low-carb high-fat (LCHF) diets 71
low-intensity rides 78–79
lower back pain 39, 41
lower body muscles 14, 16–17
   see also individual muscles
lumbar pain 25
lunges 126

M

machines, working on 152
macronutrients 60, 66
   see also individual macronutrients
maintenance, mechanical 91
massage therapy 79, 131
materials 83, 84
maximal lactate steady state (MLSS) 57
mechanical efficiency 82, 83, 89
medical supplements 68
medication, anxiety 117
medicine balls 151
   Russian Twist with Medicine Ball 188–89
meditation 78
mental performance 114–15
mental resilience 9, 111
methylhexanamine 68
metrics 9, 83, 92–93
mindfulness 111, 117
mitochondria 48, 49, 50, 52, 61
   biogenesis 51
   mitochondrial density 52, 53, 54, 55
   oxygen transport cascade 52
modern cycling 9
motion capture systems 25, 27, 34, 37
motivation, building 115
motorized assistance, ban on 87
mountain biking 9
   8-week training plan 134–35
   bike design 86
   cadence 33
   crank length and Q-factor 89
   energy systems 49
   energy use 47
   fuelling strategies 64, 65
   gearing systems 90
   muscle recruitment 22, 23
   positioning 37
   strength training 127
   tyres and pressure 90
movement economy 46
muscle oxygen sensors 92, 93, 94
muscles
   ability to oxidize fuels 46
   building endurance 47
   and cadence 32
   capillarization of 53
   creatine stores 50, 69
   fast-twitch (Type II) muscles 32, 57
   fast-twitch (Type IIa) oxidative-glycolytic muscle fibres 54–55
   fast-twitch (Type IIx) glycolytic muscle fibres 22, 23, 54, 55
   isometric contractions 23
   key muscle groups 14–15
   lactate acid 46, 47
   muscle activation patterns 34
   muscle buffering capacity 53
   muscle fibre types 54–5
   muscle recruitment 22–3
   pedalling mechanics 29
   repair of tissues 60
   skeletal muscle fibres 14, 72
   slow-twitch (Type 1) oxidative muscle fibres 22, 32, 53, 54–55
myofibrils 14
myotendinous junction 15

N

Nash Equilibrium 112, 113
near-infrared spectroscopy (NIRS) 93
neck
   neck pain 41
   posture 27
negative splits 110
neuromuscular coordination 47
nitrates (beetroot juice) 68, 69
noradrenaline 116
nutrition 8, 9, 131
   common pitfalls 70–71
   lack of nutrition planning for travel and racing 71
   managing fatigue 77
   nutrition periodization 66–67
   recovery 78
   supplements 68–69
   timing and availability of fuel sources 62–63, 70
   and training 67
   see also fuelling

O

obliques 21
one-day classics 98
ordinary bicycle 82
overtraining 122
  overtraining syndrome (OTS) 76
oxidative phosphorylation 49
oxidative system 48, 49, 51, 61, 108
oxygen
  altitude and delivery of oxygen 58, 59
  improving uptake 52
  oxygen transport cascade 52
  pressure at altitude 59

P

pacing 77, 105, 113
  pacing principles 104
  rotating pacelines 101, 103
  time trial strategy 110–11
patellar tendinopathy 25
patellofemoral pain syndrome 40
peak power 8
  achieving high 23
  aerodynamic positioning 27
pectorals 14, 19
pedal stroke
  alignment 32–33
  cadence 30–1
  coordination and technique 25
  and fatigue 30
  force application and torque production 29
  and joint kinematics 24
  mechanics of 28–29
  muscle recruitment 29
  pedal balance 34
  phases 24, 28
  Q-factor 88
pedals, power 34, 86, 92
the peloton 100–101, 112, 113
  breakaways 103, 112, 113
  dealing with crosswinds 102
  peloton dynamics 100
  structure and movement 101
  tactics 101
pelvis
  pelvic stability 17, 18, 21, 25, 34
  pelvic tilt 27

preventing rocking 25
performance
  cadence and performance strategy 33
  improving with kinematics 24–25
  mental performance 114–15
  performance anxiety 116–17
  performance metrics 83
  performance supplements 68
  performance variables 46
perineal numbness 42
periodization 8, 9, 122–23, 125, 130
  models of 122–23
phosphagen energy system 48, 49, 50, 51, 69, 106
phosphocreatine (Pcr) 47, 49, 50, 54
physics of cycling 10–11
physiological demands 46–47
  performance variables 46
  transitions, endurance, and recovery 47
physiology of cycling 44–79
Pistol Squat (with box) 159
Plank
  Alternative Foot Switch 187
  Bear Plank 187
  Low-Impact Plank 186
  Low Plank Hold 186
  Plank with Shoulder Taps 184–85
positioning 10
  aerodynamic 18, 22, 23, 25, 26–27, 52, 88, 109
  bike fits 36–39
  dropped posture 36
  extreme downhill 37
  kinematics 25
  mountain bike 37
  muscles that aid 18, 19, 22, 23
  relaxed position 36
post-event debriefing 117
postural stability 27
power 9, 10
  Critical Power (CP) 51, 105, 108
  Functional Threshold Power (FTP) 51, 56, 57, 104, 105, 108, 120, 128, 129
  impact of sweating on 76
  muscles that provide 14, 16–17

power-duration curve 50–51, 104, 121, 129
power output 83, 106, 108–9, 120
power phase 8, 24, 28
power-to-drag ratio (W/CdA) 109
power-to-weight ratio (W/kg) 11, 66, 104, 108, 109
role of hip joint 20
power meters 9, 24, 31, 83, 86, 92
  force analysis 29, 34
  monitoring training 120
  pacing with 30, 77
  tactical climbing 105
pre-performance routine 117
prebiotics 68
prefrontal cortex 114
probiotics 68
progressive adaptation 130
progressive overload 122
proprioception 23
propulsive force 108–109
  muscles that aid propulsion 17
ProTeams 98
protein 60, 61
  fuelling recovery with 63, 65, 71
  supplements 68
PST (psychological skills training) 115
psychology and cycling 114
  psychological readiness 105
  psychological resilience 77
public goods game 112, 113
Pull-Up 176–77
  Chin Up 177
pull-up bar 151
pulmonary ventilation 52
puncheur 99
punctures, self-sealing 86
Push Up 178–79

Q

Q-factor 88
quadriceps 14, 16, 17, 21, 22
  impact in different disciplines 22, 23
  pedalling mechanics 29
  Quad Stretch with Foot Raised 206–207
  strength training 126

R

races
  hydration 75
  preparation at altitude 59
  see also individual disciplines
radial force 29, 34
radiation 72
rate of perceived exertion (RPE) 105, 120, 121
recovery 9, 47, 129, 131
  active recovery 131
  fuelling with carbohydrate 64
  fuelling with protein 63, 65
  inconsistent recovery practices 71
  monitoring 79
  post-exercise rehydration 75
  recovery drinks 65
  strategies 77, 78–79
recovery phase, pedal stroke 24, 28
rectus femoris 16, 20
red blood cells 58–59
red zone 77
REDs (relative energy deficiency in sport) 67, 70
repetitive strain injuries 25
Reps in Reverse (RIR) method 152
resilience 9, 77, 111, 114–15
resistance 10–11, 100, 104, 108–109
rhomboids 15, 19
rim profiles 85, 89, 109
road captains 99
road cycling
  12-week training plan 145–47
  aerodynamic positioning 26–27
  average VO2 max 46
  bike design 82, 84, 85, 86
  biometrics 35
  crank length and Q-factor 88, 89
  energy duration 50
  energy systems 49
  energy use 47
  fuelling strategies 64, 65
  gearing systems 90
  muscle activation 23
  road racing 98–107
  strength training 126
  tyres and pressure 90

road racing 98–107
  breakaways 112, 113
  culture, commercialization and anti-doping 99
  drafting and the peloton 100–101, 112
  echelons and breakaways 102–103
  end of the race 106
  game theory 112–13
  how to race a climb 104–105
  lead-outs and sprint finishes 106–107
  muscle recruitment 22
  pelotons 112, 113
  structure and governance 98
  team dynamics and tactics 98, 106
  types and specializations 98
rocking, preventing 27
roll mats 151
rolling resistance 10, 11, 100, 108–109
Rover Safety bicycle 82
Row, Bent-Over Row with Dumbbells 180–81
RPM (revolutions per minute) 30
Russian Twist with Medicine Ball 127, 188–89

S

saddles 84, 95
  fore-aft position 38, 40
  height 24, 25, 39
  position 25, 38, 40, 42
  saddle sores and perineal numbness 39, 42
sartorius 20
scapular stabilization 27
Seated Calf Raise 170
Seated Figure 4 Stretch 203
sensors 92, 93
  integration and miniaturization 94
  personalization of equipment 95
  sensor technology 92–93
  wearable 24, 94, 95
sequential games 113
shifting systems 82, 83, 86, 88, 89
shoulder muscles 19, 22

Dumbbell Shoulder Press 182–83
shoulder exercises 127
shoulder pain 41
stabilization of 27
Single-Leg Box Jump 174
Single-Leg Calf Raise 170
skeletal muscle fibres 14, 72
skinsuit 109
sleep 79, 131
  sleep hygiene 79
  sleep optimization 9
  sleep sensors 93
social support 115
sodium 67, 75, 78, 93
  sodium intake during road races 65
  sodium levels 75
sodium bicarbonate 68, 69
soleus muscles 16, 20
somatic control training 117
speed, time trials 109
spinal erectors 18, 22
spinal extensors 15
Split Squats 156–57
  Bulgarian Split Squats 158
sponsorship 99
sportives, 12-week training plan 142–44
sports foods 68
sprinters 99, 103, 106, 107
sprints and sprinting
  altitude and 58
  cadence 33
  muscle fibres suited for 54
  muscles involved 16, 17, 19, 21, 23, 22, 23
  power output 50
  sprint finishes 106–107
  sprint training 51, 124, 125, 131
squats 126
  Barbell Front Squat 154–55
  Bulgarian Split Squats 158
  Dumbbell Goblet Squat 159
  Pistol Squat (with box) 159
  Split Squat 156–57
stability, muscles that aid 14, 18, 19, 21
stage races 98
  fuelling strategies 64, 65
Static Hamstring Stretch 208–209

step-ups 126
  Step Up with Dumbbells 162–63
steps 151
strength training 43, 122, 126–27, 129, 148–211
  best practices 152
  example session 126
  importance of 150–51
  sample workouts 153
  whole body training 127
  see also individual exercises
stress 120
  physical 116
  psychological 116
  stress inoculation 117
  stress reduction 115
stretching 127
stroke volumes 52, 53
supplements
  overuse or misuse of 71
  safety of 69
suspension 83
  maintenance 91
sustainable design 82
sweat and sweating 72, 73, 75, 76, 78
  sweat glands 73
  sweat rates 74, 77, 93
  sweat sensors 93
sweat spot intervals 124, 125
sympathetic-adrenomedullary system 116

T

tangential force 29, 34
tapering 94, 123, 131
  nutrition 66, 67
teams 9
  mental toughness 114
  sponsorship 99
  team dynamics and tactics 8, 98, 106, 107
  team time trials 111, 113
technology 80–95
temperature 105
  core body temperature 94
  core temperature sensors 93
  dehydration and 74
  thermoregulation 72–73
tempo interval training 124, 125

tendons 15
testosterone boosters 68
thermoregulation 47, 72–3, 74
  heat loss during cycling 72
  hydration and 75
  hypothalamus 72–73
  impaired 71
Thread the Needle 127, 200–201
thresholds
  functional threshold power (FTP) 51, 56, 57, 104, 105, 108, 120, 128, 129
  interval training 124, 125, 131
  lactate threshold (LT) 51, 57, 108, 122, 124
  threshold intervals 51, 124
  ventilatory thresholds 56–57
tibialis anterior 20, 23
tidal volume 52
time trials 9, 98, 108–11
  8-week training plan 138–39
  aerodynamic positioning 26–27
  bike design 83, 84, 85, 108–109
  biometrics 35
  cadence 33
  common injuries 41
  crank length and Q-factor 89
  energy use 47
  fuelling strategies 64, 65
  gearing systems 90
  key to speed 109
  mental preparation 111
  muscle recruitment 22–23
  pacing strategy 110–11
  physiology and power output 108–109
  positioning 38
  race preparation 111
  team time trials 111, 113
  tyres and pressure 90
torque 47, 92
  achieving maximal 23
  cadence and 30, 31, 32
  torque profiles 34
Tour de France 98, 102
track cycling
  8-week training plan 140–41
  bike design 84, 85
  cadence 33
  crank length and Q-factor 89
  energy duration 50
  energy use 47

217

muscle recruitment 22, 23
strength training 127
training
  how to train 118–47
  internal and external measures 120
  interval training 124–25
  load management 129
  monitoring 120–21
  and nutrition 67
  optimizing training load 130
  periodization 122–23, 125
  phases of 123
  quantification of training 120–21
  testing supplements 69
  training equipment 151
  training load metrics 92
  training phases and periodization 130
  training programmes 128–47
  training structure 125
  training volume 129
  training zones 120, 121, 128, 129, 130
  types of training 123
training programmes 128–47
  8-week criterium race plan 132–33
  8-week cross-country mountain bike race plan 134–35
  8-week cyclocross plan 136–37
  8-week time trial plan 138–39
  8-week track cycling plan 140–41
  12-week Gran Fondo/Cyclosportive plan 142–44
  12-week road cycling plan 145–47
Training Stress Balance (Training Peaks) (TSB) 121
transitions 47
transverse abdominis 18
Trap Bar Deadlift 160–61
trapezius 19, 21
triceps 19, 21, 22
trunk
  trunk angle 34
  trunk stabilization 21
tyres 8, 85
  maintenance 91
  pressure 88, 89, 90
  pressure sensors 92
  and rolling resistance 11
  self-sealing punctures 86
  tubeless 85, 86, 109

U

ultra-endurance
  crank length and Q-factor 89
  energy use 47
  fuelling strategies 64, 65
  strength training 127
Union Cycliste Internationale (UCI) 98
  regulations 87
  UCI ProSeries 98
  UCI World Tour 98
  UCI WorldTeams 98
upper body muscles 14, 19, 21, 22
  exercises 127
  mountain biking and 23
  see also individual muscles
upstroke
  force application and torque production 29
  joint actions 20
  muscles that aid 23

V

V02 max 46, 120, 122
  aerobic training 53
  altitude and 58
  assessing cardio fitness 52
  climbing performance 104
  energy duration 51
  interval training 124, 125, 131
  lactate threshold 57
  race finishes 106
  time trials 108
  ventilatory threshold 1 56
valgus alignment 32–33, 40
variation 122
varus alignment 32–33
vasodilation 72, 73
vastus medialis 16
veins 53
velocipede 82
ventilation rate 52
ventilatory thresholds 56–57
video analysis 34, 37
visualization 111
vitamins
  absorption of fat-soluble 60
  vitamin C 68
  vitamin D 68
volume, increasing cycling 130
volunteer's dilemma 113

W

W prime (W') 51
warming up 131
wattage 92
wearable devices 24, 94, 95
weight
  power-to-weight ratio (W/kg) 11, 66, 104, 108, 109
  UCI regulations 86, 87
weights 151
  choosing the right weight 152
  lifting technique 152
wheels 83, 84, 85–86, 108
  maintenance 91
  wheel selection 88, 89
  see also tyres
whole body training 127
wind 105, 107
  crosswinds 101, 102, 107
wind tunnel testing 83, 84, 87, 99
work rate, cadence and 31
wrist pain 39

XYZ

yaw angle 85
zone 1 training 56, 121, 128
zone 2 training 56, 121, 124, 125, 128, 131
zone 3 training 56, 121, 125, 128

BIBLIOGRAPHY

Asplund, C., & St Pierre, P. (2004). Knee pain and bicycling: fitting concepts for clinicians. *The Physician and Sportsmedicine*, 32(4), 23–30.

Atkinson, G., Davison, R., Jeukendrup, A., & Passfield, L. (2003). Science and cycling: Current knowledge and future directions for research. *Journal of Sports Sciences*, 21(9), 767–787.

Australian Sports Commission. (n.d.). Supplements. In AIS nutrition. Australian Institute of Sport. www.ausport.gov.au/ais/nutrition/supplements

Bailey, M. P., Maillardet, F. J., & Messenger, N. (2003). Kinematics of cycling in relation to anterior knee pain and patellar tendinitis. *Journal of Sports Sciences*, 21(8), 649–57.

Banister, E. W., Calvert, T. W., Savage, M. V., & Bach, T. (1975). A systems model of training for athletic performance. *Australian Journal of Sports Medicine*, 7, 57–61.

Barry, N., Burton, D., Sheridan, J., & Thompson, M. (2014). Aerodynamic performance and riding posture in road cycling and triathlon. *Procedia Engineering*, 72, 720–725.

Bassett, D. R., Jr., & Howley, E. T. (2000). Limiting factors for maximum oxygen uptake and determinants of endurance performance. *Medicine & Science in Sports & Exercise*, 32(1), 70–84.

Bini, R. R., & Hume, P. A. (2005) Muscle recruitment pattern in cycling: A review. Physical Therapy in Sport 6(2):89-96

Blocken, B., van Druenen, T., Toparlar, Y., Malizia, F., Mannion, P., Andrianne, T., Derome, D., Diepens, J., & Carmeliet, J. (2018). Aerodynamic drag in cycling pelotons: New insights by CFD simulation and wind tunnel testing. *Journal of Wind Engineering & Industrial Aerodynamics*, 179, 319–343.

Bompa, T. O., & Buzzichelli, C. (2019). *Periodization: Theory and Methodology of Training* (6th ed.). Human Kinetics.

Borresen, J., & Lambert, M. I. (2009). The quantification of training load, the training response and the effect on performance. *Sports Medicine*, 39(9), 779–795.

Burke, E. R. (2002). *Serious Cycling*. Human Kinetics.

Burke, L. M., & Hawley, J. A. (2018). Swifter, higher, stronger: What's on the menu? *Science*, 362(6416), 781–787.

Cheung, S. S., & Zabala, M. (Eds.). (2017). *Cycling Science* (Sport Science series). Human Kinetics.

Clarsen, B., Krosshaug, T., & Bahr, R. (2010). Overuse injuries in professional road cyclists. *The American Journal of Sports Medicine*, 38(12), 2494-501.

Clough, P., Earle, K., & Sewell, D. (2002). Mental toughness: The concept and its measurement. In I. Cockerill (Ed.), *Solutions in Sport Psychology* (pp. 32–43). Thomson Learning.

Craig, N. P., & Norton, K. I. (2001). Characteristics of track cycling. Sports Medicine, 31(7), 457–468.

Faria, E. W., Parker, D. L., & Faria, I. E. (2005). The science of cycling: Factors affecting performance. *Sports Medicine*, 35(4), 313–337

Foster, C., Florhaug, J. A., Franklin, J., Gottschall, L., Hrovatin, L. A., Parker, S., Doleshal, P., & Dodge, C. (2001). A new approach to monitoring exercise training. *Journal of Strength and Conditioning Research*, 15(1), 109-115.

Gaul, L. H., Thomson, S., & Griffiths, I. M. (2018). Optimizing the breakaway position in cycle races using mathematical modelling. *Sports Engineering*, 21(4).

Halson, S. L. (2014). Monitoring training load to understand fatigue in athletes. *Sports Medicine*, 44 Suppl 2 (Suppl 2), S139-S147.

Hopker, J., & Jobson, S. (Eds.). (2012). *Performance Cycling: The Science of Success*. Bloomsbury Sport.

Impellizzeri, F. M., Marcora, S. M., Rampinini, E., Mognoni, P., & Sassi, A. (2005). Correlations between physiological variables and performance in high level cross country off road cyclists. *British Journal of Sports Medicine*, 39(10), 747–751.

Issurin, V. (2008). Block periodization versus traditional training theory: a review. *Journal of Sports Medicine and Physical Fitness*, 48(1), 65–75.

Jeukendrup, A., & Gleeson, M. (2019). *Sport Nutrition: An Introduction to Energy Production and Performance* (3rd ed.). Human Kinetics.

Jeukendrup, A. E., & Martin, J. (2001). Improving cycling performance: How should we spend our time and money. *Sports Medicine*, 31(7), 559–569.

Joyner, M. J., & Coyle, E. F. (2008). Endurance exercise performance: the physiology of champions. *Journal of Physiology*. 1;586(1):35–44.

Kiely, J. (2012). Periodization paradigms in the 21st century: evidence-led or tradition-driven? *International Journal of Sports Physiology and Performance*, 7(3), 242–250.

Laursen, P. B., & Jenkins, D. G. (2002). The scientific basis for high-intensity interval training: optimising training programmes and maximising performance in highly trained endurance athletes. *Sports Medicine*, 32(1), 53-73.

Lucía, A., Hoyos, J., & Chicharro, J. L. (2001). Physiology of professional road cycling. *Sports Medicine*, 31(5), 325–337.

McArdle, W. D., Katch, F. I., & Katch, V. L. (2015). *Exercise Physiology: Nutrition, Energy, and Human Performance* (8th ed.). Wolters Kluwer.

Macdermid, P. W., & Edwards, A. M. (2010). Influence of crank length on cycle ergometry performance of well-trained female cross-country mountain bike athletes. *European Journal of Applied Physiology*, 108(1), 177–182.

Martin, J. C., Milliken, D. L., Cobb, J. E., McFadden, K. L., & Coggan, A. R. (1998). Validation of a mathematical model for road cycling power. *Journal of Applied Biomechanics*, 14(3), 276–291.

Martin, J. C., & Brown, N. A. T. (2009). Joint-specific power production and fatigue during maximal cycling. *Journal of Biomechanics*, 42(4), 474–9.

Marsh, A. P., Martin, P. E., & Foley, K. O. (2000). Effect of cadence, cycling experience, and aerobic power on delta efficiency during cycling. *Medicine & Science in Sports & Exercise*, 32(9), 1630–4.

Marshall, S., & Paterson, L. (2017). *The Brave Athlete: Calm the F*ck Down and Rise to the Occasion*. VeloPress.

Migliaccio, G., Padulo, M., Russo, J., & L. (2024). The Impact of Wearable Technologies on Marginal Gains in Sports Performance: An Integrative Overview on Advances in Sports, Exercise, and Health. *Applied Sciences*, 14(15), 6649.

Mountjoy, M., Ackerman, K. E., Bailey, D. M., Burke, L. M., Constantini, N., Hackney, A. C., Heikura, I. A., Melin, A., Pensgaard, A. M., Stellingwerff, T., Sundgot-Borgen, J. K., Torstveit, M. K., Jacobsen, A. U., Verhagen, E., Budgett, R., Engebretsen, L., & Erdener, U. (2023). International Olympic Committee consensus statement on Relative Energy Deficiency in Sport (RED-S). British Journal of Sports Medicine, 57(17), 1073–1097

Mujika, I., & Padilla, S. (2001). Physiological and performance characteristics of male professional road cyclists. *Sports Medicine*, 31(7), 479–487.

Mujika, I. (2017). Quantification of training and competition loads in endurance sports: Methods and applications. *International Journal of Sports Physiology and Performance*, 12(Suppl 2), S29–S217.

Mujika, I., Sharma, A. P., & Stellingwerff, T. (2019). Contemporary Periodization of Altitude Training for Elite Endurance Athletes: A Narrative Review. *Sports Medicine*, 49(11), 1651–1669.

Nash, J. F. (1950). Equilibrium points in n-person games. *Proceedings of the National Academy of Sciences*, 36(1), 48–49.

Padilla, S., Mujika, I., Orbananos, J., & Angulo, F. (2000). Exercise intensity during competition time trials in professional road cycling. *Medicine & Science in Sports & Exercise*, 32(4), 850–856.

Passfield, L., & Hopker, J. G. (2017). A mine of information: Can sports analytics provide wisdom from your data? *International Journal of Sports Physiology and Performance*, 12(7), 851–855.

Reardon, C. L., Hainline, B., Aron, C. M., Baron, D., Baum, A. L., Bindra, A., Budgett, R., Campriani, N., Castaldelli-Maia, J. M., Currie, A., Derevensky, J. L., Glick, I. D., Gorczynski, P., Gouttebarge, V., Grandner, M. A., Han, D. H., McDuff, D., Mountjoy, M., Polat, A., Engebretsen, L. (2019). Mental health in elite athletes: International Olympic Committee consensus statement. *British Journal of Sports Medicine*, 53(11), 667–699.

Sawka, M. N., Burke, L. M., Eichner, E. R., Maughan, R. J., Montain, S. J., & Stachenfeld, N. S. (2007). American College of Sports Medicine position stand: Exercise and fluid replacement. *Medicine & Science in Sports & Exercise*, 39(2), 377–390.

Selye, H. (1936). A syndrome produced by diverse nocuous agents. *Nature*, 138, 32.

Seiler, S., & Tønnessen, E. (2009). Intervals, thresholds, and long slow distance: The role of intensity and duration in endurance training. Sportscience.org, 13, 32–53.

Seiler, S. (2010). What is best practice for training intensity and duration distribution in endurance athletes? *International Journal of Sports Physiology and Performance*, 5(3), 276–291.

Skiba, P. F., & Clarke, D. C. (2021). The W' Balance Model: Mathematical and Methodological Considerations. *International Journal of Sports Physiology and Performance*, 16(11), 1561–1572.

Stellingwerff, T., & Cox, G. R. (2014). Carbohydrate supplementation on exercise performance or capacity of varying durations. *Applied Physiology, Nutrition, and Metabolism*, 39(9), 998–1011.

Stellingwerff, T., Morton, J. P., & Burke, L. M. (2019). A framework for periodized nutrition for athletics. *International Journal of Sport Nutrition and Exercise Metabolism*, 29(2), 141–151.

Sperlich, B., & Holmberg, H. C. (2017). The responses of elite athletes to exercise: An all-day, 24-h integrative view is required! *Frontiers in Physiology*, 8, 564.

Swart, J., & Holliday, W. (2019). *Cycling: Training and performance*. In Cardinale, M., Newton, R., & Nosaka, K. (Eds.), *Strength and Conditioning: Biological Principles and Practical Applications* (pp. 499–510). Wiley Blackwell.

Sunde, A., Støren, Ø., Bjerkaas, M., Larsen, M. H., Hoff, J., & Helgerud, J. (2010). Maximal strength training improves cycling economy in competitive cyclists. *Journal of Strength and Conditioning Research*, 24(8), 2157–2165.

Tarnopolsky, M. (2004). Protein requirements for endurance athletes. *Nutrition*, 20(7–8), 662–668.

Thomas, D. T., Erdman, K. A., & Burke, L. M. (2016). Position of the Academy of Nutrition and Dietetics, Dietitians of Canada, and the American College of Sports Medicine: Nutrition and Athletic Performance. *Journal of the Academy of Nutrition and Dietetics*, 116(3), 501–528.

Tucker, R., & Noakes, T. D. (2009). The physiological regulation of pacing strategy during exercise: a critical review. *British Journal of Sports Medicine*, 43(6), e1.

Union Cycliste Internationale (UCI). (2023). UCI technical regulations: Equipment. Retrieved from https://www.uci.org/

ABOUT THE AUTHOR

Dr David Bailey
PhD, Sports Scientist

David Bailey is an internationally recognized sports scientist, coach, and performance consultant with over two decades of experience in elite sport, applied research, and high-performance team leadership. He currently serves as Head of Performance for a professional cycling team, where he oversees integrated performance services spanning physiology, nutrition, coaching, psychology, and technology. David began his career in academia, earning a PhD in Exercise and Sport Nutrition from Loughborough University, a global institution in sport science. His doctoral and postdoctoral research advanced understanding of training adaptation, recovery, and sports nutrition, culminating in over 25 peer-reviewed publications. This strong scientific grounding laid the foundation for a career focused on translating evidence into real-world performance impact. He spent several years at the English Institute of Sport, delivering scientific support across three Olympic cycles, Athens, Beijing, and London, working with athletes in cycling, triathlon, and other endurance sports. These years were pivotal in shaping his interdisciplinary approach, collaborating closely with coaches, nutritionists, psychologists, and medical staff to support medal-winning performances on the world stage.

Following this, David moved into leadership roles within professional cycling teams, where he built and managed high-performance systems for some of the world's top cyclists. His responsibilities spanned coaching, race support, performance modelling, athlete monitoring, and integrating new technologies into training and competition environments.

Beyond his team roles, David is a trusted advisor to sports technology startups, elite institutions, and global organizations, including the International Olympic Committee (IOC), World Anti-Doping Agency, World Triathlon, and McLaren Applied Technologies. He is a recognized thought leader in training periodization, sports nutrition, athlete monitoring, and behaviour change. He contributed to the IOC's 2023 Consensus Statement on REDs (relative energy deficiency in sport), providing insights grounded in both clinical and coaching perspectives.

As a consultant, David plays an active role in shaping the next generation of performance solutions, contributing to AI-based coaching platforms, product development in nutrition and recovery, and innovations in wearable technology. He also continues to offer one-on-one coaching and support to a diverse group of athletes, from ambitious amateurs to seasoned professionals.

David holds credentials from organizations including British Cycling, the International Olympic Committee, the Union Cycliste Internationale, and the British Association of Sport and Exercise Sciences. He is also affiliated with professional bodies such as the American College of Sports Medicine and Professionals in Nutrition for Exercise and Sport.

Whether skiing in winter or cycling alpine passes in summer, his personal commitment to endurance sport reflects his professional ethos: that science, when practically applied, can unlock human potential.

ACKNOWLEDGEMENTS

Author's acknowledgements

I would like to acknowledge those who have contributed to my development as a sports scientist, performance specialist, and coach, enabling me to share my accumulated expertise through the writing of this book.

First and foremost, I am deeply grateful to the athletes and teams I've had the privilege to support. From Olympic champions to Grand Tour contenders, their dedication and pursuit of excellence has continually inspired me and elevated my own professional standards. Their trust, feedback, and openness to evidence-based approaches have been instrumental in shaping and evolving my practice.

I am indebted to the colleagues and collaborators I have worked with across diverse high-performance environments. I've been fortunate to be surrounded by world-class practitioners and leading sports scientists who have consistently challenged and expanded my thinking throughout my career.

I would also like to acknowledge the international organisations and partners who have entrusted me with advisory roles. Contributing to projects with the International Olympic Committee, the World Anti-Doping Agency, and other global sporting bodies has been a professional honour. I remain committed to bridging the gap between science and practice.

Finally, I would like to extend my sincere appreciation to the DK Editorial Team for their expert guidance throughout the writing of this book. Special thanks to Alastair Laing for inviting me to take on this project, and to Amy Child and Nicola Hodgson for skilfully transforming my scientific content into a visually engaging and accessible format.

My career has been guided by a simple but enduring principle: performance is a shared journey. I have been fortunate to share that journey with passionate, curious, and committed people. I look forward to continuing that path, both as a practitioner and a lifelong learner.

Publisher's acknowledgements

DK would like to thank Vanessa Bird for indexing and Susan McKeever for proofreading.

Picture credits

The publisher would like to thank the following for their kind permission to reproduce their photographs:
(Key: a-above; b-below/bottom; c-centre; f-far; l-left; r-right; t-top)

14 Science Photo Library: Professors P.M. Motta, P.M. Andrews, K.R. Porter & J. Vial (bl). **95 Adobe Stock:** Ilusiku studio (br)

All other images © **Dorling Kindersley**

Senior Editor Alastair Laing
Senior Designer Barbara Zuniga
Production Editor David Almond
Production Controller Celine MacLeod
DTP & Design Co-ordinator Heather Blagden
Senior Acquisitions Editor Zara Anvari
Art Director Maxine Pedliham
Publishing Director Stephanie Jackson

Design Amy Child
Editorial Nicola Hodgson
Illustrations Arran Lewis

First published in Great Britain in 2025 by
Dorling Kindersley Limited
DK, One Embassy Gardens, 8 Viaduct Gardens,
London, SW11 7BW

The authorised representative in the EEA is
Dorling Kindersley Verlag GmbH. Arnulfstr. 124,
80636 Munich, Germany

Copyright © 2025 Dorling Kindersley Limited
A Penguin Random House Company
10 9 8 7 6 5 4 3 2 1
001–345495–Dec/2025

All rights reserved.
No part of this publication may be reproduced, stored in or introduced
into a retrieval system, or transmitted, in any form, or by any means
(electronic, mechanical, photocopying, recording, or otherwise), without
the prior written permission of the copyright owner.

No part of this publication may be used or reproduced in any manner for
the purpose of training artificial intelligence technologies or systems. In
accordance with Article 4(3) of the DSM Directive 2019/790, DK
expressly reserves this work from the text and data mining exception.

A CIP catalogue record for this book
is available from the British Library.
ISBN: 978-0-2417-2256-5

Printed and bound in China
www.dk.com